PHYSICS ENVY

PHYSICS ENVY

*American Poetry and Science
in the Cold War and After*

PETER MIDDLETON

THE UNIVERSITY OF CHICAGO PRESS
CHICAGO & LONDON

Peter Middleton is professor of English at the University of Southampton. He is the author of three books of scholarship, most recently *Distant Reading: Performance, Readership, and Consumption in Contemporary Poetry*, and a book of poetry, *Aftermath*; and he is the coeditor of *Teaching Modernist Poetry*. He lives in Southampton.

The University of Chicago Press, Chicago 60637
The University of Chicago Press, Ltd., London
© 2015 by The University of Chicago
All rights reserved. Published 2015.
Printed in the United States of America

24 23 22 21 20 19 18 17 16 15   1 2 3 4 5

ISBN-13: 978-0-226-29000-3 (cloth)
ISBN-13: 978-0-226-29014-0 (e-book)
DOI: 10.7208/chicago/9780226290140.001.0001

"Ka 'Ba" and lines from "The Test" by Amiri Baraka reprinted here by permission of Chris Calhoun Agency, © Amiri Baraka.

Library of Congress Cataloging-in-Publication Data

Middleton, Peter, 1950- author.
Physics envy : American poetry and science
in the Cold War and after / Peter Middleton.
pages cm
Includes index.
ISBN 978-0-226-29000-3 (cloth : alk. paper) —
ISBN 978-0-226-29014-0 (e-book)
1. American poetry—20th century—History and criticism.
2. Literature and science. I. Title.
PS310.S33M53 2015
811'.50936—dc23
2015011457

♾ This paper meets the requirements of ANSI/NISO Z39.48-1992 (Permanence of Paper).

CONTENTS

ACKNOWLEDGMENTS

I HAVE FOUND the poetry world to be both friendly and keenly interested in intellectual debate about all aspects of poetry and poetics, a place where poets and researchers can mingle. For extended conversations about poetics over many years, discussions that made this project possible, I am grateful to many people, including Bruce Andrews, Rae Armantrout, Isobel Armstrong, Peter Barry, Charles Bernstein, Vincent Broqua, Olivier Brossard, Michael Davidson, Rachel Blau Du Plessis, Stephen Fredman, Robert Hampson, Carla Harryman, Lyn Hejinian, David Herd, Susan Howe, Romana Huk, Simon Jarvis, Daniel Kane, Daniel Katz, Michael Kindellan, Nicky Marsh, Tony Mitton, Will Montgomery, Miriam Nichols, the late Douglas Oliver, Maggie O'Sullivan, Geoffrey Pawling, Bob Perelman, Joan Retallack, Peter Riley, Lisa Robertson, Will Rowe, Gavin Selerie, Jonathan Skinner, Keston Sutherland, Cole Swensen, Barrett Watten, Carol Watts, John Wilkinson, and Tim Woods. But no list can adequately acknowledge the extent of my gratitude and indebtedness to the myriad conversations and publications of poets, poetry magazines, editors, and researchers on whose work I have depended.

For a long time now I have been testing out ideas about poetry and science at conferences and in published articles. I am very grateful to the many conference organizers and editors who helped make this possible, including Robert Archambeau, Carla Billiterri, Philip Coleman, Frank Davey, Steve Evans, Stephen Fredman, Kornelia Freitag, Ben Friedlander, Clare Hanson, David Herd, David Kennedy, Drew Milne, Peter Robinson, Will Rowe, Keith Tuma, John Woznicki, and John Wrighton.

I would also like to thank all those who gave time and energy to read the manuscript at various stages. Two readers for the University of Chicago Press offered extensive, insightful guidance on what worked and what needed adjustment. I would also like to thank friends and colleagues who read draft chapters of the book at various stages, including most especially Stephen Bending, Peter Boxall, and Kendrick Oliver. And many thanks to my editor Randolph Petilos, production editor Caterina MacLean, and Kevin Quach, Kathleen Raven, and Tadd Adcox, at the University of Chicago Press, as well as copyeditor Pam Bruton, for all their excellent work in producing this book.

Vital periods of research leave were funded by the Leverhulme Trust and the Arts and Humanities Research Council. I would also like to thank the School of Humanities, now the Faculty of Humanities, at the University of Southampton for its financial support and generous leave scheme and for the flexibility with which it enabled me to reschedule leave at a difficult time. My colleagues in the Department of English have created an environment in which research can flourish. I owe them a large debt of gratitude.

Finding other researchers interested in literature and science was vital to my project. The British Society for Literature and Science has been immensely inspiring for this project, partly through its activities, conferences, online reviews, and meetings and partly by the way in which it has helped to create a mutually supportive field of research that brings together researchers of all historical periods and at all stages of their career. I would particularly like to thank Daniel Cordle, John Holmes, Janine Rogers, Sharon Ruston, Michael Whitworth, and Martin Willis for their active interest and many useful suggestions.

My graduate research students were always willing to discuss poetics, and they often heard and debated the earliest and roughest form of the ideas contained in this book. Particular thanks to Ros Ambler-Alderman, Bea Bennett, Mandy Bloomfield, David Carstairs, Matthew Francis, Ross Hair, and Mark Rutter.

I would like to thank individually a few people who not only made the very idea of this project possible but also helped me to finish what sometimes appeared an insurmountable task. Conversations with Allen Fisher have inspired the project from its very earliest beginnings, indeed ever since he gently remonstrated with me about a critical remark in my review of *Unpolished Mirrors*. Marjorie Perloff has been an immensely generous and inspiring presence throughout this project, offering essential advice on every aspect, practical and intellectual. Peter Nicholls read and commented upon portions of the manuscript several times and offered crucial advice and support at many points along the way. His friendship and his unflagging commitment to high standards of literary scholarship have been enormously valuable. Rodney Livingstone has been unfailingly encouraging and offered many thoughtful comments on my emerging ideas. David Leverenz read the entire manuscript several times and gave me invaluable advice on every aspect of the project prior to my delivering it to the Press. Having his friendly voice at my shoulder was immensely helpful.

My family has been wonderfully patient and always interested in the progress of "the book." Harriet and George have lived with it for a long time and cheered me on. Kate Baker's loving support and willingness to discuss the project at every stage of what was a long process have made it all possible.

A condensed version of material from chapters 3 and 4 appeared as "Poetry, Physics, and the Scientific Attitude at Mid-century," *Modernism/Modernity* 21, no. 1 (January 2014): 147–68, copyright © 2014 The Johns Hopkins University Press. A short, earlier version of the discussion of Charles Olson's interest in the sciences and my analysis of his poem "Maximus to Gloucester, Letter 27 [Withheld]" appeared as "Discoverable Unknowns: Olson's Lifelong Preoccupation with the Sciences," in David Herd, ed., *Contemporary Olson* (Manchester: Manchester University Press, 2015): 38–51, copyright © Manchester University Press.

Grateful acknowledgment is made for permission to cite "Ka 'Ba" and an excerpt from "The Test," copyright © Amiri Baraka. Reprinted by permission of the Chris Calhoun Agency.

Research for this project was partly funded by grants from the Arts and Humanities Research Council and the Leverhulme Trust.

INTRODUCTION

J. ROBERT OPPENHEIMER introduces the September 1950 issue of the *Scientific American* with a joke about the irrelevance of poetry in a scientific age. When this most famous of American physicists was a young man, he had tried to write poetry himself. Paul Dirac took him aside and tactfully dismissed these juvenilia by reminding him of the current status of poetry: "In science one tries to say something that no one knew before in a way that everyone can understand. Whereas in poetry . . ." (his ellipses). No need to complete the sentence, everyone knows what poetry is like. Poetry's fusty, referentially challenged language can offer no reliable knowledge of the universe in which modern Americans live. And lest any of his readers be contemplating a career in poetry, Oppenheimer rubs it in: "The 10 reports here, to which these words may serve as introduction, do indeed attest that science says things that no one knew before in a way we can all understand."[1] This dismissal of poetry would have stung far more then than it does now because of the status of both this physicist and his profession, which already had what *Fortune* magazine called a "glamour" reinforced by high salaries and the enviable social standing of physicists. It was the kind of dismissal that would go on reverberating over the next decade and more as the prestige of physicists continued to grow.[2] Fifteen years later a poll showed that nuclear physicists were rated more highly than any other professionals except doctors and Supreme Court judges.[3]

Criticisms such as Oppenheimer's were all too familiar to poets. As Muriel Rukeyser begins to set out her optimistic account of what she calls "the life of poetry" in midcentury America, she first acknowledges the widespread resistance to poetry based on confused impressions of its "sexually suspect" character, its obscurity, and above all its irrelevance in a scientific era.[4] After ending the war by dropping atomic bombs on Hiroshima and Nagasaki, American physicists had further demonstrated the power of nuclear science. They revealed to the public several hitherto-secret heavy elements, including new metals such as curium and the patriotically named americium, which in the words of the *New York Times* were "artificially created out of uranium by modern alchemy."[5] Rukeyser wittily takes up this motif by suggesting that if poetry were thought to be useful, "if it were a metal," poets would not be allowed near it: "our scientists would claim their right of experiment and inquiry" over it.[6] John Ashbery later explores the consequences of the encroachment of science on poetry's domain in similar terms in "The Instruction Manual," his charming pastiche of the Romantic conversation poem. A writer of technical manuals (we might recall that this was Louis Zukofsky's job during the war) breaks off from work on a handbook about "the uses of a new metal" to daydream about visiting Guadalajara. Having by courtesy of the imagination vicariously enjoyed the hospitality of that city, the science writer regretfully returns to quotidian reality with a sigh: "I turn my gaze / Back to the instruction manual which has made me dream of Guadalajara."[7] The poem's simple realism is rendered "in a way we can all understand." The poem hints that a poetic style that invents scenes to be inhabited by means of eager readerly identification is no more than a palliative made necessary by overexposure to the rigors of science's claim to the right of experiment and inquiry.

Physics Envy is primarily devoted to Cold War poets who have also wanted to claim a right of experiment and inquiry for poetry. To explore the many implications of claiming such a right, I single out for attention mostly New American poets and Language Writers, along with a few others such as Muriel Rukeyser and George Oppen, from the many other poets who have employed scientific language or wondered at what it means to live in a scientific era. New American poets of the period between 1949 and 1989 made large demands on poetry, wanting it to have political and aesthetic resonance that would ally it with new social and political energies, while also contributing to the expansion of knowledge. These poets wrote passionate yet intellectually robust poems searching out the truth about history or mind, language or material culture, friendship or activism. With hindsight it is possible to see that these poets shared certain principles even if they did not always fully articulate them. The occasion of the poem was to be a bursting forth of language in a struggle for articulation and momentary self-definition, a process of reaching toward potentially new positive knowledge, not simply the transmission of an already-established theory, science, or fact. Poetry could have epistemological priority; it could, pace Oppenheimer, articulate things that no one knew before. Language Writing produced polemics that were often avowedly antiepistemological, yet the poetry itself was even more dedicated than its immediate predecessors to researching every aspect of language use, whether social, intimate, historical, logical, or narrative. Its signature tendency to withhold a phrase or sentence from further upward integration was almost always an invitation to the reader to subject its islanded language to the kind of scrutiny that ordinary-language philosophers, psychoanalysts, and deconstructionists habitually brought to bear on their examples of usage or the indeterminacy of language.

These poets were writing at a time when science's hegemony was pervasive. As Rukeyser says, with the explosion of the atomic bomb "the function of science was declared, loudly enough for the unborn to hear."[8] Shortly after the bombing of Hiroshima, the atomic scientist Hans Bethe told a radio audience, "The atom is the hero of the day."[9] Bethe's folksy image appears in a published collection of radio broadcasts, *The Scientists Speak*, given between 1943 and about 1946 by a roster of the leading scientists of the day. Each succinctly pictured the state of the speaker's own field in a mere fifteen minutes during the intermission of concerts broadcast to a large audience on Sunday afternoons by the New York Philharmonic-Symphony Orchestra and sponsored by U.S. Rubber. As listeners opened a beer or made a cup of coffee, leading scientists talked about oceanography, astronomy, chemistry, metallurgy, medicine, biology, genetics, social aspects of science, and, of course, physics. "The time had clearly come," says Warren Weaver, the editor of the published collection and a director of the Rockefeller Foundation, "when every man, woman, and child must recognize the role that science plays in modern society," revealingly blurring ethics and historical necessity by his insistence that everyone "must" acknowledge the importance of science.[10] The cytogeneticist and president of the University of Texas T. S. Painter sounds a common note when he begins his talk by saying, "As we turn our minds to the planning of a better world in which to live, many different groups of thoughtful people are seeking scientifically sound answers to questions of immediate human concern. One field of knowledge that may help us is genetics, the science of heredity."[11] Oppenheimer was one of the speakers and called his radio talk "The Atomic Age," with an authority that would have been hard to question since he must have seemed to embody that "age." If any science could be said to dominate the overall picture of contemporary science in *The Scientists Speak*, it was certainly nuclear physics, which appeared capable of almost anything. Two of the radio talks, "Medical Benefits from Atomic Energy" and "Atomic Energy and Medicine," even claim a direct connection between atomic energy and the clinic, because nuclear radiation can achieve "miraculous results" in the treatment of cancer and the study of human metabolism.[12] Confidence that the future belonged to nuclear science was based not only on its prospects for new, peaceful technologies but also on its recent *success* at actually converting matter into energy and, thus, theoretical equations into military victory.

The bomb had not discredited science in the eyes of the scientists—far from it.

They argued that the scientific functions to which Rukeyser alluded now covered knowledge, ethics, and even aesthetics. These functions included production of the only reliable knowledge of the fundamental constituents of all things, trustworthy self-management of their own epistemology, custodianship of a modern moral code, and, most threatening of all to poets, the transfer of creative power from the arts to the sciences. Let's consider these functions one by one. To begin with, psychophysical reductionism was in the ascendant. The material world revealed by science was all there was; true knowledge was scientific knowledge. John Dunning, a professor of physics at Columbia University who had worked with Enrico Fermi (*American Scientist* describes Dunning as "in the very top rank of modern physicists"), offers a revealing aside as he sets the scene for an account of contemporary nuclear physics: "It has long been accepted that you and I and all the world around us are made of some 92 basic types of atoms which we call elements."[13] If we are all constructed of such materials, then the science that tries to understand them must presumably be our primary and probably best source of knowledge about the world, *including ourselves*. His adverb "long" gives a patina of tradition to this very recent scientific picture of the ontology of things. Cheerleaders for science strongly supported the naturalism that increasingly dominated both its ontology and its epistemology. The logical positivist Hans Reichenbach dismissed any knowledge claim that did not arise directly from scientific study of the material world: "The philosopher who claimed to have uncovered the laws of reason rendered a bad service to the theory of knowledge: what he regarded as laws of reason was actually a conditioning of human imagination by the physical structure of the environment in which human beings live."[14] In Reichenbach's scientific world, if the human imagination wants to know the truth, it must let itself be conditioned by the actual "metal" of the universe. It is the structure of things that determines how we think.

Science could at times sound as if it were exempt from any criticism by outsiders because of its special mind-set, its self-regulation, and its ethical framework. At the height of the London Blitz, Conrad Waddington, now best known as the founder of epigenetics, recognizing that "science is already a very potent force," published *The Scientific Attitude* (1941), in which he argued that the scientific view was now dominant: "scientific thought has become the pattern for the creative activity of our age, our only mode of transport through the rough seas in front of us."[15] This pattern of creativity is evident in the "scientific attitude," which espouses "the matter-of-fact as against the romantic, the objective as against the subjective, the empirical, the unprejudiced, the ad hoc as against the a priori."[16] The idea that it was the right creative attitude that underwrote scientific rigor would be of continuing interest to poets over the next few decades, despite the gradual decline of this belief among scientists themselves. This older definition of science in terms of a "scientific attitude" continued to be endorsed by some scientists over the following decade, but they were less and less likely to equate that stance with a high degree of individual integrity, instead treating this attitude as a mode of reasoning collectively reinforced through the institutions of science, especially within the increasingly large teams of physicists and other scientific researchers.

The idea that scientists were necessarily ethical was also undergoing change in this period. At midcentury this was still a widespread assumption. In a lecture given in Chicago in 1950 and subsequently printed in the *Bulletin of the Atomic Scientists*, a journal for scientists concerned to protect the unborn from endangerment by reckless use of nuclear weapons, Michael Polanyi, a cosmopolitan Hungarian scientist from the University of Manchester in England, spoke for many scientists of his time when he asserted that the success of nuclear science highlighted the importance of recognizing the authority of scientists: a "society which wants to foster science must accept the authority of scientific opinion."[17] Some were prepared to argue that this authority was necessarily moral as well as epistemological. Mark May told attendees of "The Scientific Spirit and Democratic Faith" conference held in New York at the height of the Second World War that "the moral code practised in the fellowship of science is distinctly superior to that practised in the most civilized societies."[18] Henry Margenau went further still, optimistically telling the same conference that "no wars would be fought if the scientific attitude were to prevail generally."[19]

Such views were slow to disappear despite the gradual recognition of what Steven Shapin calls the "moral equivalence" of scientists to nonscientists. Writing in 1965, the leading sociologist David Apter could still say: "we are going to need to learn to live in a world in which the ethic of science has become the ethic of man."[20] Such assumptions had been increasingly challenged since the 1940s. Robert Merton claimed in a 1942 article that there was no evidence that scientists are "recruited from the ranks of those who exhibit an unusual degree of moral integrity" or, more importantly still, that scientific rigor derived from the "personal qualities of scientists."[21] Shapin argues that "by the 1960s, a deflationary conception of scientific knowledge and of the character of the scientist had escaped from the scientific community to—some, but not all—general cultural commentators."[22] The idea that the scientist requires a special voluntaristic commitment or "scientific attitude" over and above the orientation provided by the modes of reasoning embedded in the knowledge and social relations of science was also losing ground.

A few scientists were so confident of science's supremacy that they openly relegated the arts to the past. In a wide-ranging article published in the *Scientific Monthly* in the late 1940s, the neurophysiologist Ralph W. Gerard speaks candidly of what many scientists believed. He claims with all the authority of his profession that the arts can now be seen to be a primitive stage of human development. "Emotional behavior, and probably consciousness, depend on the activity of phylogenetically ancient brain parts, which are similar through much of the vertebrate subphylum," and therefore, because "art is concerned mainly with private feeling, science with public thinking," it can readily be inferred that the sciences are gradually superseding the more primitive cultural practices of the arts and literature that lie coiled up in the reptile brain.[23] Nothing will be lost because fortunately science is replacing art: scientific research is "a creation of man and is a work of art."[24] Gerard claims universal scope for scientific method: "I certainly cannot prove that the scientific view of the universe is the best or the final metaphysic which man will reach, or that truth is a crowning value. But I do maintain that this approach, of those man has tried, has led to great and progressive change in human affairs—mental as well as physical—and that, until an approach more satisfactory by empiric standards arises or the limitations of the present one are clearly demonstrated, man had better 'hold fast that which is good.'"[25] There would be no place for poetry in his scientific America.

Were these the scientists who would decide whether or not poetry's concerns were the sort of "metal" over which the sciences would claim rights of inquiry? In practice, most scientists were more willing to be inclusive than Gerard, and less simplistic in their naturalism than Dunning, although they too would impose a price for inclusion. Harlow Shapley, an astronomer who was then head of the American Association for the Advancement of Science, puts it characteristically simply: "fortunately, we are nearly all scientists."[26] In other words, poets need not feel excluded, because almost everyone, physicist or poet, is now a scientific American. Writing in *Harper's Magazine* in 1945, shortly after what this astronomer archly describes as the "sensational percussions in New Mexico and Japan," he makes such gestures at inclusivity because he wants the government to establish a National Science Foundation and believes that the best political strategy is to avoid the exclusiveness that can be heard in Polanyi, May, and Gerard and to argue that science embodies American democracy at work. American "science-sensitive citizens" are ready to meet the challenge of the atomic bomb. American "farmers, miners, machine operators, merchants, and housekeepers are scientific in spirit and in practice, even though not trained in schools to the scientific formula."[27] Others thought that more effort was required to understand science than simply being American, though they concurred with the idea that everyone could potentially be scientific in spirit at least.

James Conant, one of the main architects of the Manhattan Project and, as president of Harvard, in a position to know that the full spectrum of valuable modern research includes many nonsciences, tells readers of *On Understanding Science* (1947) that "a greater degree of understanding of science seems to me of importance for the welfare of the nation."[28] Unlike many of his contemporaries, Conant did see a place alongside the sciences for the humanities and even for a somewhat-idealized image of poetry. In his 1944 Franklin lecture he had cited Francis Bacon's division of learning into history, poesy, and philosophy and then argued that in modern America, "if we use the term poetry to cover all creative insights into human destiny whatever their form may be, and the word philosophy to include the whole expanse of analytical and speculative thought except for mathematics and the sciences, we see that many aspects of a scholar's labours fall within these bounds." In practice, most existing fields will involve more than one of these modes: "the humanities and the social sciences, to use our modern terms, cut across all three fields, and only rarely does the major part of a traditional subject fall within the boundaries of accumulative knowledge."[29] In Conant's mind, poets and scientists share a commitment to inquiry and a moral responsibility: "the scholar, the seeker after truth, whether he be mathematician, archeologist, scientist, philosopher, poet or theologian, must come into the court of public opinion not only with clean hands but with a consecrated heart."[30] Because of such shared commitments, the outstanding "impartial inquirer" of the future in nonscientific fields will require that they and their intellectual communities have the best possible understanding "of what science can or cannot accomplish."[31] Conant's contribution to such improvement was the development of a history of science course based on case studies of modern science that concentrated on the strategies required by scientific research, particularly the use of "conceptual schemes," using a terminology he had learned from his social science colleagues at Harvard.[32] His student Thomas Kuhn would develop such ideas into his vastly influential theory of paradigms that emerge from scientific revolutions, a theory that has had considerable influence on cultural theory. In 1947, the same year in which Conant published this call for intellectuals, in David

Hollinger's words, "to behave scientifically in social environments very different from the one in which science actually proceeds," a new journal with similar aims appeared.[33] The renascence of a long-established but then moribund periodical as the enormously successful popular-science magazine *Scientific American* was also founded on the assumption that it was now the case that to be an American was indeed to be scientific.

Such beliefs had pervasive consequences in the two decades after the war ended. Biologists, social scientists, philosophers, structuralists, and linguists all envied the intellectual authority of nuclear physics and borrowed heavily on its epistemological credit card. Even geneticists, for instance, thought physics had answers to their puzzles. After Francis Crick, James Watson, Rosalind Franklin, and Maurice Wilkins discovered the double-helix structure of DNA in 1952, scientists from many fields converged on biology convinced that they could analyze biological processes using reductionist, mathematical methods derived from physics. Historian Lily Kay explains the physicist George Gamow's important role in such strategies in terms that will become familiar as we explore the extent of intellectual interest in physics in the postwar period: "Not only did Gamow define, articulate, and attempt to solve the coding problem but also he brought the powerful culture of postwar physics with its various military linkages to bear on representations of heredity and life."[34] For many researchers during the two decades after the war, physics would be the template for scientific rigor.

Sociologists and psychologists who aspired to put their own fields on a more sound footing were drawn to the natural sciences, especially physics, as a source of methodology. Talcott Parsons's *Structure of Social Action* (1937) led the way for a generation of sociologists who called themselves social scientists. Parsons was the lead voice in a Harvard document on pedagogy that makes plain that he thought sociology could be a science on a par with physics if it emulated the use of theoretical models, or what he and his colleagues had begun in the 1930s to call conceptual schemes: "If we did not select, if we did not abstract, the writing of history would take as long as the making of history. And so it is with all the social sciences and indeed with science generally. Since the scientist cannot deal with events in all their uniqueness, the best he can do is to construct a conceptual model which reflects with a minimum of distortion certain important relationships which prevail between the phenomena."[35] The hint that even the field of historical research was becoming a social science would not have been lost on people at the time. Some sociologists made much noisier claims to scientific status. The leading advocate of scientific sociology, George Lundberg, wrote *scope of scientific influence in changing approach to humanities*

a book aimed at the general public with the blunt title *Can Science Save Us?*— Yes! was the answer—explaining that the science of sociology would be essential for understanding and solving the many problems and conflicts of modern societies. Sociology was not alone in its scientization. Psychology was increasingly dependent on what it conceived to be the fundamentals of natural-scientific method. Here too physics was the usual source of analogies and models. If you did not know that Kurt Lewin, a leading Gestalt psychologist, was talking about *people* when he sets out his methodology, you might imagine he was talking about nuclear particles. It is only when we reach the end of the following passage from a textbook that was still very influential in the 1950s that the human orientation emerges: "Direction of the Field Force. That the valence is not associated with a subjective experience of direction, but that a directed force, determinative of the behavior, must be ascribed to it, may be seen in the fact that a change in the position of the attractive object brings about (other things being equal) a change in the direction of the child's movements."[36]

Even poets were drawn to physics. Charles Olson's usage of analogies with nuclear physics, most notably in his famous manifesto essay "Projective Verse" (1950), where he proposes that the new poem should be a "high energy construct," is one of the themes of this book. That essay became a manifesto in Donald Allen's defining anthology *New American Poetry* (1960), in which the accepted ordinariness of such a strategy is evident in the manner by which the young poet Philip Whalen explains to readers how he works as a poet. Whalen borrows the image of a key piece of apparatus for the physicist: "This poetry is a picture or graph of a mind moving, which is a world body being here and now which is history . . . [his ellipses] and you. Or think about the Wilson Cloud Chamber, not ideogram, not poetic beauty: bald-faced didacticism moving, as Dr. Johnson commands all poetry should, from the particular to the general."[37] By the time this statement was published, the fiddly, temperamental Wilson cloud chamber, a sealed container of supersaturated gas that could reveal the ionizing traces of nuclear particles as trails of mist to be photographed, had been superseded by more reliable bubble chambers, which were rapidly producing a whole series of new particles whose classification was causing the physicists much difficulty.[38] The slight anachronism did not occasion any demurral. As they selectively imitated the discourse, metaphors, and images of nuclear physics, poets displayed what could be described as a pervasive physics envy. At the time, however, this eagerness to emulate the methods and models of physics was so much the norm in disciplines ranging from biology to sociology, psychology, and even

THIS IS IT

philosophy that it was rarely seen negatively. Envy of the success of physics was rational. Indeed, Olson's contemporaries thought of physics as opportunity, not as a source of envy. If there were doubts about the use of physics, they were likely to be doubts about just who had the right of inquiry, who could be called genuinely scientific.[39]

Not all scientists were as ecumenical as Conant in welcoming visitors from the arts who claimed the same right of inquiry. Shapley does what many defenders of science would do over the next two decades and reaches for an abstract stereotype of poetry as one of the most convenient signs of all that is antithetical to science. "We may hear the plaintive query, 'Does not this emphasis on science mean the death of art and poetry?'"[40] His answer to this concern could only have been published in 1945. Look to Russia, he argues unconvincingly, look at their success in integrating science into all areas of modern life without damaging the arts, which are "the flower upon the many-panicled stem we strive to cultivate and strengthen through scientific culture."[41] He soon retreats from lush metaphorical endorsement of the arts to argue that the proposed National Science Foundation should start by supporting natural sciences and then, as soon as feasible, the social sciences. "Historical, linguistic, and literary disciplines" will have to wait. Strange to think that an expanded NSF might have funded even those seekers after truth the poets. Occasionally, the cheerleaders of science threw writers a bone: "In the first place, the arts, and literature especially, can serve as a source of scientific hypotheses. In the second place, the arts are invaluable in communicating and dramatizing scientific truth and in general and through emotional appeal can motivate to action in accordance with what is scientifically sound and possible."[42] Poets and writers can usefully stimulate the imagination of scientists as long as they accept that science decides what is "sound and possible." Or they can act as publicists for the new metals, the scientific discoveries. What they cannot do is pursue rigorous inquiry or produce real knowledge on their own terms. No wonder that the eminent Harvard literary critic Douglas Bush twisted Reichenbach's argument; it was not the natural world that was necessarily conditioning human and therefore poetic imagination; it was the sciences themselves: "all modern poetry has been conditioned by science, even those areas that seem farthest removed from it."[43]

In *Physics Envy* I pursue several lines of investigation into the implications of the physics and science envy of the postwar decades. I shall argue that poets were far from alone in their envy of the status and rights of inquiry of physics and, indeed, that envy was also opportunity. Hence, one line of argument leads to a

↳ envy was widespread into other fields

discussion of the value and importance of those conceptual schemes that Parsons and Conant mention, conceptual frameworks resulting from a bricolage of metaphors, images, and concepts from physics that could be immensely effective in their new methodological location if their improvisatory character were taken into account. Such schemes were adopted in domains as different as group psychology and philosophy. A second line of discussion follows poets such as Amiri Baraka and George Oppen who become hostile to, or at least very wary of, the social scientists who have relied too heavily on the credit of physics to bankroll their own sometimes-ideological ethical and political projects. By looking hard at traces of physics envy in the social sciences, these poets are able to diagnose the pathologies of race science and urban social theory. A third, lightly sketched line of investigation encompasses a range of poets for whom physics envy might be said to be a self-diagnosis that encourages them to think harder about epistemological questions alongside their attentions to the affects. Robert Creeley, Lyn Hejinian, Jackson Mac Low, and Rae Armantrout offer differing strategies for testing what Hejinian calls a "romance with science's rigor."[44]

anti-physics envy;

pro-physics envy

"This is not an adventure in polymathy."[45] Robert Merton's disarming warning to the reader at the end of his groundbreaking essay "The Normative Structure of Science" is a useful guideline for anyone writing about poetry and science in the latter part of the twentieth century. Although I have given considerable space to the words of scientists as they talk about their work, and have learned as much as I can from both their specialist writings and histories and philosophies of science, I am not writing as a scientist or even as a historian of science (though this account may be of interest to them). My goal has been to identify the epistemological and aesthetic consequences for poetry of the enormous postwar expansion of the territory of the sciences, and shifts of public perceptions of norms of scientific method. I ask how poetic form responds to the growth of the sciences. Doing so has required close attention to the sources of poets' information about science. Poets are rarely scientists, nor do they have direct access to the nontextual hinterland of science—laboratories and fieldwork—that generates what Hilary Putnam calls the "*unformalizability* of practical knowledge" (his italics); poets have learned their science for the most part from popularizers, both scientists and laypersons, with the inevitable simplifications, distortions, and lacunae.[46] Wherever possible I have therefore tried to document the actual sources used by poets or, where that is not possible, to locate typical widely circulated accounts of scientific developments. I am convinced that close attention to the radiating discourses of scientists can provide insights into decisions made by poets who wanted to sustain their own right of inquiry. Giving attention to the voices of scientists and to the magazines and books read by poets has meant that though I do offer a number of close readings of poems, I cannot offer as many of these as I would like. My hope is that this book will send readers back to read again those and other science-oriented poems with renewed interest.

The book is divided into three main sections: "Poetry and Science," "Mid-century," and "Scientific Americans." In the first part, chapters 1 and 2 discuss interrelations between poetry and science, first from the standpoint of poets and then from the standpoint of physicists. Chapter 1 sets the scene. It starts by exploring how poets' curiosity about the world shaded into interest in the possibility of poetic research and poetic knowledge, and how such claims have been received by literary theorists. This discussion is followed by an introductory overview of different types of poetic response to science, both before and during the postwar period, by way of providing the backstory to what follows in the remainder of the book. Chapter 2 gives the physicists a voice. It shows that, despite a widespread belief that the arts were being replaced by the sciences, poetry still mattered to physicists not least because referring to an abstract ideal of poetry enabled them to talk about a tricky issue, the problems in communicating the reality of their hidden nuclear world.

Part II of the book offers a thick description of the broad intellectual culture of physics envy and opportunity around 1950. Chapter 3 starts with Charles Olson's "Projective Verse" and then shows that Olson joined a bustling crowd of many different research programs also interested in the hope that their own versions of "composition by field" would give them access to the scientific frontiers. Chapters 4 and 5 argue that both Muriel Rukeyser and Charles Olson saw radical possibilities for poetry in the uses that social scientists had made of physics. Rukeyser encountered these clever strategies through her study of the history of American science, while Olson first learned about them through friendships made at Harvard. Chapter 5 also contrasts what Lisa Jarnot calls Robert Duncan's "voracious appetite for the ideas of modern science" with Olson's self-confessed "joy" in science.[47]

Part III of the book is devoted to the dissemination of rapidly expanding American science through the *Scientific American*. Many poets read it avidly. Chapter 6 discusses well-documented encounters by Rae Armantrout, Jackson Mac Low, and Robert Duncan. Chapter 7 traces two highly critical responses to the uses and misuses of science by sociologists. I argue that analyzing the *Sci-*

entific American's presentation of technocratic sociology and race science helps us understand the intellectual context for poems by George Oppen and Amiri Baraka. Each part of the book can be read separately, although the book tells an unfolding story of intense poetic discussion about the poetic right of experiment and inquiry.

PART I
POETRY AND SCIENCE

✳ 1 ✳

THE POETIC UNIVERSE

Mapping Interrelations between
Modern American Poetry and the Sciences

POETIC INQUIRY IN SCIENTIFIC AMERICA ⟋ *no domain*

Midcentury poets faced a dilemma. Scientists owned the natural world. The elements of green shade, of rocks, and stones, and trees, of the physical pines and the red wheelbarrows that make up the thousand threads of things were being explained by new sciences, most impressively by quantum theory and by a theory of genetics known as the modern synthesis. The poet as amateur naturalist looked increasingly old-fashioned. William Carlos Williams's admission that the aim of *Paterson*—"To make a start, / out of particulars / and make them general, rolling up the sum"—relied on "defective means" looked all too true.[1] If poets wanted to talk at all knowledgeably about what George Oppen calls the "things / We live among," or the cosmos or life itself, they needed to be aware of what the scientists were saying about the intangible entities of atoms and genes.[2] At first glance, this seemed to be a manageable dilemma for poets willing to stay off the turf of the natural scientists. Unfortunately, however, for any strategy of retreat into the inner life, scientific naturalism also extended into the green thought that accompanied the green shade, turned spirit into neurons, eliminated metaphysical pines, and provided scientifically materialist reductions of every aspect of the body, mind, language, and society of the poet and poem. Wittgenstein's insistence at the end of the *Tractatus* that "we feel that even when all *possible* scientific questions have been answered, the problems of life remain completely untouched" (his italics) was no longer reassuring.[3] Morality, meaning, values, relationships, memory, history, languages, concepts, desires, and moods were all claimed by one science or another. And when the reductionism reached its limits, all these features of human culture turned out to be the product of atoms and their forces.

These pressures from the sciences were especially powerful in the United States after a war that was widely believed to have been won by physics, with the assistance of the achievements of other sciences, such as medical advances in antibiotics, cybernetic developments in weapons design, and chemical products with valuable military uses. Nuclear physics was seen as America's special achievement since many of the leading nuclear physicists from Europe had fled there from persecution, and more had been persuaded to come by the enormous investments in the new science. American social science was also rapidly growing in prestige and put the humanities and the arts under pressure to justify themselves. The claim that only scientific method could be relied upon to produce reliable knowledge of the world, coupled with the idea that anyone, farmer or poet, could be a "scientific American," created a dilemma for poets. Should they insist that they were not part of the new America, or—and why not?—should they become *scientific* American poets? I believe that the response of poets in the late-modernist lineage to this challenge to produce a poetics capable of full participation in a scientific America was to create some of the most ambitious Anglophone poetry of the twentieth century. These poets of the second half of the twentieth century wanted to claim their own right of experiment and inquiry as scientific Americans, and they could be vocal about this desire.

In this introductory chapter I shall sketch out how these poets conceptualized their interest in sharing the responsibilities and rewards of inquiry with the sciences. I shall address several questions. Is it possible to schematize the main kinds of poetic interaction with new scientific knowledge of quantum fields, genetic codes, electronics, astrophysics, and biology? What did literary theorists think of the use of experimentation and of reasoning with propositions in poetry and, above all, of poems that engaged with scientific knowledge and methodology? Could poems claim legitimacy as inquiry? Did poems have any epistemological traction as they engaged in what Ronald Bush aptly calls "negotiations between reason and desire"?[4] The postwar poets that I shall be discussing saw themselves as inheritors of a modernist tradition that they wanted to renew. What were the main legacies of high-modernist poetics for poetry's active pursuit of interrelations with the sciences? What new directions did poets take? The majority of late twentieth-century poems about science borrow scientific ideas and discoveries for metaphoric uses. This chapter maps out a broad theoretical and historical context for the postwar poetic ventures onto the terrain of the sciences, as a preliminary for the more detailed studies that follow in later chapters.

Alice Fulton speaks for many contemporary poets when she says of her poem

"Cascade Experiment": "I often lift scientific language for my own wayward purposes. That isn't to say I play fast and loose with denoted meanings. I'm as true to the intentions of science as my knowledge allows. But my appropriations from science are entwined with other discourses, other ideas, so that a term such as 'cascade experiment' comes to stand for more than the laboratory event it is."[5] "Cascade Experiment" borrows a popular version of quantum theory to talk about the importance, if a relationship is to flourish, of having faith in the existence of unobserved emotions: "faith in facts can help create those facts, / the way electrons exist only when they're measured, / or shy people stand alone at parties." The poem appropriates a popular version of the uncertainty principle to suggest that the lover is like an atomic particle because intense interest about who and what this entity is both makes it knowable and at the same time sets limits to that knowledge: "Because believing a thing's true / can bring about that truth, / and you might be the shy one, lizard or electron / known only through advances / presuming your existence, let my glance be passional / toward the universe and you."[6] Ronald Johnson's poem "The Invaders" wittily finds scientific analogies to represent the aesthetic achievements of writers, artists, and musicians: "Mahler is a nova of / virus," while Blake "descends in mystic / 7's / the stairs / from Quasar to Orange to Atom." In a note to the poem he criticizes the unquestioned facticity of science, saying that we live in a time when "science is no longer seeing it is fiction, or poetry, and makes its own solutions with intuition as stepping stones."[7] Fulton's and Johnson's scientific metaphors suggest that the poet and reader must now recognize that their everyday living takes place inside an organism genetically driven by molecular processes and living amid a phenomenal whirl of particles in electromagnetic fields. The image of the lover as shy electron also typifies a tendency in this kind of poetry. Metaphorization of science disembeds the new scientific entities from their epistemological contexts and often results in an ironic stance toward the expansion of scientific method into all areas of human experience. Not all poets treat the sciences as sources of fertile metaphors. Some treat them as inspiration for a poetics of inquiry.

Robert Creeley liked to quote William Carlos Williams's "contention that 'the poet thinks with his poem, in that lies his thought, and that in itself is the profundity.'"[8] "We," said Creeley of his generation of poets—"that unimaginable *plural* of I!—want our lives to be known to us," a knowledge created in poems that form "a record, a composite fact of the experience of living in time and space."[9] Oppen argued that "the emotion which creates art is the emotion that seeks to know and to disclose." He believed that the modernists had demonstrated that poetry could attain a "skill of accuracy, or precision, a test of truth." The poetic image represents "an account of the poet's perception," and its effectiveness is "a test of sincerity, a test of conviction, the rare poetic quality of truthfulness."[10] Oppen uses his terms with great care. Accuracy and sincerity are what Bernard Williams calls the "virtues of truth," dispositions essential for a discourse to be truthful and capable of producing knowledge.[11] In effect, Oppen is arguing that poetry is as capable of rigorous inquiry as any scientific or scholarly discipline.

Robert Duncan did not want poets to give away any rights to define epistemic values: "We work as poets and take seriously what seems to most men the one ground surely not to be taken seriously—the play-reality of imagined religions, philosophies, sciences."[12] Poetry is an art of the hypothetical, an aesthetics of theories. Charles Olson told poets that "the human universe is as discoverable as that other," the material universe, if only they would compose their poems as high-energy fields.[13] And in a memo to the Black Mountain College faculty written with a copy of Lincoln Barnett's 1949 *Life* magazine profile of Oppenheimer at his elbow, Olson echoed an aspiration attributed to the physicist: "I am, then, concerned as any scientist is, with penetrating the unknown."[14] Denise Levertov's well-known definition of organic form reveals in a quieter terminology a similar interest in inquiry: "For me, back of the idea of organic form is the concept that there is a form in all things (and in our experience) which the poet can discover and reveal."[15] Gerard Manley Hopkins's "inscape" was now reframed in a more science-friendly manner.

The generation of avant-garde poets who followed them were even more willing to affirm explicitly their commitments to experiment and inquiry. Lyn Hejinian entitled a selection of her essays *The Language of Inquiry* because "the language of poetry is a language of inquiry, not the language of a genre. It is that language in which a writer (or reader) both perceives and is conscious of the perception."[16] Charles Bernstein also endorsed poetic inquiry and linked it directly to knowledge: "It is just my insistence / that poetry be understood as epistemological / inquiry." In place of Duncan's mythic language of play and imagination, Bernstein knowingly makes a specimen out of the jargon of Continental philosophy in this free-verse essay first performed at the MLA: poetry resists the "hegemony of restricted / epistemological economies."[17] When Ron Silliman explained to readers of the first major anthology of Language Writing, *In the American Tree* (1986), that what the poets share is not a style or an ideology but "a perception as to what the issues might be," his choice of these issues closely echoed the domains that define the scope of different, mostly scientific disciplines: "The nature of reality. The nature of the individual. The function of language in the constitution of either realm. The nature of meaning. The substantiality of language. The shape and value of literature itself. The function of method. The relation between writer and reader."[18] Each item in this list can be paralleled with a natural, social, or human science: physics, psychology, linguistics, philosophy, literary theory, and perhaps even the philosophy of science.

Anyone familiar with modern American poetry is likely to recall many other similar bold claims. Poems, say the New American poets, can think and inquire. Thinking about composite facts such as time and space with accuracy and tests of truth, working with imagined sciences in field compositions, and penetrating the unknown, discovering laws or forms in all things—these and other related modes of poetic research shape the poetry and poetics of the New American poets. Language Writers argue that poems can test the workings of public and personal language. Inquiring into the role of discourse, sentences, words, and phonemes in constituting subjectivity, politics, history, power, and knowledge—these are potentials of an expanded poetry. They and other postwar poets in the modernist lineage repeatedly argue that poems need not concede epistemological primacy to any other domain of research, whether a natural, social, or human science.

Puzzlement about the implications of such claims was the origin of this book. I say "puzzlement" deliberately. The conceptual resources to reformulate as research questions the mix of curiosity, uncertainty, and fascination that generated this puzzlement have been hard to find in literary and cultural theory. What are we now to make of such late-modernist aspirations for poetry? Are they at all plausible? These poets, whether associated with the generations of the objectivists, the New American poets, the Language Writers, or the less defined poetic movements of feminist or black activist networks, believed that poetry (like other arts) can be a site of inquiry and investigation, that poetry is more than entertainment, instruction, Orphic revelation, or self-realization, though in the course of their experiments and inquiries they might practice some or all of these other modes. Are such beliefs no more than boosterism for poetry? How might literary theory help us weigh up such beliefs?

We clearly need to approach cautiously such claims that the poem can be a site of inquiry, not least to look to the poetry for confirmation of the poetics. Bryan Walpert's study of recent contemporary American poetry provides evidence that poets known for their interest in science can talk in their essays about the importance of science and yet manifest in their poems a steady resistance to scientific authority, "specifically claims that science holds a monopoly on knowledge."[19]

Although three of the four poets whom Walpert chooses to represent the diversity of recent poetry—Pattiann Rogers, Alison Hawthorne Deming, and Albert Goldbarth—each approach the rendering of subjectivity differently, their poems often belie their authors' discursive enthusiasms for scientific ideas and methods and display "different kinds of resistance to the authority of scientific knowledge."[20] Deming may say that "a mode of questioning we associate with science can become a nest of poetic delight," but in the well-known poem "The Woman Painting Crates," Walpert thinks she rejects the world revealed by physics "in favor of her common perception of her surroundings."[21] Only the fourth poet he discusses, Joan Retallack, manages to align her commitment to science (as in *The Poethical Wager*) with her poetic practice, and she does so by writing performative poems that are as experimental as good scientific research. Her poem "AID/I/SAPPEARANCE," by "enacting linguistic critiques of scientific language, shows a concern for the human subject."[22] Although her poetic theory is of interest to anyone writing about poetry and science, Retallack's poetic engagements with science are outside the historical frame of this book—Walpert's insightful discussion of her is a reminder that poetry continues to develop new versions of Rukeyser's "methods of science."

In *Physics Envy* I argue that resistance is, in practice, usually mingled with admiration and sometimes, as in the cases of the poets on whom I concentrate, with a searching cognitive and affective negotiation between the poem and specific scientific epistemology. Deming herself might be better described not as resistant to science but as disputatious, in a scientific sense, because she is concerned to counter the assumption that one specific science can explain everything. Science is nothing if not intense argument, as Deming knows. In "The Woman Painting Crates," the poet does not directly reject physics in favor of her own perception, but instead she challenges physicists to develop an even better model than is provided by particles or by the new halo theory (presumably a version of string theory) about which she has just heard a lecture. Although the poet reaches a point where she insists that "knowing is what I try / to train myself out of," this implies that she herself wants to be as scientific as the physicist, and so she is training herself not to think that she already knows what she hopes to find out.[23]

METHODOLOGY

By midcentury the sciences were making strong claims to the right of inquiry over the outside and the inside places, over the green shade and the green thought.

⌐ intellectual
superiority
(concrete results)

Poets who wanted their work to stand up to intellectual scrutiny as more than narcissistic fantasy had to find ways to negotiate with the epistemologies of science. When reading across the full range of postwar American poems, especially poems in the long modernist lineage, antifoundationalist dogmas can appear unconvincing guides to analysis. Even the most "languagey" poems can resist dissolution into the pure play of signifiers. Pound's allusions, H.D.'s etymologies, Stein's language experiments, Olson's ships, Duncan's dreams, Hejinian's memories, and Howe's surfaces of Puritan discourse have an epistemological robustness, a dependency on accurate relations to verifiable information, that in each case asks of the reader a commitment to what the poem is thinking. Should we really treat poetic language as frictionlessly moving within the world?[24]

Literary theory has not always appeared helpful in studying these poetic negotiations. Poets, as well as literary critics, work within what John Guillory calls an "epistemic hierarchy."[25] In my view, literary theory, despite its indispensible achievements, has sometimes avoided confrontation with this hierarchy by bringing theory in-house. Instead, for example, of negotiating with the actively researching social-scientific discipline of psychology, we mostly work with our tailored modes of psychoanalysis sympathetic to literary imagination. Similar strategies of selective containment have been applied to sociology (cultural materialism), linguistics (structuralism), and philosophy (deconstruction), enabling literary theory to largely avoid confronting other refractory major developments in these fields, and especially to avoid difficult trade-offs around rights of inquiry and claims to knowledge.

Guillory believes that the abandonment of knowledge as a goal and a value has been intellectually disastrous: "Abdicating the responsibility for defending claims to knowledge in favor of provocative antirealism and a self-congratulatory skepticism, we critics cease thereby to question the beliefs that function as truth, common sense, or consensus for literary and cultural studies." What then is the answer? Guillory hopes that literary and cultural studies will come to recognize themselves as a part of the range of human sciences located in the field of historical studies and develop an interdisciplinarity responsive to cultural, historical, and social science methodologies, recognizing the likely need to negotiate new lines of exchange between them. Above all, literary studies should recover interpretive methods that can produce "positive knowledge." Guillory offers prescriptions for the steps needed. Literary theorists should renounce the strategy, evident in much postmodernist criticism, of claiming that interpretation could be a mode of "skeptical critique of knowledge-claims" justified on the grounds that other disciplines rely on social constructions of the world and human life whose constructedness is treated as "natural." In the 1980s, literary and cultural studies committed themselves to a strong "analogy between literary representation and social construction" that made it possible to "equate the political and the epistemological." He gives as an example the widely promulgated idea that most supposed truths about human life were "really" social constructions: "If race was once a category of biological science (it still is for some scientists, though very few), might not the manifest constructedness of this category, which resembles a fiction, be indistinguishable from the constructedness of any object of scientific theory, including genes or quarks?"[26] Theory aimed at the wrong target. Constructs, models, theoretical frameworks, and conceptual schemes have been integral to the development of many areas of modern science. They are not automatic clues to hidden ideology. Although I think this overstates things—some conceptual schemes do embody ideological assumptions—Guillory is surely right that the very *fact* of constructedness is not itself evidence of false naturalism. As I show later, constructs and conceptual schemes have played a valuable role as agents of inquiry both in the development of recent scientific method and in the poetics of a number of poets.

There are two further reasons why literary theory has had little to offer by way of methodology. The first is that by its tendency to refuse to engage with epistemological questions as other than false consciousness, literary theory has not fully recognized the diversity, messiness, and incompleteness of actual scientific practice. This need to recognize the discursive improvisations and sometimes-dissonant heterogeneity of the sciences is one reason why I give extensive space to the words and ideas of many different modern scientists, both famous and little known. Paisley Livingston, in his unjustly neglected study *Literary Knowledge*, argues that literary and cultural studies need to recognize that literary knowledge has to operate in "the institutional and conceptual space of 'knowledge,'" because the "human sciences" (employing a usage more familiar in the French term *sciences humaines* and the German *Geistwissenschaften*) ultimately share "an object domain and epistemic principles" with the natural sciences.[27] He believes that because of the lack of forthright engagement with the sciences, literary theorists often have distorted images of what science actual does. They accept uncritically a scientistic vision "of a perfectly deterministic natural science that grasps the movements of a timeless nature in terms of a set of universal and perfectly necessary laws" or that reduces everything down to a few fundamental units.[28] They think science always claims it is the only path to truth. In practice, many scientists, as well as philosophers of science, now argue that, in John Dupré's words, "there are surely paths to knowledge very different from those currently sanctioned by the leading scientific academies."[29] Peter Burke's major study of knowledges from the early-modern period to the present makes a similar case.[30]

The second reason for the difficulty of finding help from literary theory is that although the sciences generally avoid epistemological dialogue with humanities research disciplines, science has been incorporated into the DNA of modern literary theory. The New Criticism developed out of an anxiety about the place of poetry in a scientific age. New Criticism's founding thinker, I. A. Richards, challenged Matthew Arnold's rousing statement that "the future of poetry is immense, because in poetry, where it is worthy of its high destinies, our race, as time goes on, will find an ever surer and surer stay,"[31] with the riposte that "a more representative modern view would be that the future of poetry is *nil*" (his italics).[32] The "principles" (or axioms) from which this comprehensive basis was constructed can be traced to the series of books on language, literature, and poetry that I. A. Richards wrote between 1923 and 1930 in which the issue of science and literature is central. One of these books, Richards's short study *Science and Poetry* (1926), remains one of the most influential of all modern theoretical discussions of the interrelations between science and literature.[33]

The groundwork for this account of science and poetry was laid in two earlier books: the philosophical study Richards wrote with C. K. Ogden, *The Meaning of Meaning* (1923), and *Principles of Literary Criticism* (1924), his own attempt to emulate the system builders in other disciplines such as mathematics and psychology by setting out axiomatic principles for a theory of literature. *The Meaning of Meaning* (whose title's opaque circularity ironically hints at the limitations of definition) ambitiously attempts to synthesize many recent developments in philosophy, linguistics, and psychology concerned with the shaping influence of language upon thought. Although *The Meaning of Meaning* is more usually thought of as a philosophical than a psychological text, the authors insist throughout the book that theirs is a *scientific* approach to the system of signs by which language operates. This is no Kantian account of judgment or Saussurean system of entities with only loose mental definition; signs are precise psychological phenomena: "more accurate knowledge of psychological laws will enable relations such as 'meaning,' 'knowing,' 'being the object of,' 'awareness,' and 'cognition' to be treated as linguistic phantoms also, their place being taken by observable correlations."[34] Philosophers must stop thinking of "words as containers of power" and recognize that words gain their force from the interactive context of use in which reference is often only one factor.[35] This new clarity about the function of signs can be demonstrated by thinking of different uses of the word "energy": "Thus for the physicist 'energy' is a wider term than for the schoolmaster, since the pupil whose report is marked 'without energy' is known to the physicist as possessing it in a variety of forms. Whenever a term is thus taken outside the universe of discourse for which it has been defined, it becomes a metaphor, and may be in need of fresh definition. Though there is more in metaphor than this, we have here an essential feature of symbolic metaphorical language."[36] Ogden and Richards aim to construct a science of meaning based on the axial mapping of allegedly orthogonal functions of language: the referential (or, as they usually call it, the symbolic) and the attitudinal or emotive.

This theory of language becomes the foundation for *Principles of Literary Criticism*, an extraordinary text that anticipates much of the literary theory of the rest of the century. In the spirit of Nietzsche, Richards consequently deploys a flexible team of scientific metaphors, some of which would resurface in postwar poetic usages of science. Poetry and poetic drama are his touchstones as he sets out a materialist psychology of "streams of impulses" that mediate between the visual stimuli of writing and reading and the eventual "affective-volitional attitudes" the signs produce.[37] Few later critics found this part of the theory convincing; they talked less rigorously of emotions and tone. The power of *Principles of Literary Criticism* as a theory of literature largely derives from the punctilious self-consciousness that leads Richards to reflect constantly on his own conceptual strategies of explanation and in doing so to elicit a series of descriptive metaphors that would prove endlessly fruitful for future literary theorists. The intense experience of a work of art cannot be simply the result of the mere addition of isolable components: "if we regard this as an affair of mere summation of effects it may seem impossible that the effect of the form can be the result of the effects of the elements, and thus it is easy to invent ultimate properties of 'forms' by way of pseudo-explanation." Having made this point he quickly signals dissatisfaction with its simple atomism and elaborates on parallels between the physics of matter and aesthetic experience to create what is explicitly said to be a better model: "Our more intense experiences are not built up of less intense experiences as a wall is built up of bricks. The metaphor of addition is utterly misleading. That of the resolution of forces would be better, but even this does not adequately represent the behaviour of the mind. . . . The intricacies of chemical reactions come nearer to being what we need. The great quantities of latent energy which may be released by quite slight changes in conditions suggest better what happens when stimuli are combined."[38] By setting up and knocking down these metaphors—elements, construction, forces, and so on—Richards tacitly conveys the idea that the formulation of analytic metaphors is itself a site of inquiry. In this instance, having halfheartedly offered the metaphor of a "resolution of forces" to describe the mind in the process of aesthetic composition, he tries to complicate it further by adding energy to the picture. He even alludes to the concept of a "field," though this is a "field of attention," a concept taken from psychology, for its multidiscursive resonances: "The parts of a visual field exert what amounts to a simultaneous influence over one another."[39] His argument shows that he fears that the reductionism of scientific analogies will too readily be accepted as incontrovertible and may also hide supervenience, or emergent properties of the constituent elements. Aesthetic synthesis creates new forms whose characteristics cannot be predicted just by adding together properties intrinsic to the raw materials. Richards makes his readers aware of how literary theory (or poetics) investigates literary phenomena through a process of checking and discarding metaphors because they come with existing attachments and epistemic debts.

Energy, construction, forces, and fields: the kernel metaphors of Olson's "Projective Verse" are all here. It might seem that Olson could have constructed his poetics essay simply by reaching back beyond the New Critical interpreters of Richards's literary theory to this foundational text. Although Olson may have known *Principles of Literary Criticism*, it would not, however, have furnished all that he needed, because these kernel metaphors are in the service of a psychological model of poetic composition that treats them solely as names for the participants in an inner psychodrama of conflicting impulses. All the epistemological action is suspended, leaving behind only the pursuit of "the equilibrium of opposed impulses."[40] This model of poetry assumes that poetic art is the management of the stresses and strains of mental life rather than an inquiry into the nature of the world, mind, or society. Art is not a process of inquiry. "So far as any body of references is undistorted it belongs to Science," he writes, and then goes on to distinguish between the referential use of language, which can be judged true or false, and the emotive use of language, which cannot be so judged because it works to affect attitudes independently of its truth-values.[41] His theory of the mind in terms of impulses would even seem ill-equipped to explain the normative rationality of his own discourse in which the theory is set forth.

The logic of his thought led Richards to confront poetry with science directly in his next book *Science and Poetry*. Much of the lasting power of the book stems from the apocalyptic tone in which he announces that humanity itself faces a "biological crisis"; new certainties offered by scientific knowledge are destroying the "magical view" of the world as a place shaped by "Spirits and Powers which control events" and may take poetry down with it.[42] Now "all the varied answers which have for ages been regarded as the keys of wisdom are dissolving together," as old beliefs about God, the universe, human nature, relationships, and the soul lose their credibility.[43] No one has more thoroughly and persuasively set out the challenge that the sciences represent for poetry. On the page facing his opening chapter he places as an epigraph Arnold's rousing statement that "the future of poetry is immense," though he doesn't quote Arnold's added Wordsworthian avowal that "without poetry, our science will appear incomplete," because such confidence would now seem risible to scientists. Richards tells a quite-different story in which the new sciences have no need of poetry now that *they* are the future: extending human life, rapidly increasing both knowledge of the cosmos and the power to control the material world, and developing scientific methods that should soon give a much "better understanding of human nature," one traditional domain of the arts.[44] "In its use of words most poetry is the reverse of science" to the extent that poetry's idea of "'truth' is so opposed to scientific 'truth' that it is a pity to use so similar a word."[45] Those who might have become poets once upon a time are now much more likely to become biochemists, while he knows mathematicians who cannot read poetry because "they find the alleged statements to be *false*."[46]

What can poetry do? In an inspired move that would have far-reaching consequences for literary theory, Richards transforms the scientific view that poetic statements are wrong into the fertile idea that propositions presented in poems might be potentially useful simulacra or models of real propositions. He arrives there by a process of reasoning that first maps the opposition between referential and emotive onto the difference between science and poetry. For this step he finds precedent in Kant, who separated scientific thought from practical reasoning, because although science incontrovertibly demonstrates its power to make available "genuine knowledge on a large scale," such knowledge has "no *direct* bearing" (his italics) on what we ought to feel or do.[47] Then he borrows from the emerging philosophy of logical positivism its interest in how to establish the validity of the proposition given the sometimes-deceptive linguistic forms it can take. A few years later, Moritz Schlick, one of the leaders of the logical-positivist Vienna Circle, would famously articulate the necessary criterion for winnowing the wheat from the chaff of propositions in the formula "the meaning of a proposition is the method of its verification."[48] Richards offers a workaday version of positivism in *Science and Poetry*: in the scientific approach, "if any of the consequences of a statement conflicts with acknowledged fact then so much the worse for the statement."[49] Richards's use of the term "statement" is carefully judged. The German term employed to refer to the statement or proposition was often *satz*, or "sentence," but ambiguities arise in English because some sentences can be either true or false depending upon a changing referent and would therefore not qualify as genuine propositions, which for logical positivism are by definition always either true or false. Even if a proposition does dress itself in the proper syntax and logical forms, it may still not be admitted to the realm of truth. Rudolf Carnap defined a whole category of these failed propositions: "if it only seems to have a meaning while it really does not, we speak of a 'pseudo-concept,'" and in particular, "logical analysis reveals the alleged statements of metaphysics to be pseudo-statements."[50] This is why Richards insists that "the poet is not writing as a scientist"—the value of a poetic statement does not depend on any correspondence with the actual world.[51]

Richards's chosen terminology for propositions that lack propositionality is notorious. Later critics would assume that when he calls these poetic propositions "pseudo-statements," he means that poems offer only fake propositions. Perhaps it was this misprision that led Richards retrospectively to observe of the book: "soon after its appearance I took a dislike to it without being very sure why. What seemed to me its best and most clearly stated points were, I found, understood in ways which turned them into indefensible nonsense."[52] Yet the original version of the book does appear at times to give support for this interpretation of the pseudo-proposition as a fake. Some thinkers, he tells us, have suggested that the validity of poetic statements depends on whether they cohere with a "universe of discourse," the imaginative space belonging to the poem.[53] The pseudo-statement is true for this poem, or for this speaker in the poem. He rejects this attempt to redeem the truth-value of poetic propositions. We have no criteria for identifying the specific universe in question, and the coherence that is claimed for this relation of poetic statement to the universe of discourse is not dependent on logic. This does not mean, however, that the pseudo-propositions are mere fantasy statements. Richards has a more complex view of the matter, calling these poetic propositions "pseudo" (a favorite term of intellectual abuse in that period) because their truth is not dependent on verification, as scientific statements are, but on "acceptability *by* some attitude" (his italics). The pseudo-statement "is a form of words which is justified entirely by its effect in releasing or organising our impulses and attitudes."[54] If we look back at the development of his thinking, we find further evidence that he did not believe that a pseudo-statement is a nonfunctional, plastic replica of the real thing. In *The Meaning of Meaning* Richards and Ogden cite Wittgenstein's dogmatic assertion in the *Tractatus* that "the propositions of mathematics are equations, and therefore pseudo-propositions."[55] Mathematical equations can be used as idealized models of complex phenomena, but they are not therefore merely toy replicas, any more than a theory of gravity is a cartoon version of some real force. Is it possible that poetic propositions are more like mathematical formulations than neutered propositions? If so, what are the implications for our understanding of the proposition in a poem?

Science and Poetry remains a challenging head-on engagement with the problem of how to defend poetry in an age when all reliable knowledge is assumed to be produced by scientific methods of inquiry. Richards was one of the first modern thinkers to explain to poets and literary critics why poetry might have to reconceive its relation to the linguistic proposition in the face of the claims by the sciences to offer the only valid form of verifiable propositions about the truth of things. Many subsequent literary critics and writers, even if they disagreed with his conclusions, would follow his lead and try to find a new role for poetry when confronted by the challenge from science. His skepticism about poetic statement would persist, often unexamined, as successive literary theories followed one another.[56] The next generation of critics, especially the New Critics, would try to make sense of Richards's electrical theory of psychology and, in the spirit of philosophical debate in the 1940s, treat the "pseudo-statement" as a form of emotivism. Later generations of critics would try to turn the tables on the sciences and treat them as "universes of discourse," texts whose claims to truth and knowledge are social constructions driven by power and desire. The idea that the statements in poetry are superficial copies of propositions has unfortunately lingered on, despite the overthrow of its positivist underpinnings by many philosophers, including Wittgenstein himself.

Poets may not all have read Richards but they heard the message. The most common response was to create an unchallengeable version of the universe of discourse: instead of claiming that the poetic universe was its own closed system, the poet could write a poem whose propositions depended solely on the author's warrant, the author's insider knowledge of her or his own inner world. You could in other words write confessionally, write a personal lyric. A more risky, more ambitious way to deal with the accusation of epistemological nullity would be to alter the linguistic presentation of the poetic statement so that it would no longer be measured against the real (scientific) thing at all, rather than letting it hover in its shabby pseudo-propositional dress. Strip away the sentence elements, the verb or the subject, or present only a participial phrase or some other verbal fragment, as Ezra Pound did, and you could leave the reader to imagine possible propositions constructed from these underaffirmed "piths and gists." You might present your statement between inferred scare quotes so that the reader not only grasps that this is only a pseudo-statement but is encouraged to engage in a metapoetic reflection on the implications that it would have *if* it were to be used in one context or another, making its pretense a virtue. You might even be able to have your cake and eat it if you could invite the reader to ask what sorts of truth claims would be entailed if this statement were to be made directly (though *of course* the poem is not doing so). Such nuances were to become widespread in the following two decades of American avant-garde poetry.

Richards's influence stretched beyond New Criticism to the literary theory that emerged in the 1960s, including structuralist criticism, which was open about its affinities with the human sciences. Roman Jakobson argued that because "poetics deals with problems of verbal structure" and "linguistics is the global science of verbal structure, poetics may be regarded as an integral part of linguistics."[57] Today, cognitive theory looks to neuroscience for assistance. Other areas of literary theory less obviously entangled with the sciences still have significant intellectual debts. Tillotama Rajan offers a persuasive case that "deconstruction tries to renew philosophy in the face of the challenge, even more the menace, of the 'human sciences.'"[58] Historicism has always been more receptive to scientific influence, which has resulted in an enormous expansion in the field of historical studies of literature and science led by George Rousseau, Gillian Beer, George Levine, and, recently, many others. Even a historicist critic writing about the early-modern era can find himself relying on a series of metaphors whose significance depends on contemporary physics. New Historicist Stephen Greenblatt uses scientific metaphors to lend conceptual authority to this argument: "Plays are made up of multiple exchanges, and the exchanges are multiplied over time, since to the transactions through which the work first acquired social energy are added supplementary transactions through which the work renews its power in changed circumstances."[59] The underlying model—of an entity composed of an accrual of energy exchanges from which its aesthetic power derives—encourages Fredric Jameson to talk of Greenblatt's *Renaissance Self-Fashioning* in Kuhnian terms of scientific revolutions: "in hindsight [it] looks like one of those classical and paradigmatic scientific discoveries achieved by triumphant accident in the process of solving a false problem."[60] Greenblatt himself harks back to the inspiration he gained from Clifford Geertz, pointing out that "Geertz's account of the project of social science rebounded with force upon literary critics" because Geertz himself acknowledged that the methods of reading cultural codes depended upon skills "of the literary critic" for the "determining of their social ground and import." In a telling aside for our understanding of the relation between New Historicism and the sciences, Greenblatt adds: "I perhaps did not wholly appreciate the scientific ambition lurking in the word 'determining.'"[61]

Literary theory has played a *fort-da* game with science, alternately repelling it and clutching it. We need to reflect on the implications of Peter Galison's question to historians of science: "can there be a history with no transcendental 'theory package' that escapes historicization?" Theorists and theories can themselves be historicized, and perhaps should be, leading to Galison's further question: "What if we can no longer invoke Wittgenstein or Kuhn or Peirce as the scaffolding on which historical detail is to be pitched? Such a fully historicized project would no doubt be hard to write. Philosophy, small group sociology, semiotics, anthropological 'culture'—these would not be 'givens' but instead would play out as part of the historical field, though with different rhythms and breakpoints."[62] We could replace his names with Derrida, Foucault, Lacan, Bourdieu, Butler, Deleuze, Badiou, and others. Much literary theory is based on historically sedimented attempts to respond to the sciences, so that if we try to employ New Critical, structuralist, poststructuralist, or psychoanalytic ideas, we risk circularity, since these developed as historically specific responses to science, sometimes oppositional, sometimes opportunistically emulating a salient feature of what appeared to be the latest scientific method. Literary criticism has also had great difficulty engaging with discourses in which truthfulness and accuracy are paramount values. It is therefore important to historicize what counts as knowledge in the natural sciences and in literature, rather than assuming that one or another standpoint, whether that of a scientist, a philosopher of science, a deconstructionist, or a historicist, can provide a fixed point of departure beyond question. But as Galison concedes, historicism cannot and should not be total. We do need to start somewhere. In what follows, I have employed methods derived from provisional uses of literary historicism, the new pragmatism, and the ideas of philosophers of science such as Hilary Putnam and Bas van Fraassen. I do not try to justify my use of their ideas; that would take another book. What I attempt is to map out a landscape, not to derive all its features from one theory or process. This is more geography than geomorphology or, to be more literal, a study of what

Charles Taylor calls the "background of available meanings" for poetry based on a gathering of the voices, writings, and institutions through which poetry and the sciences have encountered one another.[63]

One central theme of this book is therefore the development from the 1940s to the early 1970s of new ideas about poetic epistemology.[64] In the 1950s the idea that poetry could offer knowledge was respectable far beyond avant-garde circles. "Poetry gives us knowledge," affirm Cleanth Brooks and Robert Penn Warren in the first sentence of their preface to the third, 1960 edition of *Understanding Poetry*, the most influential poetry textbook in America over several decades, and itself an extended defense of poetry.[65] They anticipate readers more familiar with science than poetry, readers who are prone to the "confusion that causes people to judge formal poetry as if it were science" and who fail to realize how different they are.[66] For Brooks and Warren, poetic knowledge is admittedly a somewhat-different animal from scientific knowledge; it is a knowledge based on "attitudes, feelings and interpretations" found in everyday speech, not a knowledge based on the precisions of scientific attention to the material world.[67] Defending poetry in the scientific fifties requires that they promote it as a valuable source of knowledge, but they think of it as what John Crowe Ransom nearly two decades earlier had called "a kind of knowledge which is radically or ontologically distinct" from scientific knowledge.[68]

For the poets that *Physics Envy* studies closely, this ontological divide is either unstable or nonexistent. While none of these poets believe that their poems will produce knowledge that can then be directly assimilated into physics or biology, they do believe that they can do more than hold the sciences to ethical or aesthetic account. When Creeley writes about the uncertainty principle, when Duncan ponders the genetic code, or when Armantrout puts the metaphors of sociobiology under pressure, their aims are similar to those discerned by Stephen Wilson in the work of some visual artists who deal with scientific ideas and materials. He believes that artists who inform themselves about science and exchange ideas with scientists can produce an art that "explores technological and scientific frontiers." Anyone who talks to scientific researchers will have heard as much about their doubts and uncertainties, their struggles with ever-expanding probability arrays, as about their confident knowledge. Wilson encourages artists to engage with this zone of emergent knowledge, to try to supplement scientific research by pursuing "different inquiry pathways, conceptual frameworks, and cultural associations than those investigated by scientists and engineers."[69] Many of the poets I discuss are also extremely interested in the whole process of how knowledge is produced, whether by scientists, scholars, or ordinary citizens. Wilson's careful phrase "different inquiry pathways" is especially apt. For many of the poets discussed here, there are other pathways than science to knowledge, which might eventually at least be recognizable as such by scientists. Or to put this differently, these poets are sharply aware of the tensions between what Philip Kitcher calls "naturalistic epistemologies"—approaches to questions about what we can know that assume the answers will depend on the use of scientific knowledge about human capacities—and other philosophical and existential approaches to knowledge.[70]

"Epistemology" is a highly controversial term in itself. In this study I shall use "epistemology" in the broad sense of the study of any kind of human knowledge rather than in the narrower senses of naturalized epistemology or philosophical epistemology. Most philosophers now accept that it is a mistake to think that skepticism can ever be satisfactorily opposed by means of traditional epistemology. It is not possible to treat our direct sensory knowledge of the world as an incontrovertible foundation of knowledge. For postmodernists, epistemology has been a bête noire embodying everything they mistrust: power, dogmatism, ideology, autonomy, and a belief in the instrumentality of language. Although he understands this animus, Charles Taylor cautions against dismissing epistemology on such grounds. Traditional epistemology has been closely bound up with the growth of science since the early-modern period, while also being rooted in core moral and spiritual beliefs. Since Hegel, philosophers like Heidegger, Merleau-Ponty, and Wittgenstein have repeatedly felt the need to question what they see as a dubious "interpenetration of the scientific and the moral." But Taylor does not believe that these thinkers have made a complete break with the epistemological tradition and simply abandoned knowledge altogether. They continue to make a "demand for self-clarity about our nature as knowing agents" because they "conceive this self-understanding as an awareness about the limits and conditions of our knowing."[71] Taylor's account of the lingering interest in questions about knowledge and our responsibilities to it could also apply to the work of most of the poets in this study.

SCIENCE ENVY OR SCIENCE EMULATION?

The poets that *Physics Envy* studies were all too aware that the very phrase "poetic knowledge" could sound like an evasive way of saying "fanciful knowledge" or at best "ornamented knowledge," more akin to the "world building" of science fiction than to informed speculation. Or more bluntly, it can sound like the "science envy" that Lisa Steinman diagnoses in American modernist poetry of the early twentieth century. She argues that envy of the socioeconomic authority of the natural sciences, especially physics, was the driver of an interest in pairing up poetry and science among leading modernist poets of the first half of the twentieth century who felt that America marginalized poetry. They followed two strategies: to claim authority for poetry as an American technology or to liken poetic vision to the vision of the new physics of Einstein and quantum theory. The first of these strategies led poets to "borrow the prestige of science and technology" for an art desperately needing cultural esteem, whose absence poets felt keenly. This appropriative strategy was risky, according to Steinman, because these modernist poets were defending poetry "by comparing it to science and by appealing implicitly to values they did not endorse."[72]

The second strategy proved more enduringly effective. In the late 1930s and 1940s poets began to construct "analogies with physics" that enabled them "to overcome or overlook their commitments to potentially opposed aesthetics." Wallace Stevens "rested his case for poetry's importance on the apparently scientific nature of the reality poetry captured and enacted." Steinman exaggerates when she says that Stevens was the poet "who made the most serious use of the new physics," unless we do not understand this seriousness as a claim that he was well informed about quantum physics—his knowledge, as she admits, was mainly derived from mass circulation magazines or letters from friends. But in the late 1940s and the 1950s (i.e., in the early part of the period with which this book is concerned), he certainly was keen to articulate connections between the underlying epistemological purposes of physics and poetry.[73] The argument that poets were straightforwardly envious of the prestige of science in a very utilitarian society does not, however, hold for the situation after the war. As I shall show, many fields of research were now imitating physics because it was widely believed that the use of theoretical models and conceptual frameworks in physics was the key to its success. The use of such models, and even sometimes transposed versions of the structural constituents of matter (the elementary particles, forces, and fields), could enable other research disciplines, especially in the social and human sciences, to achieve similar scientific rigor. Postwar poets as different as Rukeyser, Oppen, Olson, Duncan, and Baraka were aware of such developments. Their complicated relationship to physics included envy but also encompassed a mix of admiration, hostility, and aesthetic appreciation. Sometimes they went further still and respectfully or skeptically borrowed, tested, and experimented with its methodological strategies.

Yet we can also recognize some truth in the idea of a pervasive science envy among poets. David Antin's accusation that Olson and his contemporaries experienced "physics envy" does stick. The confident invocations in Olson's "Projective Verse" of "high-energy" poems and an exciting new method of "composition by field," however much they were in line with many other similar attempts at appropriation, do need to be defended against the charge that this is more rhetoric than method, more envy than reason. Antin makes his charge in an interview with Larry McCaffery, speaking for a generation that often felt consigned to a disregarded mezzanine between the New American poets and the Language Writers. Antin disowns any intellectual debt of his own to Heisenberg's uncertainty principle: "It might have demoralized sociologists in the earlier part of the century because they were still suffering from physics envy . . . but the questions surrounding the discourse of human experience raise completely different issues—issues of the capabilities of language itself, the meaningfulness of notions of agency, intentionality, desire, or need, of their relation to the meaning of causality, the semiotic force of metaphor or narrative and their relation to some notion of truth."[74] As he well knew, Olson, Duncan, and notably Creeley, plus several others among their contemporaries, repeatedly invoked the authority of Heisenberg. Antin's clever post-Freudian insult not only ridicules the desire for borrowed epistemic authority as itself as obsolete as the psychoanalytic simplicity of "penis envy" compared with the more sophisticated model of phallocentrism but also simultaneously manages to cast aspersions on the potency of the thinker who feels the need for the stimulant of physics. In fact, Antin himself was not the originator of this clever insult. Historians of science believe it was the invention of, or at least first used in print by, Joel Cohen, a molecular biologist fed up with hearing about the supposed superiority of the methodology of physics.

"Physics envy" appears in Cohen's 1971 review of Robert Rosen's dauntingly titled *Dynamical System Theory in Biology*. Cohen criticizes the book's interdisciplinary promiscuity and makes the damning pronouncement that "physics-envy is the curse of biology."[75] Only a few years earlier a newspaper story on molecular biology used physics as a term of praise: "the science of biology has reached a new frontier said to be leading to 'a revolution far greater in its potential significance than the atomic and hydrogen bomb.'"[76] Now in the 1970s, molecular biologists were beginning to resist such analogies. Rosen had written that "dynamical analogies can formally unite two apparently totally dissimilar areas, and thus conceptually enrich the whole of biology."[77] Dynamical analogies uniting several apparently dissimilar areas—this might be the defense used by Olson or any number of social scientists, philosophers, and others who used such analogies to drive inquiry in their own fields. Cohen objects less to the use of analogy than to its misuse to create what sounds oddly like a stereotypical view of poetry: "Because the elaboration of language is, for some people, much easier than the labor of establishing concordance or tension between general principles and experience, there is an enormous temptation to spin out language that has the sound and syntax of general principles and to declare it scientifically satisfying without scrupulous regard to its relations to reality."[78]

Cohen's quarrel with Rosen is an old argument that goes back at least as far as the founding of the Royal Society and flares up regularly as disciplinary boundaries and modes of scientific discourse change. Perfect timing made it such a successful put-down. The coinage of "physics envy" occurred at a turning point in the fortunes of high-energy physics, when it stopped being the obvious paradigm in public understanding of all that was scientific, to be replaced by molecular biology, the disciplinary standpoint from which Cohen speaks. Now that physics was losing its potency it was safe to mock those who tried to appropriate its power. Government funding for science was shifting rapidly away from high-energy physics—a shift marked by the cancellation of a high-profile linear accelerator, for instance—to molecular biology, and with that shift came another. Instead of the epistemic metaphors of fundamental particles and the fields that they generate, a new discourse treated the fundamental units of life as a language, and the genes as an archive. This new paradigm was fertile soil for structuralism and other linguistically oriented cultural theories. Physics, we might say, was losing its phallic status, and so the charge of "physics envy" could now seem convincingly derisory. Who wants to envy a loser?

Complaints about physics envy in the social sciences have a history going back to the birth of modernism, as Philip Mirowski shows in his histories of economic theory. Steinman's high-modernist poets were in good company. But Mirowski cautions against dismissing the results of emulating physics just because such imitation may be driven in part by envy. In *More Heat Than Light* he traces criticisms of the "relentless, remorseless, and unremitting" attempts of economists to be scientific: "The neoclassicals opted to become scientific by ignoring what the physicists and the philosophers *preached* and to cut the Gordian knot by directly copying what the physicists *did*."[79] He cites Norbert Wiener's withering judgment on such excessive emulation: "The success of mathematical physics led the social scientist to be jealous of its power without quite understanding the intellectual attitudes that had contributed to the power."[80] Mirowski's conclusions are more nuanced. Given that science is a human activity deeply embedded in our society, "a social process intertwined with all social organization," then recourse to physical metaphors such as energy and field may be inescapable.[81] The challenge, for poets as well as economists, is to avoid drawing on out-of-date physics and employing terms from physics without due cautionary reflection. Commentators critical of science envy need to look behind the envy at the achievements of the supposedly envious borrowers as they construct valuable new theories or new poetics in their own disciplines and arts, theories that may be both rigorous and epistemologically accurate.

Poets were among good company in their envy and emulations of physics, but we do need to acknowledge the negative connotations of envy. The psychology of envy is complex. We readily associate envy with unjustified and self-diminishing greedy desires for what others have. It seems, as Sianne Ngai says, an "ugly feeling." We can discern something of this ugliness in W. H. Auden's memorable reaction to what he perceived as scientists' disdain for poetry: "When I find myself in the company of scientists, I feel like a shabby curate who has strayed by mistake into a drawing room full of dukes."[82] Auden's class-based metaphor jars with the American experience. American poets are more likely to be responding, as Ngai suggests is often the case when envy manifests itself, to a perceived inequality that may then elicit a "polemical mode of engagement with the world."[83] Calling your poems "high-energy constructs" is partly a riposte to those who too readily promote their constructions as high-energy contributions to an America that many politicians liked to describe as a high-energy civilization.[84] Ngai argues that by tracing the work of envy in cinematic representations it is possible to reveal a "critical interrogation" of gender norms, but we can find other possibilities for critical interrogation too.[85] John Rawls, for instance, in his discussion of the problem of envy for his theory of justice, insists that envy need not be rancorous.[86] It may be benign envy for the good fortune of others that does not diminish one's own. The more difficult cases occur where inequality does diminish one's own self-respect or capacity to live fully, and in such cases envy may be a rational, even transformative response. Physics envy may therefore be a symptom of social inequities in the way the status of knowledge in physics is compared with knowledge in other fields, including the arts, and provide a useful tracer for interrogating how knowledge and inquiry are valued—indeed, how they are conceptualized and how they change over time.

MAPPING INTERRELATIONS
BETWEEN POETRY AND SCIENCE

In her biography of Josiah Willard Gibbs, Rukeyser identifies what she calls "the deep error that so many great writers have made in their dealings with scientific thought; the error that is the weakest link in the marvellous chain of which science and imagination form the alloy. It is the error of a rigid analogy, of using the *discoveries* of science instead of the *methods* themselves in dealing with other material" (her italics).[87] This insight that if poets want to keep up with the sciences they must find ways to adopt scientific methods is the basis of her claim in *The Life of Poetry* that "the trap is the use of the discoveries of science instead of the methods of science."[88] George Levine advises anyone trying to understand the history of science and literature to recognize "the variousness and incompleteness of writers' and scientists' interrelations."[89] This is extremely good advice for anyone studying American poetry in the second half of the twentieth century, when some engagements between poetry and science are neither self-evident nor look much like earlier types of poetry-science interrelation and require extensive rethinking of methodology. In *Physics Envy* I concentrate on poetic interrelations that are based on the belief that poetry can still be a mode of inquiry commensurate with scientific methods of research if, as Rukeyser proposes, the poet finds means to engage with scientific methodology. Interesting and important as allusions to the scientific news can be, especially when exploring unforeseen or damaging consequences of a scientific discovery, they work differently from other sorts of poetic allusion. They assume that, in relation to the sciences, poetry plays a secondary role, as if the poet were a columnist responding to the discoveries publicly announced by scientists, adding his or her own reflections on the likely social and aesthetic impact of those findings, whose scientific and epistemological veracity has to be conceded to the sciences.

In the remainder of this book I concentrate on poets who have been interested in going further and engaging with the epistemologies, methods, and modes of inquiry by which the sciences were expanding. In this section, however, I want to look briefly at the whole landscape of late twentieth-century poetry-science interrelations and also point to their earlier twentieth-century modernist antecedents. My sketch of the modernists is intended only as brief backstory to the following detailed accounts of Rukeyser, Olson, and the New American poets and their legacy. Although I too shall be primarily interested in poems that use or resist the methods of science, I think Rukeyser is too hard on what she calls "error." Historically poetry has had many different valid types of encounter with science. Many of these remain active today.[90] Therefore, I shall now offer a short outline of the range of modes of poetic engagement with science that have been most salient for the long modernist tradition.

Here is a partial list of these many poetry-science interrelations. Poems have treated science and its social consequences as a theme for wonder, critique, or didactic transmission. Many of A. R. Ammons's earlier poems combine wonder at science with speculation about its consequences. The Nobel Prize–winning chemist Roald Hoffmann's poems "What We Have Learned about the Pineal" and "Jerry-Built Forever" offer accurate lessons in human biochemistry along-side their speculations about human frailty. Poets have taken advantage of a series of new technologies made possible by scientific advances as they compose and circulate their poems—the use of tape recorders, electric typewriters, and then computers has stimulated many innovations. Poets' bodies and minds have been modified by medicine and new pharmaceuticals: Robert Lowell and Allen Ginsberg shared this propensity, though Lowell used Miltown while Ginsberg tried new and traditional hallucinogens. Poets have protested about the troubling social consequences of applied scientific research, whether because it appeared to turn the universe into a machine, threatened human life with nuclear extinction, or was polluting the planet and destroying species diversity. In *Cold War Poetry*, Edward Brunner gives a detailed account of heterogeneous poetic responses to the nuclear threat in the 1950s, in which he observes that poems about the science or technology of the bomb itself are rare and that instead poets tend to treat the atomic bomb as a regrettable part of the new cultural landscape. Don Gordon's poetry takes this to the extreme of assuming, in Brunner's words, "that the Bomb is the single most important mechanism by which the cold war institutionalized itself throughout every level of society."[91] Gary Snyder's *Turtle Island* is a striking example of this genre not least because it exhorts poets to seize scientific knowledge for new, more utopian ends. As we saw earlier, poetry-science interrelations have been a rallying point for several different aesthetic and literary theories, including the New Criticism and structuralist poetics. Poetry continues to provide an abstract image of a type of thought contrasted both favorably and unfavorably with science—one theme of chapter 2. Later chapters explore in detail how specific poetic practices have been sites of envy for scientific prowess and campaigns of protest against the encroachment of allegedly scientific methods that claim rights to authoritative knowledge into domains where poetry has flourished or politics has been displaced.

During the Cold War and after, the most salient interrelations between poetry and science have been evident in references to the changing scientific picture of human beings and their universe, expressing awe, disgust, concern, critique, or a desire to make new scientific ideas tangible. The great majority of poems in the various anthologies of science and poetry that have appeared since midcentury follow the line set out by Kurt Brown, the editor of one of the best of these, *Verse and Universe* (1998): "A poem's connection to science and mathematics is largely utilitarian: different fields of knowledge are explored for themes, images, metaphors, and language that might be used in the making of new and unique poems."[92] This exploration is playful self-instruction in an existing knowledge leading to the use of fascinating up-to-date metaphors that can help renew poetic language. Brown's poets represent a vast, largely unmapped, stylistically hetero-geneous archive of late twentieth- and twenty-first-century poems that express the curiosity of nonscientists about the consequences for lived experience of the rapid scientization of their society. Mapping that archive is beyond the scope of this study. One example will have to suffice. William Bronk's "The Thinker Left Looking Out the Window" is one of a number of poems in which he reflects on what it is like to live in the universe revealed by modern physics. Here he con-trasts the view from nuclear physics of "his grossness" as an unthinkably large number of what, from his perspective, are necessarily invisible motes with the astronomer's picture of the cosmological distance he would move if he were to leave his desk and go into the yard outside his window, a shift so tiny from the perspective of the whole universe as to appear as infinitesimal as the nuclear par-ticles.[93] Such existential vertigo is acutely described in Thomas Nagel's study of the aporiae of objectivity, *The View from Nowhere*: "How can I, who am thinking about the entire, centerless universe, be anything so specific as this: this measly, gratuitous creature existing in a tiny morsel of space-time, with a definite and by no means universal mental and physical organization?"[94] Bronk's answer is to end his poem with a line that refuses the specificity of exact nouns and verbs for the openness of the small, tethering deictic words "there," "this," and "it": "No, there was only this, and what was there for it?" No more-precise nouns would suffice.

Several modernists, notably Hart Crane and Archibald MacLeish, had also asked whether the new universe was habitable. MacLeish's mock portrait of Einstein, "Einstein" (1929), is an example of such modernist questioning of what it meant to live in a relativistic universe where "Still withstands / The dust his penetration and flings back / Himself to answer him" and deserves to be better known.[95] With subtlety and some unexpected comedy it depicts an Einstein puzzled by the failings of Darwinian biology and of his own relativistic physics to account fully for experience, as he walks around Geneva. As is evident in MacLeish's poem, the fascination with the universalizing claims and technical language of the sciences can tempt a poet to ridicule. Marianne Moore offers mild mockery of claims to precision in poems such as "Four Quartz Crystal Clocks." Postwar poets tend to be more savage. In *Gunslinger*, Edward Dorn exposes the epistemological paradoxes of contemporary physics to skeptical hilarity. Dr. Jean Flamboyant, first encountered laughing as he reads the *Scientific American*, is the author of a PhD dissertation entitled "*The Tensile Strength of Last Winter's Icicles*." "Like the star whose ray / announces the disappearance / of its master by the presence of itself," comments the narrator.[96] As several critics have observed, Dorn literalizes, personifies, and allegorizes metaphors.[97] One source of inspiration for such knowing poetic ridicule of science is 'Pataphysics, that eccentric, intermittent, twentieth-century literary movement with strong affinities to poetry that originated in France and later found support in America and Canada. 'Pataphysics traces its roots to the ideas of Alfred Jarry, who coined the term and argued, as the poet Christian Bök explains, that 'Pataphysics questions the authority of science "by becoming its hyperbolic extreme."[98] The May–June 1960 issue of *Evergreen Review*, which carried a full-page advertisement announcing that *The New American Poetry*, "the first comprehensive anthology of the new poetic voices since World War II," was now in bookstores, was devoted entirely to 'Pataphysics. Its essays argued the case for the magazine's subtitle: '*Pataphysics Is the Only Science*. The magazine printed a translation of Jarry's work in which he calls 'Pataphysics the "science of the particular" that extends "as far beyond metaphysics as the latter extends beyond physics," and insists that it is a challenge to "the laws that govern exceptions."[99] In his introduction to the *Evergreen* issue, Roger Shattuck, one of the editors of the poetry magazine that first published Olson's "Projective Verse," explains the relevance of 'Pataphysics to that cultural moment: "Like the sorcerer's apprentice, we have become victims of our own knowledge—principally of our scientific and technological knowledge. In 'Pataphysics resides our only defense against ourselves."[100] The 'Pataphysical strategy is inherently paradoxical: to be as informed as possible about science and then deform its metaphors and concepts until they altogether lose their power to condition us.

Behind the comedy and ridicule of scientific pretensions is often a suspicion of the relentlessness of scientific novelty. In "A-12" Louis Zukofsky says emphatically, "A poet is not at all surprised by science." The vernacular idiom "not at all surprised" undercuts the literal meaning—poets are well informed about scientific developments and perhaps anticipate them—with an additional shrug. We say we are not at all surprised with a sense of slightly disapproving superiority toward a person who may appear to have surprised everyone and yet in doing so has revealed herself or himself to be just as we thought them to be, and paradoxically therefore not at all surprising.[101] Zukofsky was as justified in making such a comment as any modernist poet, given his scientific competence. Mark Scroggins underlines Zukofsky's "respect for—even envy of—the vocabulary and methods of science."[102] Like MacLeish, Zukofsky had a lasting interest in Einstein, and even more justification for it. His very first book, a 1931 translation of Rudolf Kayser's *Albert Einstein: Ein biographisches Porträt*, may have played a larger part in his influence on poetry-science interrelations than has been appreciated. This biography was published in English (without mentioning Zukofsky's work as translator)[103] under the pseudonym Anton Reiser.[104] Kayser was Einstein's son-in-law and presumably wanted to avoid the impression that he was Einstein's public-relations agent, and moreover, Kayser was still living in an increasingly anti-Semitic Germany at the time, an additional reason for not using his name. As a novice translator, Zukofsky does reasonably well, displaying a wide vocabulary and a care for precision, although his style can be stiff and occasionally awkward: "To make theory the object of our reflection evolve physical concepts, so that events and our experiences of them are connected in intelligent order, is a task of such beauty and magnitude that the significance and force of Einstein's theoretical physics can easily be understood from this standpoint as well."[105] Some of this stiffness was Kayser's. Zukofsky complained about the author's poor style in a letter to Pound, yet was evidently impressed enough with the portrait of Einstein to find inspiration for passages of "A."[106]

Einstein's "entire working procedure is surprisingly analogous to that of the artist," writes Kayser.[107] This idea that the physicist is akin to the artist, which runs through the whole biography, helped shape the legend of Einstein the genius and rubbed off on other physicists, notably Oppenheimer. More worrying for poets was the thought that perhaps physics was now the art of their time and that other visual, sonic, or verbal arts could no longer expect to express the epoch.

Zukofsky, of course, did believe that poetry was still possible, but he was certainly very aware of this conjunction of physicist and artist. In "A-6" he uses an anecdote from Kayser's book in which Einstein responds irritably to a questionnaire about his interest in Bach with the words: "In reference to Bach's life and work: listen, play, love, revere, and—keep your mouth shut!"[108] Zukofsky writes:

> Asked Albert who introduced relativity—
> "And what is the formula for success?"
> "X = Work, y = play, Z = keep your mouth shut."
> "What about Johann Sebastian? The same formula."[109]

Kayser conflates Einstein the theorist with Einstein the moralist: not only is he fundamentally a humble man (an idea repeated several times), but his ideas are a deductive "simplification" of existing laws of nature. Einstein's art lay in his ability to construct astonishing new theoretical models of the universe, especially a new concept of the field. Even as a boy, Einstein "experienced great aesthetic pleasure at the thought that the world could be built up by conceptions—products of human understanding." After his success at creating the "constructive systems of thought" that resulted in the theory of general relativity, he concluded, according to Kayser, that "the greatest task of the physicist is the search for those general elementary laws, out of which by pure deduction his picture of the world is formed." At the heart of Einstein's artful construction of concepts is a reconfigured idea of the field: "the concept of the 'field' is to him the most profound idea which theoretical physics has developed since the classical age of Newton." His theory expands the concept of the electromagnetic field into "a natural representation of the physical properties of space as framed by a field theory [which] should logically be the result of a fusion of the Riemannian metric and the postulate of the convergence of parallel lines in infinity."

In the same period that he was translating Kayser's *Albert Einstein*, Zukofsky wrote a long review of Charles Reznikoff's entire body of work for *Poetry* magazine. Similarities between the way Einstein's work is described and Zukofsky's conception of objectification suggest that objectivism had roots in physics as well as philosophy. In this essay now simply known as "Sincerity and Objectification," he first broached the core ideas of objectivism, in language that has close affinities to the language of Kayser's biography. For example, after Kayser cites a passage about the "restful lines which seem created for infinity" to be seen in a mountain landscape, a passage in which Einstein paraphrases Schopenhauer, the biographer explains that this desire to escape the ego is paramount in Einstein's life: "His work not only frees him of his ego, inasmuch as it carries over all energies and experiences into an objective world, but also simplifies and clarifies the picture of this world. This too means the fulfilment of a human desire, since only a simplified world of forms discloses a spiritual life of rest and harmony."[110] Zukofsky retains much of the metaphorical, as well as conceptual, content when he writes: "In sincerity . . . writing occurs which is the detail, not mirage, of seeing, of thinking with the things as they exist, and of directing them along a line of melody. . . . Presented with sincerity, the mind even tends to supply, in further suggestion which does not attain rested totality, the totality not always found in sincerity and necessary only for perfect rest, complete appreciation. This rested totality may be called objectification—the apprehension satisfied completely as to the appearance of the art form as an object."[111] Objectification is a key concept in Schopenhauer's *The World as Will and Idea*, but these affinities between the language of Zukofsky's translation of the account of Einstein's overcoming of subjective egoism in his work and the theory of objectivism suggest that it is Kayser's somewhat-casual use of Schopenhauer's ideas to describe Einstein's special objectivity that captured Zukofsky's attention. Borrowing Kayser's formulation enabled Zukofsky to hint that the objectivist poem somehow balanced a constructive system of thought with detailed observation in the manner of the new physics.

Zukofsky himself was as cautious about conceptual schemes as his insistence on detail would suggest. He took to heart the idea that for the thinker of this time "his difficulty is now simple," the idea that the process of creation involves the transformation of the difficult complexity of observation into aesthetically satisfying simplicity of conceptual form. In his brilliant poem about the problem of understanding quantum physics as a layperson which starts "It's hard to see but think of a sea" (*Anew* 12, 1946), he depicts the writing of a poem in terms of a similarly painful but, in the poem's words, "desired path" toward a solution.[112] He asks tacitly what quantum physics does to subjectivity, to the perspectival character of individual experience, and to the reliance of cognitive certainty on the senses. This is a question that Daniel Tiffany addresses in his study of modern poetry as a modeling of representation, where he argues that quantum realities triggered a "crisis of the equation of materialism and realism" because "as long as quantum mechanics failed to provide pictures of an invisible material world, it failed to constitute a new reality." Tiffany argues that the poetry most aligned to the new physical realities was "a symbolic mode in which literal images of ordinary experience betray, but do not represent, an invisible world of impossible bodies and events."[113]

Einstein was not the only physicist celebrated by the modernist poets. Wallace Stevens insisted in the early 1950s that the great scientist of the century was Max Planck, rather than Einstein, revealing a significant difference between his relation to physics and that of Zukofsky. By this time the quantum physics that Planck inaugurated had become as prominent or more so than Einstein's general relativity. Stevens would have understood why Zukofsky might want to translate a biography of Einstein, and why MacLeish tried to write a poem about such an abstract figure, but his candidate for top physicist was the originator of quantum theory because, as Tiffany says, quantum physics makes much greater demands on the imagination than relativity theory. Physicists construct speculative analogies, images, and conceptual frameworks in order to understand the empirical and nomological features of this invisible yet material world. Stevens treats such acts of scientific conceptualization as extrapolations of poetic imagination and therefore comparable to poetry: "It is as if in a study of modern man we predicated the greatness of poetry as the final measure of his stature."[114] This symbolic scientist implicitly recognizes the need that physics continues to have for poetry, a need that Stevens's friend Jean Paulhan finds painfully evident in the deficit of words and images for the understanding and communication of "quantic phenomena."[115] Paulhan cites the very architect of wave mechanics, Louis-Victor de Broglie, who commented that quantum physics was stalled for lack of cognitively fertile representations. Stevens is drawn to quantum physics for those very reasons that Einstein mistrusted it—that is, because it requires a poetic imagining of contradictions and alien phenomena. Stevens's supreme fictions are simulations of scientific realism, whose reliance on mathematics, hypotheses, theories, and models of supersensible realities places it, in the eyes of many poets, into the category of new and not-always-persuasive mythmaking, replacing the pantheon of gods with a zoo of subatomic particles. Stevens's practice resembles the critique of pure scientific realism offered by philosophers of science such as the empirical realist Bas van Fraassen: "According to the realist, when someone proposes a theory he is asserting it to be true. But according to the anti-realist, the proposer does not assert the theory: *he displays it*, and claims certain virtues for it" (his italics).[116] Stevens displays his fictions in poetic arguments in which he considers their virtues and failings.

Part II of *Physics Envy* primarily concentrates on the poetics of such fictions, models, and conceptual schemes. One further type of modernist engagement with the sciences also influenced the poets I discuss, especially in part III: the idea of the poem as itself a site of experimentation. Ezra Pound, Gertrude Stein, and other poets that I cannot discuss here, such as William Carlos Williams, treated the poet as experimenter testing out an immense variety of poetic structures, modes of allusion, conceptual frameworks, prosodies, and syntactical and semantic arrangements. For some postwar modernist poets, experimentation would become the most trusted manifestation of poetic inquiry, and they would regard their predecessors as having taught them to recognize the value and scope of possible experimentation, a legacy that was far more important than specific observations about social credit, urban life, or new technologies.

The importance of experimentation for modernist poetry was one of the touchstones in Seldon Rodman's widely circulated anthology *One Hundred Modern Poems* (1949), a pedagogically conceived genealogy of international modernism where readers could encounter Rilke, Marinetti, Blok, Valéry, Neruda, Mayakovsky, Brecht, Lorca, and others, alongside Americans such as Stein, Williams, Stevens, and Pound. Although he overrepresents poetry that satirizes modernity and at times chooses what we would now consider marginal work (Wallace Stevens is represented by "The Pleasures of Merely Circulating"), the strength of the collection depends on its perceptive choice of representative poets and its thoughtful preface that foregrounds science as one of the key forces at work. Rodman looks back to a time when "the poet was expected to be (and often was) as learned in science, religion, psychology, politics and morals as he was in metrics" and evaluates modern poets accordingly, advising poets to heed Robert Graves's warning that they need to pay more attention to dangerously "capricious experiments in philosophy, science and industry."[117] Above all, the poets need to master experiment for themselves by heeding Rimbaud's praise for Baudelaire: "Unknown discoveries demand new literary forms."[118]

Did the unknowns welcomed into the new physics demand new poetic forms? A midcentury poet reading this injunction could readily convert it to a timely conditional: if you want to discover what is currently unknown, you must experiment; you will need literary forms that body forth that unknown. And for such poets, Ezra Pound and Gertrude Stein were exemplary figures. Stein made linguistic experimentation a defining characteristic of her writing. As Steven Meyer says of her work: "Writing, ordinarily treated in science as a means whereby experiments are reported and analyzed, itself becomes the medium for experimentation."[119] Unlike Pound, she was wary of calling her writing "scientific," because for her science was represented by the positivism of the nineteenth-century methodology of the physicist Ernst Mach.[120] She did, however, believe the writer should recognize the new scientifically conceived universe and try to work within it: "So far then the progress of my conceptions was the natural progress entirely in accordance with my epoch."[121] Pound was much more willing to say openly that modern poetry should look to the new sciences, asserting as early as 1913 that "the arts, literature, poesy, are a science, just as chemistry is a science."[122] Yet although he talked a fair amount about science and began his *ABC of Reading* (1934) dramatically with the assertion "We live in an age of science and abundance," signs of this interest in contemporary science are scarce among the abundant allusions to world history in his poetry. His advice to readers of his book to emulate the comparative "method of contemporary biologists" and treat his examples as "measuring rods and voltmeters" pointed later writers to his main contribution to understanding how poetry could engage with science: by adoption of an experimental approach to form.[123]

Pound's archive of poetic experimentation in the *Cantos* simulates the accumulation of modern knowledge as an open-ended set of poetic experiments resulting in statements, findings, reports, speculations, and allusions to other authorities. What interests Pound almost as much as the material he finds is a problem that also haunts the modern scientist, which is how to authorize the knowledge an inquiry generates. Pound has two answers: either present an idea as affirmed by a significant individual whose standing underwrites the idea's legitimacy (the humanities vision of research) or intimate its position on a spectrum of certainty to uncertainty (the scientific method of communicating results). This latter strategy is also a response to assumptions that poetic statements are pseudo-statements. Pound marks out the assertoric force of his propositions by carefully muting the full propositional force of his poetic statements through a variety of means—often by omitting a verb—so that the force of the reliability ascribed to what is being said will depend on a whole network of inferred contexts. He showed later poets how to find a method similar to that nuanced rhetoric used by scientists to measure the degree of certainty of their findings. Scientific publications evolved what by today has become a highly codified rhetoric for the precise calibration of certainty and uncertainty in assertions about research results. Ronald Bush argues that in "Pound's late poetics, poetic vision and scientific investigation become one and the same. . . . [P]oetry in this formulation is nothing but observation founded in the rigor—the care and sharpness—that guides the scientist as he shapes data into hypothesis."[124] Many postwar poets would take up Pound's strategy of creating fractal propositions and extend such experimentation in different directions, ranging from the use of isolated words in an open-field poem, to the use of chance procedures to eliminate the warrant of propositional intention, and to the decontextualized sentence of Language Writing. As I shall show later in the book, latent in much of this experimentation is the tacit principle that poetic manipulation of propositional affirmation is a form of experimental inquiry commensurate with the researches of the natural and social sciences.

My aim in this book is to show that key developments in late twentieth-century American poetry can be understood as responses to the cultural impact of changes in the epistemic authority of the natural sciences. The next step is to look more closely at the part that poetry plays in the rhetoric of the scientists themselves as they go about their business in the two decades after the war. Physicists of the period were very conscious of the tensions around their attempts to communicate the implications of their experiments with invisible, intangible entities that were best described by mathematical formalizations. They wanted to defend their priority as architects of scientific knowledge and distinguish their imagined objects from those of mere speculators and fantasists. Poetry proved a useful paradigm for explaining the protocols of their linguistic communications. I therefore give considerable space to the actual words of the scientists as they venture their phenomenological and theoretical accounts of the quantum world. Knowledge of physics was rapidly altering, usually in the direction of more fine-grained phenomenological knowledge accompanied by greater explanatory and theoretical uncertainty. Physicists' relations with a public who after more than a decade of nuclear fear had begun to look elsewhere to other sciences for paradigmatic methods of inquiry also began to alter. At the time of physics' ascendancy, however, ideas about the role and limits of poetry that circulated in the scientific world were hard facts that poets themselves had to reckon with, especially if they wanted to make any claims about their own right of inquiry into the possibilities of both Rukeyser's new metal and those other new metals that were playing an increasing role in every aspect of American culture.

WHAT THE PHYSICIST
SAID TO THE POET

*How Physicists Used the Ideal
of Poetry to Talk about Uncertainty*

THE STRUGGLE FOR DIALOGUE

"Come out and talk to me," shouts the "poet to physicist in his laboratory" in David Ignatow's well-known poem of that title. Why might a poet want a physicist to come out from behind the institutional barriers and talk openly? Alice Fulton speaks for many poets when she confidently asserts that the quantum mechanics developed by Niels Bohr, Werner Heisenberg, and others is essential reading for poets, and "a truly engaged and contemporary poetry must reflect this knowledge."[1] But why this sense of urgency, of needing to get the physicist *out* of the shelter of the laboratory *now*? The allegorical encounter that Ignatow imagines is a sharp reminder that the actual world of the laboratory practice of physics is largely inaccessible to outsiders, however well informed they try to be. This inaccessibility matters because, coupled with the discipline's epistemic ascendancy, physics appears to be an esoteric knowledge that devalues the poet's own knowledge.

Ignatow spells out the danger:

I give up on thought
as I see your mind
leading into a mystery
deepening about you.

Because the enigmas of the new physics can overwhelm outsiders to the point where they give up on their own approach to knowledge, Ignatow's poem asks what it is that justifies this obscurity. What is the physicist "trying to discover / beyond the zone of habit / and enforced convention" in the "patterns / of research"? Discovery, the poem continues, is also the aspiration of poetry, though the poet tries to avoid being bound, as the physicist is, "to laws within your knowledge." Finally, the poem turns back to the physicist with a request that the scientist "will find yourself patient / when you are questioned."[2] "Find yourself patient" can be read slightly against the grain to suggest "When you yourself are a patient," as if the physicist's physics is the sign of a pathology. If, in addition to a call for physicists to make a greater effort to communicate their findings to the public, we also hear both this hint that physics is a pathology and a threat of interrogation in the final words of the poem—"when you are questioned"—we are probably right to do so. Poets of the postwar generation found the conceptual depths of the new bomb-making physics pathological, threatening, even dangerous, and yet unignorably significant.

In this chapter I shall show that some physicists did think they were making an effort to step outside the closed linguistic community of the laboratory to talk to poets and that this talk created cultural expectations about the interrelations between poetry and science that poets had to negotiate. The scientific idea of what constituted poetry could be a barrier or opportunity for poets, but it could not be ignored by poets who wanted to engage with the sciences. In the simplest terms, alluding to poetry had become one means for scientists, especially but not exclusively physicists, to manage the boundaries of inquiry. They (and other natural scientists too) used the abstract idea of poetry for rhetorical purposes, they read and quoted canonical poems and some even mentioned the avant-garde, they wrote doggerel of their own, they taught "physics for poets" courses, and they read debates about poetry in their professional journals. Sometimes the journals even ventured to encourage scientists to come out of the laboratory and talk to poets. General-science journals, notably *Science*, a journal aimed at both professional scientists and educated nonscientists, intermittently published thoughtful articles about poetry and science. The cumulative effect of all this was to create a strong expectation of what poetry was and was not, an expectation that deeply influenced poets who were receptive to science. Although individual poets only rarely directly encountered physicists' allusions to poetry in specialist journals, they were familiar with similar examples written by the physicist popularizers. Moreover, the poets did chat informally in person with physicists—who sometimes lived on the same street—and the poets did register acutely how scientific issues emerged into public discourse.

There is sufficient evidence of such activity to make George Rousseau's hypothesis that influence is a two-way street sound particularly applicable to this period:

> There is no reason to disbelieve on logical or epistemological grounds that literature and science affect each other reciprocally. That is, that each influences the other in just about the same degree, although conceivably in different ways. It is also probably valid to assume, although it would be practically impossible to prove, that science shapes literature to the same degree that imaginative literature shapes science.[3]

Yet reciprocity is not one's first impression. Instead, what one finds within science discourse is a constant low hum of dismissal of the relevance of poetry for the sciences and modern America.

Muriel Rukeyser and Charles Olson may have wanted poets to be scientifically literate and take advantage of scientific methods of inquiry, but they knew they could not expect scientists in the second half of the twentieth century to show openly a reciprocal interest in the potentials of poetry for scientific research. Although Rukeyser's biography of the physicist and chemist Josiah Willard Gibbs elicited considerable praise from literary readers, scientists were wary of it. "Reading the judgments of Willard Gibbs by scientific and academic reviewers makes one marvel all the more at Rukeyser's sheer courage in undertaking the book," writes her literary biographer.[4] An elderly mathematician, who had known Gibbs, irritably demurred: "The author is a literary woman rather than a historian or scientist; she states as facts a great many things she cannot possibly know, such as what someone felt or thought on a given occasion, even though there be no record to indicate it."[5] Charles Kraus, reviewing the biography in the *Journal of the American Chemical Society*, also feels the book is "replete with inaccuracies" that no scientist would make.[6] The scientists particularly dislike her poetic style. Kraus thinks it "borders on the cryptic"; another reviewer advises that "readers who are accustomed to technical accuracy and complete sentences will not enjoy Miss Rukeyser's repetitious style, fragmentary sentences, and round-robin chapters."[7] Who are these readers expecting accuracy and only comfortable with complete sentences? Certainly not poets.

Olson claimed personal experience of the physicists' suspicion of approaches from poets. In "The Gate and the Center" in which he observed that American physicists increasingly worked in large teams, Olson wryly describes the reaction when he suggested that a poet might be a useful addition:

> It is one of the last acts of liberation that science has to offer, that is, modern science stemming from the Arabs, that all the real boys, today, are spending their time no longer alone but in teams, because they have found out that the problem now is not what things are so much as it is what happens BETWEEN things, in other words:
>
> COMMUNICATION (why we are at ripe, live center)—and the joker? that from stockpile Szilard on down, what the hot lads are after (under him at Chicago, Merritt at Columbia, Theodore Vann at the Univ. of Paris, and at the Princeton Institute) is, what is it in the *human* organism, what is the wave (is it H-mu) that makes communication possible! It kills me. And I made one physicist run, when I sd, quite quietly, the only thing wrong with yr teams is, you have left out the one professional who has been busy abt this problem all the time the rest of you and yr predecessors have been fingering that powerful solid, but useless when abstraction, Nature.[8]

The poet may be a masculine enough "professional," one of the boys, yet the physicist flees the suggestion that poets might contribute to physics. Michael Davidson notes this tendency in Olson's writing, even in "Projective Verse," where his exhortations take the homosocial form of "go by it, boys" and "there it is, brothers, sitting there for USE."[9] Olson's use of "professional" neatly hints at what he finds hard to justify—that the poet may be a poet-researcher, an intellectual, a writer committed to ideals of inquiry that underwrite a profession.

Instances where postwar scientists and their advocates deplore the irrational, emotive, or illogical thinking of poets of all kinds are easy to find. These and other instances of scientific resistance to their attempts at dialogue with scientists would appear to show that Rukeyser's and Olson's ambitions to create a joint laboratory space where both scientists and poets could work alongside each other were futile. Certainly, poets would have encountered many dismissals of poetry by scientists and philosophers of the sort that George Lundberg made: poets, he suggests, are like children who smash their toys when their desires are thwarted, and they use their adult powers only to rebuild them "nearer to the heart's desire."[10] Scientists also knew that any hint in their research publications that they had been influenced by fiction or poetry might damage their reputation. Only a few extremely confident scientists have been able to do this. One exception to this rule will help us see why this was the case. *exception*

We can see how the seemingly merely illustrative use of poetry can become integral to the theorizing of physics in one of the most celebrated modern papers in physics, Murray Gell-Mann's "A Schematic Model of Baryons and Mesons," published in *Physics Letters* in early 1964. Gell-Mann could make an allusion to the poetic prose of James Joyce's *Finnegan's Wake* in his famous paper not only because he was exceptionally well read in modern literature and had the stature as a physicist to be able to present his work as he wished but because this allusion was integral to his strategy for introducing his hypothetical elementary particle the "quark." Physicists were finding it increasingly difficult to bring conceptual order to the rapidly growing list of supposedly fundamental particles, which were, in Olson's words, "isolated, observed, picked over, measured, raised / as though a word, an accuracy were a pincer."[11] The sorting into the small electron-like particles, or leptons (the term is simply the Greek for the smallest coin), and the larger hadrons (from the Greek for "stout" and "strong"), which were subject to strong, interactive forces much greater than gravity, had still not managed to yield a convincing pattern. The eventual solution proposed by Gell-Mann (and Yuval Ne'eman) was that all these fundamental particles are actually not quite so fundamental after all but are made up of a small number (initially three) of even tinier, more basic bits of stuff, called quarks. These sub-subatomic particles brought order at the cost of raising even more strongly than ever the question of just how phenomenological they were.

Gell-Mann's paper is one of the most striking instances of the use of poetry to mark this question. Indeed, so cleverly did Gell-Mann employ the rhetorical device of locating poetry at the point of a communication knot that "quark" was soon universally adopted as the name for a new, more fundamental type of particle. He chose the name from one of the most unclassifiable literary texts of the twentieth century, *Finnegan's Wake*, a text that exists at the crossroads of poetry, fiction, and prose. Gell-Mann's paper has eight endnotes, most of them to the *Physical Review* and *Physical Review Letters*. Note 6, however, notoriously reads: "James Joyce, *Finnegan's Wake* (Viking Press, New York, 1939) p.383."[12] If we look up the reference, we find a song beginning with the line about the three quarks:

— Three quarks for Muster Mark!
Sure he hasn't got much of a bark
And sure any he has it's all beside the mark.
But O, Wreneagle Almighty, wouldn't un be a sky of a lark
To see that old buzzard whooping about for uns shirt in the dark
And he hunting round for uns speckled trousers around by Palmer-stown Park?
Hohohoho, moulty Mark!
You're the rummest old rooster ever flopped out of a Noah's ark
And you think you're cock of the wark.

Apart from the appropriateness of the reference to "three" to name the "unitary triplets," as Gell-Mann calls them, and the hint of puns on quirks, quarts, and corks, there is no particularly obvious crux to the source.[13] Perhaps the physicist himself, faced with the confusions of the subatomic world, identifies with the unhappy King Mark ridiculed in the song, or at least with the image of Mark receiving three bird calls, three quarks, and flying off in all directions, beset with cosmological confusions brought about by the Creator. There may well be such a mischievous hint of such personal identification, but the main reason for alluding to the poetry of Joyce's text is to negotiate a problem of communication.

Gell-Mann's endnote is attached to a passage halfway through the paper, a passage in which mathematical entities that have up to this point been referred to as "members" of a symmetry group (an enclosed mathematical system of hypothetical entities that need not have any correspondence to actual entities) are abruptly named quarks and so cross over from the imaginary world of Joyce into the material universe. "Quark" signals the moment when sign and referent begin to link up. What happens almost subliminally in the quark paper is that quarks are distinguished from particles. Particles are reasonably real, observable, known entities, whereas the quarks are tenuous, speculative signs that have only a wisp of referentiality clinging to them. In a controversial account of Gell-Mann's research and its consequences, provocatively entitled *Constructing Quarks*, the historian of science Andrew Pickering argues that "throughout the 1960s and into the 1970s, papers, reviews and books on quarks were replete with caveats concerning their reality."[14] In my view the allusion to *Finnegan's Wake* helps Gell-Mann avoid unwanted metaphysical speculation arising from these warnings. The historian of physics Helge Kragh reports that Gell-Mann insisted on calling quarks "'real,' as opposed to 'mathematical,' not to deny that quarks existed but to avoid philosophical discussions concerning the meaning of existence of permanently confined objects."[15] Gell-Mann himself concludes his famous article by saying disarmingly that "it is fun to speculate about the way quarks would behave if they were physical particles of finite mass," a remark which further displays the sheer speculative fizz of physics at this time.[16] *Fun* to speculate? *If* they were physical particles? Quarks were weird—somewhat like the black holes that dominated the astronomy of the decade after their announcement in December 1963 at a Dallas hotel a few blocks from the site of the Kennedy assassination— not material, not energy, and thanks to complementarity, not wholly bound by the house rules of space-time.

Although Lundberg's attitude to poetry was widely shared, Gell-Mann's use of *Finnegan's Wake* in a specialist article points us to another contrasting use of poetry in physics. Rukeyser herself, for instance, was invited a few years after the appearance of the biography and its mixed reviews to write about Gibbs for an audience of physicists in *Physics Today*. The journal is happy to introduce her profile "Josiah Willard Gibbs" as having originated in a poem: "First coming upon Gibbs through the writing of a poem, Muriel Rukeyser was drawn into a study of his work and later wrote his biography."[17] And the editors apparently didn't flinch at her opening claim that Gibbs displayed "an imagination which for me, more than that of any other figure in American thought, any poet, or political, or religious figure, stands for imagination at its essential points," nor when she went still further and insisted that because they are both "workers in systems," scientists and poets will gradually converge as "the method of science approaches the method of poetry."[18] Even if they did squirm a little, the editors at *Physics Today* were still willing to give space to such a claim alongside reports of physics research. Physicists were interested in what poets had to say.

Rukeyser's essay and Gell-Mann's playful allusion to Joyce are instances of a phenomenon that has not been studied. Scientists made references to "poetry" when they needed to illustrate knots of difficulty in their attempts to communicate the character of esoteric mathematical or experimental data. In doing so, they were usually talking not about specific poems or poets but about a concept as abstract as "government," "family," or "machine," a concept whose usefulness in their rhetorical portfolio depended on its accepted social currency, which their usage helped reinforce at the same time that such usage inflected the normative idea of poetry with their own concerns. At times "poetry" was not much more than simple metaphor, but to call it metaphor risks diminishing its significance, which depended on the dense conceptual infrastructure of this stereotype. This complexity, however, remained largely out of sight, and for the most part the scientists did not bother to analyze what they meant by "poetry." Its usefulness largely depended on not examining its use too closely, and not considering the actual phenomenon of poetry, whose diversity could quickly undermine the viability of this useful stereotype.

Poetry as an idealized abstraction provided rhetorical resources of several kinds. Talking about poetry could, for instance, be a means to illustrate elusive problems concerning the questionable reality of representations of nuclear phenomena without having to expand riskily on the ontological and epistemological issues underlying these representations. Poetry often performed within a negative theology of science, acting as a placeholder for the very antithesis of good science: archaic, emotional, undisciplined, and not answerable to the factual nature of the world. Poetry was the ugly twin who makes science look good. Or it could stand for a form of wild inquiry that might anticipate later scientific inquiry by indicating paths that science could follow. It could even point back to a golden age of unified thought, to earlier forms of wisdom now largely superseded, and thereby act as a reminder of the humanity that was assumed ought to orientate scientific research. Occasionally, poetry stood in for ambivalence about the new secular scientific metaphysics of the cosmos.

Physicists were the main users of the poetry abstraction. What united their illustrative usages of poetry was a need to reflect on difficulties of communication. The idealized image of the poet could provide an image of science's other, the nonscientist whom the scientist needed to enlighten and persuade and who appeared frustratingly uncomprehending. Poetry as a fine art of language could stand for the heights of expression apparently demanded by the impossible stresses put on language by quantum strangeness. As the zoo of particles became ever larger, poetry could illustrate the need for new paths through the troubled borders between theory and experiment. For researchers, poetry's illustrative role was therefore not to picture the weirdness of the invisible quantum world but to point up the effects of this weirdness on discourse. Allusions to poetry enabled physicists to represent the difficulties of talking about the status of their research speculations, about their reality and their conceptual provisionality, or what they tantalizingly called the "semiphenomenological" character of some particle physics. Allusions to the abstraction of poetry offered momentary rhetorical assistance as physicists awkwardly maneuvered their conceptual schemes into closer contact with the elusive irregularity of actual phenomena.

PHYSICISTS READING AND WRITING POETRY

Scientists did read poetry. In 1948 a team of journalists at *Fortune* magazine used extensive interviews to offer a collective biography of the contemporary American scientist. They found unequivocally that "physicist-mathematicians occupy the top cultural group," which meant not only were they the highest-paid group of scientists, but they also "read voraciously" and "tend to be the most roundly literate and politically conscious."[19] Other evidence suggests that this roundedness sometimes extended to poetry. Robert Oppenheimer's biographer reports that he liked reading John Donne, while Murray Gell-Mann took the title of his autobiography, *The Quark and the Jaguar*, from a poem by Arthur Sze. At the Smithsonian Astrophysical Observatory during the winding-up discussion for the third symposium on gas dynamics in astrophysics in June 1957, Leo Spitzer reviews the achievements of three earlier symposia on the same topic and jokes, "When I look at these three dates, I am reminded of the poem by Browning on the philosopher Cleon, who, as you remember, had written three books on the soul, 'proving absurd all written hitherto and putting us to ignorance again.'"[20] The actual allusion matters little. Its purpose is to amuse the audience and associate their work with a cultural capital already respected by physicists. They would like to think that they too are sometimes reminded of canonical poems when considering the wider implications of their research.

Physicists also wrote poetry. Around the edges of some science journals doggerel flourished. Here is "Practitioner's Lament," an amusing example from Martha J. Schwartzmann and Martin D. Turner at General Electric, which was published in the Phimsy column of *Physics Today*. These are the first two and the last stanzas:

> They've transformed cps and mho.
> Nothing stays the same.
> My head now swims with synonyms
> And quantities renamed.
>
> As farad is to daraf,
> So once was ohm to mho.
> Sieman came to take mho's place.
> Now mho is mho no mo'.
>
> . . . .

I read through every journal.
I dare not sleep nor eat
For fear the latest jargon
Will make me obsolete.[21]

The poem asks why the language of physics is changing so fast. Is it that scientists want memorials: "To honor all the scientists / We make the language grow"? Whatever the reason, and the poem does not try seriously to find it, poetry is felt to be an appropriate medium for reflecting on the uncertainty as to whether terminological instability is nominalism run wild or linguistic barnacles accreted over time. This poem is typical of the poetry written by scientists: witty, crudely metrical, signaling its lack of seriousness, while also tacitly suggesting that their knowledge and professional lives can be the subject of poetic art.

It was not all physics whimsy. *Science*, *Physics Today*, and other journals that were aimed at a range of professionals (though strikingly not the most widely read of all, *Scientific American*) intermittently published essays that called the scientists' attention to the value of poets as cultural brokers for science. In the early 1930s, *Scientific Monthly*, which was published, like *Science*, by the American Association for the Advancement of Science but offered science journalism rather than specialist papers, presented a series of articles on science and poetry. These covered such topics as color and insects in poetry, followed later in the decade by pieces entitled "Poetry of the Rocks," "Poetry and Astronomy," and (in *Science*) "Mathematicians, and Poetry and Drama." Others on entomology, geology, and astronomy were less original, following a well-worn track that continues today: the writer lists instances where poets refer perceptively to natural phenomena that later become the subject matter of scientific inquiry, while also drawing attention to poetic misapprehensions about the scientific principles behind these phenomena. In her essay on insects in poetry, Pearl Faulkner Eddy cites Walter De La Mare's poem "The Fly," which begins, "How large unto the tiny fly / Must little things appear," and then comments: "the scientist suggests that a more accurate picture of the actual appearance of things to a fly might be given in a mosaic, their many tiny ocelli overcasting all things with the diagrammatic outlines of a mosaic into which pictures are fitted."[22] The scientist corrects the poet. Her terms of praise for poetry—"charming," "fascinating," and "beauty" are repeated throughout—implicitly contrast with the rational objectivity of the sciences, though she does not try to address the implications of this tension between the aesthetic and the scientific.

Implicit in Eddy's comment on De La Mare's poem is the idea that he would have been a better poet if he had paid more attention to the discoveries of the scientists, to those ocelli and compound eyes. Frederick W. Grover, an electrical engineer, makes the point explicitly in another article in *Scientific Monthly*, a point that would be repeated by other scientific commentators over the following decades. He had been collecting poetic references to astronomy for some years and had found "many striking and beautiful passages," yet he believes that something important is missing. Where are the new discoveries of the sciences? He cannot find a poet who has tried "to picture the whirl of the spectroscopic binaries, and the rhythmic oscillations of the Cepheid variables, to describe the individualities of the dwarf and giant stars, or to soar in imagination to the confines of an expanding universe."[23] Where, one might ask from the scientist's perspective, have those poets been soaring, or have their poetic imaginations lacked the escape velocity to follow the scientific imaginations that have made the latest scientific discoveries? This was a criticism that would often be repeated in the following decades.

These essays of the 1930s and early 1940s were not much more than field notes cataloging random sightings of science in poems, whereas essays on science and poetry published in the 1950s were, on the whole, more sophisticated, more interested in probing the underlying methodological, cultural, and epistemological differences between science and poetry. The journal *Science* was in the forefront of this discussion and was quite willing to give a platform to a writer claiming that poetry itself could be scientific. The April 1953 issue published a forum on communicating science to the nonspecialist in which most of the articles were written from the standpoint of the scientist, who needs this sort of advice: "By taking poetic license it is often possible to describe a difficult concept 'as if' it were something simpler, but of the same order of function."[24]

One science writer, an editor at *Harper's* with a literary training, Eric Larrabee, in an article entitled "Science, Poetry, and Politics," wants more than an extension of the license. He calls for a new model of research that would bring scientists, humanities researchers, and poets together in an expanded vision of science. Scientists think of poetry and what it represents as "an unrelated avenue of experience" just because it handles intuitions. Larrabee is particularly aware of the expansion of the social sciences and wants other new sciences to learn from this success to travel those routes of experience previously mapped only by poetry: "If sociology is part of the scientific continuum, if only in principle, then so also must be the humanistic studies of behavior, which draw on poetry and politics, among other resources, for their factual evidence." Science can be poetic, and poetry can be scientific: "The materials on which the poetic intuition works are no less factual because they are not statistically handled, nor is the intuitive process less accurate because it is rapid and deals with probabilities."[25] The importance of this article lies less in the effectiveness of its arguments than their location: Larrabee is telling a readership of natural scientists that they should do what Olson advocated and bring poets onto their teams. He is saying this in a journal devoted to the natural sciences, whose previous issue was dedicated largely to technical articles in chemistry and whose next issue would be dedicated largely to medical articles on such topics as corticosteroids, penicillin, and hematology, all of them requiring an advanced readerly expertise not usually thought of as compatible with expertise in poetry.

In the 1950s one scientist did come all the way out of the physics laboratory not only to talk to poets about contemporary poetry and science but to encourage other scientists to do the same. June Zimmerman Fullmer was researching radioactivity in metals when in June 1954 she published in *Science* the first of a series of three articles on poetry and science. In "Contemporary Science and the Poets" not only does she show herself to be well informed about modern poetry, but she also displays an impressively wide knowledge of the books and articles that had been written on the subject of science and poetry over the previous decade. She is aware of the illustrative potentials of poetry for scientists struggling to explain the knots in their communications to nonspecialists but interested in the same kind of question that George Rousseau would later ask. She wants to know "who are the poets that might be examined by a scientist for their scientific content?" But she is quick to add that she is not looking merely for the scientific furnishings provided by electrons and test tubes; instead, "what is meant is the appearance in poems of the newer theoretical concepts and broad points of view; what is meant, too, is the scientific spirit, the scientific attitude." She undertakes this task because, like Rousseau, she believes that "poetic usage of science, of the scientific attitudes and spirit, performs an 'inestimable service' for scientists too." Tentatively, she suggests that this is because scientists and poets share an interest in abstraction. The use of abstraction in poetry can be evidence of a scientific stance because "the subject matter of science, abstractions, also occupies a dominant role in poetry" and "the essential nature of the abstracting process is the same, whether it be used to extract the properties of a collection of selenium atoms or the characteristics of a man's philosophic dilemma." By saying that science takes abstractions as its domain, Fullmer is using the term "abstraction" to mean something close to "model" or "conceptual scheme." Yet despite this hopeful picture of poets and scientists sharing an expanded laboratory, she concludes with an explicit reservation about the future of interrelations between science and poetry: "Poets as such will probably never suggest the direction of future scientific inquiries, but they will always provide a fairly reliable index of the extent of popularization of major scientific advances."[26] Poets are more focus group than research group.

Not surprisingly her essay provided plenty of provocation to critics and scientists. A young literary critic, John Hagopian, wrote a bad-tempered riposte. There can be no real relation between science and poetry: "For the poet, then, the ideas of science are nothing more than waves on the surface of the sea of his mind; his significant act is the plunge beneath that surface and the return to it with a new symbolic linguistic form in which we may perceive the conative-affective nature of human experience."[27] The first scientist to write a reply was no more sympathetic to efforts to reconcile science and poetry. "Scientific poetry is a bore," writes Herbert Hirsch, a researcher in the Division of Cancer Biology at the University of Minnesota.[28] Poets lack sufficient knowledge to write intelligently about new scientific developments; they freshen up their poetry with a dab of scientific vocabulary and a spritz of philosophy, and then think they can "interpret man's place in nature."[29] Such attitudes are archaic. Poets can no longer be allowed to repudiate, along with D. H. Lawrence, the scientific picture of the sun as "a ball of flaming gas" in favor of a poetic vision.

Fullmer's argument that poems can be scientific in attitude and method clearly went too far for the readership. Her defense, in a reply to her critics, was to go back into the laboratory from which Ignatow's poet wanted the scientist to emerge.

> A scientist goes daily to the laboratory to carry out certain operations, some manipulative, some paper and pencil calculations. From the operations he reaches certain limited conclusions whose validity is subject to check, usually most efficiently by another scientist or group of scientists. On the other hand, the nonscientist does not enter into the operations; for him the conclusions may frequently appear only as Venus of miraculous birth.[30]

Authority to affirm scientific knowledge (rather than just to transmit it) depends on the trust invested in oneself through membership in the scientific community.

WTF
?

Poets, unless they are also scientists, will be excluded. Her faltering hope at this end of the debate is that poets and scientists might meet not in abstractions but in the aesthetic. She adduces the authority of the chemist Cyril Hinshelwood to claim that science and poetry can converge on invention and creation: "the imposition of design on nature is in fact an act of creation on the part of the man of science, though it is subject to a discipline more exacting than that of poetry or painting."[31] We can, however, still hear in that word "discipline" the point of view of the laboratory. Although the debate about poetry launched by Fullmer had no obvious legacy, Fullmer's essays in *Science* remain important, both for their insights and for the vivid display of the tensions between laboratory science and poetry. They give us a strong sense of the climate of scientific opinion about poetry in which the poets of the time had to work. The issues she raises still remain current.

Scientists might be readers of poetry, as well as readers of articles about poetry in their house journals, but they tended to be skeptical about claims like Larrabee's that poetry and science could travel the same routes toward a unity of knowledge. In one of his many beguiling anecdotes, Richard Feynman recalls attending a talk at Princeton Graduate College on poetry at which the speaker concluded his classification of the elements of a poem by comparing the critical process to mathematics. This bold claim prompted a mathematician in the audience to ask Feynman to discuss the connections between physics and poetry.

> I got up and said, "Yes, it's very closely related. In theoretical physics, the analog of the word is the mathematical formula, the analog of the structure of the poem is the interrelationship of the theoretical bling-bling with the so-and-so"—and I went through the whole thing making a perfect analogy. The speaker's eyes were beaming with happiness.
>
> Then I said, "It seems to me that no matter what you say about poetry, I could find a way of making up an analog with any subject, just as I did for theoretical physics. I don't consider such analogs meaningful."[32]

Underneath the typical mixture of glee and slight cruelty with which Feynman deals with scientific ignorance is the belief that poetry is utterly different from physics. He also tacitly assumes that poetry lacks the intellectual rigor to produce knowledge. Poetry can say anything and therefore says nothing. Feynman's anecdote represents a wide swathe of scientific (and philosophical) opinion that poetry as a form of intellectual inquiry has been completely superseded by sciences such as physics. Even if we suspect that the intensity of this physicist's recoil from the promiscuous ambiguities of poetry hints at a forbidden fascination with whatever it is that poetry evokes, his judgment is damning.

It was not just physicists who believed that poetry lagged far behind. Even literary critics wondered at poetry's resistance to science. In 1957 *Science* published an impassioned plea by a New York high school English teacher, Joseph Gallant, calling for poets, writers, and teachers of literature to become modern. As he says in a later response to his critics: "the humanities sweepingly ignore the role played by scientific insight and thinking in the ideology of our times and disdainfully march on their archaic way as though the atomic and electronic age had not yet arrived."[33] Modern poets have been remiss about keeping up with the sciences, especially when compared with their predecessors. Even "the work of so dreamy a romantic as Shelley is interlaced with the new scientific perspectives of his age," whereas now in the 1950s there is almost no poetry of such stature engaged with the sciences.[34]

Gallant calls in evidence Helen Plotz's recent anthology of poems about science, *Imagination's Other Place: Poems of Science and Mathematics* (1955), dismissing it as a "thin compilation" that has to "lean heavily on Emerson, Masefield, Hardy, and Shelley" and a few "living poets [who] are flippant about flying saucers and the like."[35] Gallant is unfair to Plotz. She is vulnerable to his dismissal largely because she is aiming her collection at schoolchildren and has understandably, though perhaps mistakenly, undersold modern poems about science, either by truncating major early twentieth-century science poems such as Rukeyser's "Gibbs" and MacLeish's "Einstein" so much that they lose their complexity or by choosing poems on a scientific theme that aim solely to be entertaining. This latter point was made by William Newberry, a chemist working at the Olin Mathieson Chemical Corporation. Newberry broadly agrees with Gallant about the need for better humanities education in the sciences and notes several other poets, unmentioned by Gallant, who also write about science, such as "Max Eastman, James Franklin Lewis, Alfred Noyes, Selden Rodman, A. M. Sullivan and William Carlos Williams."[36] Poetry was not as distant from the new sciences as Gallant and Feynman feared: even industrial chemists were profitably reading modernist poetry.

SCHRÖDINGER'S NEW LAWS

In *The Night Sky: Writings on the Poetics of Experience*, Ann Lauterbach reflects at one point on a widespread disenchantment with academic thought. Even "the model of science, which has held our faith in reason, and by which the real is aligned to the true through test and experiment, has not so far shaped our moral imaginations or informed our spiritual quests."[37] What she says may be generally true, but there were important exceptions. There were models of science that engaged with moral imagination and spiritual aspiration, notably the models offered by the two physicists whose writings most inspired the New American poets: Erwin Schrödinger and Werner Heisenberg. Both were adept at intimating how quantum physics might shape moral and spiritual imaginations while not straying into what Elizabeth Leane calls the "holistic and mystical implications of quantum mechanics" that attracted later successful popularizers such as Fritjof Capra and Gary Zukav.[38] Both Heisenberg and Schrödinger were philosophically inclined physicists of the earlier, Solvay generation that pioneered quantum physics.[39] A wide liberal education gave them, like many of their generation, a respect for literature that encouraged them to mention poetry prominently in their writings on quantum physics for the general public. Indeed, Schrödinger even wrote poetry himself, though his friend Stefan Zweig's comment, "I hope your physics is better than your poetry," tartly indicates its limitations. Schrödinger's *What Is Life?* (1944) and Heisenberg's two books *Philosophic Problems of Nuclear Science* (1952) and *Physics and Philosophy* (1958) gave many poets their most authoritative idea of the potential interrelations of science and poetry.

"Everybody read Schrödinger," writes Horace Freeland Judson in his magisterial history of molecular biology, and that everybody included the New American poets Olson, Duncan, and Creeley.[40] Olson recommends Schrödinger to his students in *The Special View of History*, a set of posthumously published classroom notes that Olson wrote in 1956 at Black Mountain College shortly before it closed. He interprets *What Is Life?* as demonstrating that "a methodology becomes the object of its attention." Olson's strong interest in the conceptual engines of scientific methodology leads him to assimilate Schrödinger to Whitehead and claim that Schrödinger's book demonstrates that "there is no difference between process and reality," just as energy and matter are simply polarities "of a greater movement."[41] By contrast, Robert Duncan was not so interested in questions of method; he was captivated by the idea that genetics depended on a code inscribed within the cell.[42] In his 1964 Voice of America lecture, "Towards an Open Universe," which became one of his most widely read statements on poetry when it was reprinted in Donald Allen and Warren Tallman's defining collection, *The Poetics of the New American Poetry* (1973), Duncan writes with awed fascination at the molecular "code-script" that enables living organisms to resist entropy: "What interests me here is that this picture of an intricately articulated structure, a form that maintains a disequilibrium or lifetime—what it means to the biophysicist—to the poet means that life is by its nature orderly and that the poem might follow the primary processes of thought and feeling, the immediate impulse of psychic life."[43]

The current Cambridge edition of *What Is Life?* describes it as "one of the great science classics of the twentieth century," and few readers have doubted that if there can be said to be a canon of modern scientific texts, this is one of them. *What Is Life?* began as a series of seven public lectures given at Trinity College Dublin at the height of the Second World War, between February and April 1943, to a large audience of over four hundred people, including appropriately enough the taoiseach, Eamon de Valera, who had rescued Schrödinger by bringing him to Dublin when he had to flee Austria because of his anti-Nazi views. The lectures answer the question in their title by drawing on theories of entropy from both statistical and quantum mechanics and applying these insights to recent biological research into the mechanisms of heredity. Their tone is always hesitantly respectful of disciplinary expertise, and when Schrödinger moves on from physics to biology, he repeatedly reminds his audience that he is a physicist whose scientific authority to discuss biology is at best very limited.

Leah Ceccarelli argues that such tact was strategic. Schrödinger worked hard for a "negotiation of interests" between physicists and biologists by emphasizing shared values and goals and by the use of linguistic and conceptual ambiguity.[44] In making sure that he "did not obviously privilege one discipline over the other, nor limit the power of scientists who chose to embark on the new path of inquiry," he also left room for others, even poets, to go down that path. No one could have been more qualified to lead such negotiations between quantum physics and biology. Schrödinger was one of the most famous living physicists because of his work on the vexed question of whether the subatomic world was made of particles or waves, or some hard-to-imagine combination of both. His fame rested on a series of papers he wrote in 1926 that established a new mathematical model of the behavior of elementary particles, the so-called wave equation, a differential equation that made it far more straightforward than previously to calculate the wave behavior of particles such as electrons. This equation has been described as "the fundamental dynamical equation of quantum theory" and was foundational for the field theories of quantum electrodynamics (QED) that gave the field concept new currency at midcentury.[45] Mathematical accuracy was not its only benefit. The wave equation helped physicists visualize the atomic processes because as the elderly theoretical physicist Hendrik Lorentz put it, the wave equation had "greater *Anschaulichkeit*" (concreteness, vividness) than such theories as those of Heisenberg's matrix mechanics.[46] *Anschaulichkeit* was arguably Schrödinger's greatest gift as a scientist, and it helps account for his influence far beyond the two sciences he wished to see work together on his big question.

At the start of the first lecture he tells his audience that his aim is to ask whether the new physics that he represents can help us understand the workings of living organisms: "How can the events in space and time which take place within the spatial boundary of a living organism be accounted for by physics and chemistry?"[47] Instead of laying claim to complete knowledge over these events, he concedes that they cannot yet be fully explained by the methods of physics, though importantly he believes that they eventually will be. This confidence leads him later in the lectures to make a startling claim that over the years would intrigue readers and provoke much debate: a full explanation of the workings of the organism may require new laws of physics. Recent genetics research by Max Delbrück and others strongly suggests the validity of Schrödinger's suspicion that "living matter, while not eluding the 'laws of physics' as established up to date, is likely to involve 'other laws of physics' hitherto unknown, which, however, once they have been revealed, will form just as integral a part of this science as the former."[48] In a passage that displays his gift for *Anschaulichkeit*, Schrödinger explains that living organisms are able to do something that appears to run counter to the laws of physics: "An organism's astonishing gift of concentrating a 'stream of order' on itself and thus escaping the decay into atomic chaos—of 'drinking orderliness' from a suitable environment—seems to be connected with the presence of the 'aperiodic solids,' the chromosome molecules, which doubtless represent the highest degree of well-ordered atomic association we know of."[49] These new laws of physics may emerge from research into the role of the gene, which safeguards its future with what he calls a "code-script." Perhaps too the central insight of quantum mechanics, that energy is discontinuous, may help foster better understanding of mutations in the code-script, since they also are jumps from one state to another.

Once the series of lectures was finished, he revised them, added a preface and an epilogue, and gave each chapter a distinctive epigraph with a philosophical twist: one each from Spinoza and Descartes, three from poems by Goethe, and an aphorism about the necessity of contradiction from the poet, novelist, and essayist Miguel de Unamuno. Goethe represents the poet-sage of an earlier era, and the first of these epigraphs, for instance, explicitly talks about physical law in aesthetic terms:

Das Sein ist ewig: denn Gesetze
Bewahren die lebend'gen Schätze,
Aus welchen sich das All geschmückt.[50]

The tacit effect of these epigraphs from Goethe is to remind readers that poetry, at least canonical poetry from an earlier century, can still, just as much as the logical arguments of the great philosophers, be a source of wisdom and enlightenment for present-day scientists. Poets reading *What Is Life?* would feel that poetry could still make a contribution to human knowledge alongside the sciences. These epigraphs were not the only appeal to poets; their tone was carried over into the remarkable epilogue. Schrödinger speculates that consciousness, though we experience it individually, is actually "only one thing," a collective mind that is best described by the Upanishads and whose power is visible in the experience of free will, the inner experience that each of us is "the person, if any, who controls the 'motion of the atoms' according to the Laws of Nature."[51] Here in this vision of a spiritual intersubjectivity appears to be the sort of science that Lauterbach wishes for.

What Is Life? was an instant classic. Yet as Francis Crick later observed, Schrödinger's biology was "embarrassingly gauche," perhaps because of his mistake about the number of chromosomes (Schrödinger reports the erroneous figure of forty-eight), the idiosyncratic description of the gene molecule as an aperiodic crystal, and the inaccurate though exciting claim that a living organism keeps alive "by continually drawing from its environment negative entropy."[52] The book transcends such egregious errors because of what Ceccarelli calls Schrödinger's skillful use of ambiguity, thanks to which readers with different commitments can take from it different yet equally satisfying interpretations. When there was disagreement between readers, Schrödinger's rhetorical strategies encouraged disputes about how to interpret his arguments, not disavowals of his aim of uniting physicists and biologists in a program of research.[53] His phrase "other laws of physics" appealed to both reductionist and antireductionist camps, depending on whether they heard it to mean more laws like the laws of gravity or thermodynamics or to mean entirely new types of nondeterminist, perhaps normative laws. Although Ceccarelli doesn't say this, the rhetorical skill at reaching out to other disciplines clearly worked on poets too. The epigraphs, the vivid images and metaphors, not to mention the mystical epilogue, all helped enlist poets in the speculative inquiry about the physics (and metaphysics) of life.

This was not all that the book had to say to poets. Schrödinger also makes a rhetorical move that had the potential to provoke poets reading his account by reminding them of their now-marginal status, and this too probably added to its power to fascinate them. As he reaches his conclusion that a biological code or "aperiodic structure" embedded in every cell provides a blueprint that enables an organism to reproduce itself and to manage the remarkable feat of resisting the universal fate of entropy, Schrödinger alludes disparagingly to any claims poetic imagination might have as a means of inquiry to compare with scientific method. He compares the dissemination of genetic instructions within the organism to "stations of local government dispersed through the body, communicating with each other with great ease, thanks to the code that is common to all of them." In a clever rhetorical maneuver, like an attorney slipping inadmissible speculation into a courtroom statement without seeming to, Schrödinger immediately disowns his elaborate metaphor, saying: "Well, this is a fantastic description, perhaps less becoming a scientist than a poet. However it needs no poetical imagination but only clear and sober scientific reflection to recognize that we are here obviously faced with events whose regular and lawful unfolding is guided by a 'mechanism' entirely different from the 'probability mechanism.'"[54] The polarization is sharp: scientists are objective observers whose reasoned analysis reveals the underlying laws of the universe, while poets are fantasists whose excessively subjective imagination is drunkenly lacking in rational self-examination. And that might seem to be that. Poets are to be excluded from serious inquiry. This is not quite how the rhetorical swerve works, however. By making this analogy the physicist himself momentarily comes onstage as a poet, and though he then compares the relation between science and poetry to that between sobriety and inebriation, like the lawyer introducing inadmissible evidence Schrödinger leaves the implicit impression that poetic imagination *can* contribute to scientific inquiry if the poets sober up sometimes and develop rigorous poetic methods capable of taking on epistemic tasks suggested by their wild, fertile imaginings.

Schrödinger enabled poets to claim epistemological authority for poetic imagination. Speaking on Voice of America, Robert Duncan told his audience: "Atomic physics has brought us to the threshold of such a—I know not whether to call it certainty or doubt."[55] This performative statement of undecidability from his lecture "Towards an Open Universe" helps clarify further why Ignatow was pleading with the physicist to cross the threshold between laboratory and public life. Duncan is acutely conscious he is not in San Francisco now; he has an international radio audience and therefore feels the need to justify his claims for poetic imagination by proving that it stems from the same authoritative, prestigious sources as scientific discovery. Schrödinger speaks for an atomic physics that has led Duncan and other poets to a threshold (a very potent image for Duncan, who entitled his best-known collection of poems *The Opening of the Field*) opening out on new possibilities, though the reliability of what is revealed is unclear.

Duncan is also exceptionally frank about the poet's response to what the physicist has to say: "What interests me here is that this picture of an intricately articulated structure, a form that maintains a disequilibrium or lifetime—what it means to the biophysicist—to the poet means that life is by its nature orderly and that the poem might follow the primary processes of thought and feeling, the immediate impulse of psychic life." And Duncan is very interested indeed. His lecture is close to being an extended commentary on Schrödinger's "biophysics," several times mentioning Schrödinger explicitly and extensively discussing his ideas, including negative entropy or the intake of "orderliness," the code-script, and the epilogue's inner creator who mysteriously controls the motion of the atoms. Duncan is even moved enough by Schrödinger's allusions to Goethe to follow them up himself: "Remembering Schrödinger's sense that the principle of life lies in its evasion of equilibrium, I think too of Goethe's Faust, whose principle lies in his discontent, not only in his search but also in his search beyond whatever answer he can know."[56]

Duncan's main aim is to argue that poems are sensitive instruments for registering the cosmos and its effects on living organisms, sensitive as any scientific device. His argument rests on the construction of his own poetic modern synthesis from Schrödinger's ideas. Because "we are all the many expressions of living matter," cellular beings responsive to the physics of the sun and the biochemistry of planetary life, poets can reach down into the roots of their organic, cellular being and draw up new insights into the world. Duncan also recalls Dirac's notorious suggestion in an article in the *Scientific American* that "it is more important to have beauty in one's equations than to have them fit experiment."[57] Duncan infers that "the truth does not lie outside the art," or in other words that Keats was right: beauty in poetry is a signature of truthfulness. Today it requires a physicist to corroborate the older wisdom of a Romantic poet. With Schrödinger's authority, Duncan can infer that the poet's body can itself be as accurate as a scientific instrument in its registration of the "most real," whether it is the "immediate pulse of psychic life" or "the natural order" the poet "may discover" in the environment.[58]

Duncan spells out this poetic modern synthesis: "The most real, the truth, the beauty of the poem is a configuration, but also a happening in language, that leads back into or on towards the beauty of the universe itself."[59] In a subsequent essay, "Rites of Participation," Duncan makes even more explicit his assumption that Schrödinger gives the poet warrant for considering his own body to be a site of truthful apperception capable of sustaining scientifically legitimate inquiry. Schrödinger, Duncan claims in a theosophical flourish, has shown that "[o]ur secret Adam is written now in the script of the primal cell" or, in the less exalted language of *What Is Life?*, that a "code-script" is at the heart of all life.[60] Poetry itself is an art of scripts and codes and therefore intimately connected to both life and language. Since the code-script is a language used by all living things, poetry can then claim to be a higher-order manifestation of this universal cellular code-script.

Creeley's interpretations of Schrödinger were less direct. One stands out, an interpretation that is close to Duncan's, though expressed in an entirely different discourse. Creeley concludes his introduction to *The New Writing in the USA* with a resounding endorsement of the new poetry expressed in what appear at first to be remarkably airless abstractions: "That undertaking most useful to writing as an art is, for me, the attempt to *sound* in the nature of the language those particulars of time and place of which one is a given instance, equally present. I find it here" (his italics).[61] "Particulars of time and place of which one is a given instance"— what could he possibly mean? Given the insistence on personal authenticity of Creeley's affirmation ("That undertaking most useful to writing as an art is, *for me*"), it is surprising Creeley might be paraphrasing Schrödinger's statement cited earlier: "The large and important and ever much discussed question is: How can the events in *space and time* which take place within the spatial boundary of a living organism be accounted for by physics and chemistry?" (my italics). Creeley's formulation speaks not of "events in space and time" but of "particulars of time and place," and talks not of "accounting" for them but of "the attempt to *sound*"— which neatly conflates sonics and analysis. In place of Schrödinger's "living organism" is "one" (oneself). Particulars (a term that for a poet easily evokes the "particles" of nuclear physics) are the opposite of universals, and in some philosophies they are instances of those universals and the laws governing them. The accounting that is Schrödinger's theme, the theory derived from the encounter between quantum physics and biochemistry, may as we saw require "new laws" of some unspecified kind. Creeley's allusive formulation intimates that the poetry he admires will similarly be searching or sounding for the laws of these particulars.

Perhaps Creeley was wondering how to justify his own new experiments in poetry that would become *Pieces* (1969). This collection of short poems, fragments, epigrams, anecdotes, and philosophical quibbles marked such a substantial break with his previous practice that it must have seemed badly in need of critical and conceptual self-justification. In the second section, Creeley merges Schrödinger's idea of events needing new laws with Heisenberg's theory:

No forms less
than activity.

All words—
days—or
eyes—

or happening
is an event only
for the observer?

No one
there. Everyone
here.[62]

Trying to observe the entirety of the event of the sentence, trying to know its position and momentum, appears to be as impossible as Heisenberg explains is the case for quantum measurements. In this section, and in other opening sections of *Pieces*, Creeley tacitly assumes his audience will be aware of the speculative physics of Schrödinger and Heisenberg, as he isolates words and syntactical moves for attentive meditation. His final stanza points to a troubling aporia within the idea of events that are not events unless they have an observer: the risk that the intervening stage of mediation disappears, so that "everyone / here" is all that there is, and the perspectivalism of individual subjectivity disappears.

HEISENBERG'S UNCERTAINTIES

When we come to consider Heisenberg's influence on the poets we have to start by acknowledging that although "everybody read Schrödinger," everybody did not read Heisenberg.[63] To work with Schrödinger's ideas was to be part of a general scientific ferment around the possibilities of extending the ideas of physics into biology and of bringing to bear the skills of textual decipherment to the study of the genetic code. To read Heisenberg did not bring poets into contact with current thought to the same degree. Heisenberg's collections of lectures on the philosophical implications of quantum physics were popular, but unlike Schrödinger's lectures they did not introduce a new idea that then transformed a science, nor were they essential reading for the intellectual.[64] The uncertainty principle had long been an established if unstable pillar of modern physics. Poets who read Heisenberg probably did so because the uncertainty principle offered all sorts of imaginative possibilities of extension, and because Heisenberg directly discussed poetry's relation to science. Nonspecialist readers of Heisenberg had no more grasp of what the actual science portended than readers of Einstein had understood the fourth dimension. At best the uncertainty principle became the half-understood idea that a residual imprecision was unavoidable in all atomic measurements because the act of observing the atomic world necessarily altered its behavior. For poets, the uncertainty principle appeared to provide a naturalized epistemology justifying the perspectival and affective character of poetic judgments.

Heisenberg's *Philosophic Problems of Nuclear Science* was assembled mainly from lectures given in Germany and Austria in the 1930s and 1940s, including a talk comparing the color theories of Newton and Goethe given in Budapest in May 1941 under the auspices of the sinisterly named "Society for Cultural Collaboration."[65] This lecture sets up a comparison of theories of color as a discussion of the interrelations between science and poetry. Initially, Heisenberg tries to argue that there is less distance between scientist and poet than we might think. Kepler's admiration for the harmony of planetary orbits shows he had a "poetic sensibility," while Newton himself keenly pursued "philosophical and religious investigations." It would be truer to conclude that "the world of poetry has been familiar to all really great scientists."[66] Heisenberg employs the idea of poetry for mainly rhetorical purposes. The first is to make conventional references to poetry as a constellation of unique insights into realms outside science — to harmony, spirituality, and metaphysics. But it is his second usage that is likely to have caught the interest of poets. By talking about poetry he is able to allude to a problem central to quantum physics: the difficult art of achieving reliable *discursive* insights into material domains inaccessible to the senses.

He directs attention to the conceptual and cultural work of poetry, arguing that it is closer to the search for "real understanding" characteristic of science than is often assumed: "Every genuinely great work of creative writing transmits real understanding of all aspects of life otherwise difficult to grasp." In communicating such understanding of what other forms of inquiry find hard to portray, poetry offers something that modern science needs. Physics "no longer deals with the world of direct experience but with a dark background of this world brought to light by our experiments." Poetry can provide assistance. In a lecture entitled "On the Unity of the Scientific Outlook on Nature" he argues that quantum theory offers a far more subtle understanding of the way our minds necessarily conceptualize experience than its Newtonian predecessor, but in doing so, this theory creates new problems for perception and language: "We are now more conscious that there is no definite initial point of view from which radiate routes into all fields of the perceptible, but that all perception must, so to speak, be suspended over an unfathomable depth."[67] This troublingly inchoate image of an observer hovering over "an unfathomable depth" in which the quantum mysteries hide requires assistance to make it more tangible and intelligible, so where better to find such assistance than poetry's insights into point of view and perception and its famed ability to see down into the ghostlier demarcations of unfathomable depths?

Casting poetry rhetorically in the role of a guide to the labyrinthine worlds produced by active intervention at a scale far below the capacity of human perception helped make Heisenberg's meditations on the uncertainty principle compelling to poets of the 1950s and 1960s who saw themselves as searching for "real understanding." Having been tacitly invited to think of physics and poetry as sharing a similar challenge, Creeley and others were ready to extrapolate to their own concerns when they read Heisenberg saying that in quantum physics "there was good reason to disbelieve that the course of an event was objective and independent of the observer."[68] Poets could transpose to their own situation the idea that "the effect of the means of observation on the observed body has to be conceived as a disturbance, partly uncontrolled."[69]

This is why Creeley, who was notoriously insistent on firsthand knowledge, is willing to invoke Heisenberg's epistemic authority when discussing his own poetics in the 1967 talk "I'm Given to Write Poems." Trying to justify his reluctance "to speak directly of the writing itself," Creeley says, "I am persuaded by Heisenberg that 'observation impedes function.'"[70] Although Creeley puts the statement attributed to Heisenberg in quotation marks, it appears to be a product of Creeley's own bricolage. Heisenberg himself repeatedly insists that observation does *not* impede the functioning of the measurement; instead, observation is constitutive of the act of measurement, which itself depends on a measuring device that becomes part of what is measured. In a revealing moment during an interview ten years later with William Spanos, Creeley actually concedes that his formula misrepresents Heisenberg but insists that this doesn't matter. The passage is worth quoting at some length because it shows how the poet views the authority of physics and paints a picture of the poet having actual conversations with physicists:

> Again, in that long and extraordinary friendship with Olson I was very, very aware in his writing but equally in his conversations or all else, that he really disliked an abstraction that wanted to be apart from that which is observed. I think that's why those persons in my generation, friends as Allen Ginsberg, were delighted by Heisenberg. We would chant almost as a mantra, "Observation impedes function!" I was told many times by friends who were physicists that that wasn't what Heisenberg had in mind. But we didn't care. We loved the sense of "Aha! You see, if you attempt that discrete distance from the thing happening, you will literally not only not get such an actual view of it or actualizing view of it—you very possibly won't get into it at all!" Then too in college I recall taking very briefly courses in cultural anthropology and coming upon such delicious terms as "participant observation" and then being told by Clyde Kluckhohn, who was teaching such a class, that his sister Jane had got into an extraordinary dilemma by "participating" and "observing" in a "cultural rite," I think, of the Navajos in which she discovered she was momently to be married if she didn't take more care, so to speak.[71]

Jane Kluckhohn needs to take more care; why don't the poets take more care that they don't misrepresent the physics? Might they not find themselves entering into some equally unwanted alliance with physics that they had not intended? Is Creeley just demonstrating an epistemological irresponsibility, or at least an indifference to epistemic authority, typical of poets? Or is this response that the precise details of the uncertainty principle don't matter encouraged by Heisenberg, however unintentionally?

What Creeley takes from Heisenberg is first of all permission to treat poetry as a means of inquiry into problems of understanding created by unfathomable depths opened up by modern encounters with the world, whether quantum or quotidian. He and other poets treat the uncertainty principle as a kind of rhetorical maneuver. They listen to the rhetorical appeals to the idea of poetry in order to illustrate the tricky issue of distinguishing the perspectival uncertainty of observation from purely subjective responses, and they think of the uncertainty principle in similar terms. Creeley's poem "I Keep to Myself Such Measures" presents the difficulty of treating thought as capable of full self-knowledge, by using the comic image of wholly unsuitable rocks being placed by the poet along a path as reminders of thoughts that he hopes to revisit. Creeley and his contemporaries treat the uncertainty principle as an image of poetic interaction. Heisenberg says to them that the virtues of poetry can help physicists illustrate the difficulties of returning from their gaze into the unfathomable depths of the quantum world with any sort of communicable understanding. They concur. Poetry is an art of transmission engaged in the project of finding a "real understanding" of "aspects of life otherwise difficult to grasp."

Creeley's own explanation of his willingness to customize the uncertainty principle takes a second step. His seemingly random anecdote about Clyde Kluckhohn, the Harvard anthropologist, gestures at another reason for interpreting the uncertainty principle as he does. Historian of science Joel Isaac shows (in a history of social science at Harvard from which I shall be drawing later for a discussion of Olson's interest in conceptual schemes) that Harvard social scientists such as Kluckhohn developed a pedagogic and "artisanal conception of epistemology."[72] During his brief stay at Harvard, Creeley encountered this conviction that epistemology is inseparable from pedagogy and the actual practices of research inquiry. Kluckhohn was a strong advocate of a theoretical approach to anthropology coupled with a more interdisciplinary training. He echoed Whitehead's warning about the "celibacy of intellect" resulting from the disciplinary narrowness of a "single limited set of abstractions."[73] Through Kluckhohn as well as other teachers, Creeley would have met one of the most striking forms of the new Harvard social science, the "case study" method of research based on the use of a conceptual scheme. Kluckhohn argued, according to Isaac, that the Rimrock Navajo "offered the perfect case study for a theoretical exploration of the functioning of non-logical elements in the constitution of a social order."[74] For Creeley, Heisenberg's willingness to enlist poetry to understand observation licenses conflation of the uncertainty principle with a scientific, practice-based epistemology in which understanding emerges from immersive inquiry. It therefore doesn't matter that physicists disagree with the poet's interpretation of Heisenberg's uncertainty principle.

POETRY'S ILLUSTRATION OF THE DIFFICULTIES
OF SCIENTIFIC COMMUNICATION

Heisenberg and Schrödinger were not only exceptional popularizers of science but also eminent physicists, philosophically trained thinkers capable of producing highly illuminating images of very difficult concepts. The problem of communicating what the physicists saw in the nuclear depths also troubled less philosophically minded, and less literary, physicists. Victor Weisskopf, a Manhattan Project scientist who was later head of MIT before becoming director general of the European Organization for Nuclear Research (CERN), gave a presidential address to the American Physical Society in 1961 outlining the difficulty of picturing the quantum world. "Most problems of nuclear structure cannot be dealt with by direct approach in solving the Schrödinger equation for all nuclear particles involved," and therefore, "new concepts and pictures are introduced" that "are useful for the description of the observed facts, but their connection with the fundamental forces is vague and only qualitatively understood."[75] One way to keep this vagueness under control was to call it "semiphenomenological." Michael Moravcsik's account of the Rochester conference on high-energy physics in 1959, for example, reports that "a complete (although not necessarily unique) semiphenomenological description of the proton-proton interaction has been given up to 400Mev in terms of phase shifts and the one-pion exchange contribution."[76] But what does "semiphenomenological" mean when used in the "external relations" of physics? Robert Marshak, writing for the *Scientific American* in 1960, was forced to find a translation intelligible to the layperson and came up with a very revealing one.

His essay about the "nuclear force" outlines the result of many years of nuclear research: a model physical system composed of a few types of particle, most of which can be pictured as having spins, trajectories, and forces of mutual attraction and repulsion. Yet this picture is misleading. In this "quantum world" ordinary expectations of solid objects do not apply, and particles can pass directly through one another or turn out not to be particles at all but energy waves. It is as if a baseball could also be a flash of light. Marshak tries to be scrupulous about the epistemological and the consequent ontological unease felt by the experimenter:

Now throughout this detailed description of the dynamics of nucleon-nucleon interactions we have been speaking as if the actual particles were visible to us, and we could see them spinning, curving, diffracting and so on. It should be remembered, however, that all we can really see are the recording devices of a particle-counter, or the dark spots on a photographic emulsion, representing relative intensities of a particle beam scattered in various directions. From these data the experimenter draws a curve, which he then presents to the theoretical physicist and asks: What kind of force produced such a graph?[77]

Poets might well ask themselves what kind of imagination produced such a force? Marshak has an answer of a kind. These are provisional diagrams of the quantum world where "not only are the particles far too small to see and trace, but also the uncertainty principle of quantum physics tells us that sharp trajectories do not even exist except at energies much higher than those we shall deal with." Therefore, the physicist talking to the nonphysicist has to rely on what Marshak revealingly calls a "semifictional particle language."[78] His elegant defense of the specialist word "semiphenomenological," carefully hidden from *Scientific American* readers it should be noted, appears as a poetic practice, the creation of a language of semifictions, "semi" because these images have some relation to actuality, but no one is clear just how. The "semi" tries to provide a solid bridge between objectivity and interpretation but only draws attention to the rift between them. Not surprisingly, historians and philosophers of science since the 1960s have increasingly argued for various forms of what has become known as "antirealism."

Faced with this semifictional world, other scientists reached for the abstract idea of poetry just as Heisenberg and Schrödinger had done. Sometimes poetry lent a little heritage magic to the science, but sometimes there were the same hints that the language art of poetry made it a useful junior partner, if a very junior backroom partner, in certain kinds of scientific thought. Speaking to an audience of scientists at the Centennial Celebration of the National Academy of Sciences, the leading Manhattan Project scientist Isidore Rabi explains the "aspiration for discovery" as something that unites the "true poet or scientist": "The scientist, the experimental scientist, shares with the poet and artist a feeling for the value for the immediate and the empirical face of nature."[79] Scientists might learn from poetry's capacity to discover the feelings and values embedded in the phenomenological world. Rabi has been credited with having encouraged C. P. Snow to write about the two cultures, though the situation in America was somewhat different from that in the United Kingdom.[80] A pervasive lack of scientific expertise in government circles in the United Kingdom contrasted unfavorably with the more active contribution of science advisers to American government policy. Both men lamented a general lack of scientific literacy, which meant, as Rabi complained in his 1955 Loeb lecture, that "the non-scientist cannot listen to the scientist with pleasure and understanding." Rabi longed for a collaborative enterprise of scientists and artists, based on mutual recognition and understanding, because "only with a unified effort of science and the humanities can we hope to succeed in discovering a community of thought, which can lead us out of the darkness, and the confusion, which oppress all mankind."[81]

Such longings were probably shared by many physicists. Writing a memoir rueful of his acceptance of the necessity for large scientific teams, Robert Wilson, director of Fermilab, describes individual scientists who resist co-optation into "big science" as retaining the virtues of poets: "these men are doing creative, poetic, and enduring work—true intellectuals they, not bureaucrats enslaved by a computer."[82] Creative, poetic, enduring work? This is what poets had thought *they* were doing. Physicists even taught courses known as "physics for poets," a title that was a physicists' joke, since the students who took them were not normally poets but simply nonscience majors. The joke tells us that to physicists everyone who didn't understand physics might seem to be a poet. Adolph Baker's textbook for such an introductory course in physics stages a dialogue between a "scientist" and a "poet." At one point the scientist compliments the poet on a growing understanding of the true character of physics: "You are beginning to discover the shallowness of the simplification which dismisses physics by calling it an 'exact science,' and assumes that the gift of subjective vision is the exclusive province of the poet."[83] Despite the poet's resistance to physics, and the suspicion that the poet relies on faith rather than reason to understand the world, Baker needs this poet to speak about "observation and interpretation" in physics. Recourse to the abstract idea of poetry was a device for illustrating communication difficulties associated with the quantum world.

Self-aware analyses of this strategy of bringing poetry momentarily onto the scientific stage are hard to find. The effectiveness of this strategy depends on treating poetry as a known, taken-for-granted quality, and any self-consciousness would bring up unwanted questions about just what sort of poems and poets (Victorian moralists, Romantic pantheists, metaphysical rationalists, or avant-garde abstractionists) were intended. There are a few partial exceptions to this guardedness, however. One of the most striking occurs in a 1963 lecture by J. Robert Oppenheimer for the Centennial Celebration of the National Academy of Sciences. By this late stage in his life, Oppenheimer felt fairly free to say what he liked, perhaps because nothing could damage his reputation any further than it had been sabotaged already by the hearings that confiscated his security clearance and by the sense that as a result he had ceased to be part of the inner circle of nuclear scientists. This freedom enabled him to reflect expansively on the role of language, concepts, and imagination in scientific research, laying bare much of what is happening backstage when physicists allude to the abstraction of poetry.

The topic of the National Academy of Sciences centennial, "The Scientific Endeavor," was suitably open-ended, although speakers were assigned themes, and Oppenheimer was asked to address the theme of scientific communication. Oppenheimer was a perfect choice. He was good at communicating with the public and, by all accounts, quite brilliant at helping physicists communicate with one another. His biographer Ray Monk cites two eyewitness reports of Oppenheimer in action at the 1947 Shelter Island conference on the latest developments in nuclear physics, including the discovery of new versions of the troublesome meson. Oppenheimer showed himself "simply masterful" at "directing a group of physicists during their scientific deliberations." Karl Darrow recalls with some awe Oppenheimer's "analysis (often caustic) of nearly every argument, that magnificent English never marred by hesitation or groping for words (I never heard 'catharsis' used in a discourse on [physics], or the clever word 'mesoniferous,' which is probably O's invention), the dry humor, the perpetually recurring comment that one idea or another was certainly wrong, and the respect with which he was heard."[84]

At the centennial celebration Oppenheimer and the other speakers were introduced by George Kistiakowsky, a key Manhattan Project colleague of Oppenheimer's who had been special assistant to President Eisenhower and was now a member of President Kennedy's Science Advisory Committee. Until the Renaissance, says Kistiakowsky, the scientist shared "a public position next to the philosopher, artist, and poet," but then the new experimental methods of science enabled the scientist to "project a new, more exclusive public image."[85] Having placed poets firmly in the past, Kistiakowsky hands over to Oppenheimer, who begins with the anecdote that he deployed back in 1950 in the *Scientific American* to boost science at the expense of poetry, an anecdote that at first appears to confirm the chairman's low opinion of poetry. "In an important sense," starts Oppenheimer, "the sciences have solved the problem of communicating within and with one another more completely than has any human enterprise."[86] He then repeats the Paul Dirac anecdote, of which he was very fond, about the apparent incompatibility of science and poetry, which I cited at the start of this book. If

Oppenheimer had left it there, the opposition would be damning: poetry would belong to the past while its latest incarnations would be mere fashionable dress on an aging body of ideas. But the Oppenheimer of the 1960s is different from the colossus of physics of the 1940s. Now he is far more aware of the need for physics to come out of the laboratory and experience the consequences of imperfect communication with the public. In a wily manner, almost imperceptibly, he begins to question complacencies behind Dirac's contempt for poetry and by implication behind Kistiakowsky's tacit relegation of poetry to the past. He starts with the observation that scientists must recognize that there are "things of which we cannot talk without some ambiguity, and in which the objective structure of the sciences will play what is often a very minor part, but sometimes an essential one." Such issues are not trivial, for they include "the arts, the good life, the good society."[87] On its own moreover, communication is not enough; it must be accompanied by the effort to comprehend what is communicated, "to question, to try, to apply, to adapt, to ask new questions." Without this effort, "communication would provide the fuel pipes, the electrical wiring, the transmission of a car, but not the combustion which gives it power and life."[88]

Oppenheimer has now established two important points: that scientific method may be inherently unable to be precise about important aspects of human life; and that communication or dialogue is not in itself understanding but only a vehicle for conveying it. He can now take a crucial step toward dismantling the remainder of Dirac's (and perhaps Kistiakowsky's) foundational assumptions about language. Instead of the poets being the offenders when it comes to hanging on to comfortably familiar ways of thinking, Oppenheimer is about to find scientists just as guilty. Not only do they underestimate the difficulties of reception and interpretation, and the demands these tasks make on our "imagination, play, curiosity, invention, action," but they cling to the past. They "love the old words, the old imagery, and the old analogies" so much that clear understanding itself is impaired.[89] This is a harsh judgment, especially on physicists who think of themselves as being at the forefront of knowledge. If scientists often fail to grasp that scientific communication requires the ability to handle ambiguity with imagination and drive, and to be open to new uses of language, what we might call, in the spirit of Oppenheimer, its *mesoniferousness*, what are the implications for attempts to improve scientific communication?

Here too he challenges the assumptions revealed in Dirac's dismissal of poetry. The specialist discourse of internal scientific communications should not be a private language confined to the laboratory; to be methodologically effective it needs to have a similar texture to that used for the "external relations" with non-scientists. Oppenheimer's complex foldings of thought are not easy to excerpt. Here is a passage that at least captures the way he reasons about this issue of going public:

> Though I do not suppose that a thorough knowledge of science, which is essentially unavailable to all of us, would really be helpful to our friends in other ways of life in acting with insight and courage in the contemporary world, it would perhaps be good if in talking with them we could count on a greater recognition of the quality of our certitudes, where we are dealing with scientific knowledge that really exists, and the corresponding quality of hesitancy and doubt when we are assessing the probable course of events, the way in which men will choose and act, to ignore or to apply, or make hypertrophic or nugatory the technological possibilities recently opened. I think that some honest and remembered experience of the exploration of nature, of discovery, and of the way in which we talk to one another about these things might indeed be helpful; but that is because it would remove barriers and encourage an effective and trusting converse between us, and make more fruitful the indispensable role of friendship.[90]

Oppenheimer is saying, in the midst of these almost Jamesian sentences freighted with allusions to further possible topics of reflection, that scientists need to be more open about their uncertainties, about the quality or degree of their certitudes, when talking to nonspecialists outside the laboratory. Scientists also need to be more honest in talking to outsiders about the "partly accidental quality to the effectiveness of [their] converse with one another, and thus to the effective unity of [their] view of the world."[91] Scientists should remember that they come to "new problems full of old ideas and old words, not only the inevitable words of daily life, but those which experience has shown fruitful over the years."[92]

When talking to nonscientists, the specialists should try to retain the improvisatory, uncertain qualities of their internal, specialist conversations and not just rely on dogmatic affirmations of scientific fact. Doing so could have great benefits. Just as scientists sometimes take up at least an amateur interest in the arts, nonscientists may thereby be encouraged to a similar skilled amateur engagement with scientific enterprises, but only if "we do what we can to open the life of science at least as wide as that of song and the arts."[93] Oppenheimer now has one more rhetorical rabbit to pull out of the hat. Affirming his confidence that in future scientists will embrace such openness because they will want to share pleasures they derive from the practice of science, he turns song into an extended metaphor and in doing so invites poetry back into physics: "We all know this, and all share it; but each of us, I think, must be free to use his own words to sing its praise, even to describe it."[94] To sing the praise of science? He has just implied that the arts are metaphorically a form of song. Isn't this final statement then a way of arguing the exact opposite of Dirac? Scientists will sometimes require the assistance of lyric forms of expression, at least behind the scenes, if they want to communicate fully. All this is typically dry and playful Oppenheimer. Appealing tacitly to the abstract idea of poetry enables him to illustrate his reasoning about the significance of different sorts of interdependent scientific communication with images of self-expression, verbal ambiguity, affect, human values, and other, even less tangible concepts, without appearing to leave the scientific space of reasons. Although the idea of poetry starts out as being purely illustrative, it begins to inflect the very theorizing that employs it. Poetry is, Dirac and Kistiakowsky notwithstanding, a necessary ally in the conceptualization required by the scientific enterprise.

Interrelations between science and poetry viewed from the perspective of the scientists, especially the physicists, were not simply dismissive. There was plenty of ridicule then as now directed at the primitive, obscurantist, fantasizing verbal arts; there was also a recognition that poetry remained an important part of an advanced education, at least a sentimental education. Physicists in particular found in poetry a handy abstraction as a placeholder for difficult epistemological questions around linguistic representation of the strange, nonsensuous subatomic domain. Oppenheimer's insistence that scientists need to provide "combustion" in their communications echoes with Olson's insistence that poets need to provide some high-energy discharge in their poems. Oppenheimer's belief that scientists need to convey to nonspecialists the "accidental quality," the mix of everyday and specialist language, in their discourse has affinities with Olson's interest in constructing poetic texts that enact the process of halting, uncertain inquiry. Listening to the physicists, a poet might well conclude that they were inviting other researchers to likewise acknowledge their own uncertainties. *Not*

knowing need not be a sign of error or failure. Perhaps the composition of the poem could also be a self-reflexive process capable of making as explicit as possible its own "semifictionality" or "semiphenomenologicality." This need not be a license for surreal flights of imagination. Simulating the rigor of physics might similarly require an empirical approach to observation and action that precede or are coeval with the composition of a semifictional language or, as the poets would say, a poetic language of diminished referentiality.

PART II
MIDCENTURY

* 3 *

PROJECTIVE VERSE

///

Fields in Science and Poetics at Midcentury

OLSON'S "JOY OF SCIENCE"

"My joy of science is such, I am apt to forget most people have a double-trouble: they are either captive of its mechanisms (unable to see how Heisenberg restored science to man) or they are full of the old religion-art suspicion of it as robber of the luster of the day-dreams of man." Charles Olson confessed to this excitement with science in his "Lectures in the New Sciences of Man," which he gave at Black Mountain College in 1953. He tells his audience that his aim is the construction of "the crucial 'human' science" or, more ambitiously still, "the ultimate science of man."[1] The language and metaphors that he enthusiastically displays in "Projective Verse" are full of this joy at the possibilities for a new poetry to be found in scientific understanding of energy, fields, matter, and forces. So too is one of his earliest major poems, "The Kingfishers," although here the joy is more evidently the exuberant skepticism of Nietzsche's *fröhliche Wissenschaft* than an uncritical pleasure in scientific achievement.[2]

From start to finish of his career as a poet Olson wanted other poets to share his excitement about the repurposing of scientific methodologies for poetic ends. One of his first publications after leaving his job with the Office of War Information in Washington, "This Is Yeats Speaking," takes Ezra Pound to task for scientific illiteracy. If any poet living in the next world were to try to make contact with the living it would be Yeats. Yeats/Olson criticizes Pound for the "obsession to draw all things up into the pattern of art" that led him to defer to the authority of the "brawlers and poets" of the past and present, which Yeats himself knew well could result in a dangerous submissiveness.[3] Olson's share in this criticism might sound hypocritical coming from another arch-synthesizer, but it is not the search for *pattern* that Olson mistrusts; it is the insistence on a solely *aesthetic* pattern rather than a pattern informed by the protocols of contemporary knowledge. If only Pound had recognized the importance of other knowledge, especially the sciences: "He was ignorant of science and he will be surprised, as Goethe will not be, to find a physicist come on as Stage Manager of the tragedy."[4] Stage-managing scientists such as Oppenheimer were doing cultural work just as important, perhaps more so, than any poet. Olson was determined not to make Pound's mistake, and he would continue throughout his life to broker scientific ideas. One of the last poems Olson wrote, "The Heart is a clock," subtly amends the physicist Hermann Weyl's relativistic picture of lived time implicit in his suggestion that "the life-processes of mankind may well be compared to a clock."[5] Olson depicts a more Whiteheadian picture of life as a cluster (a variant on Whitehead's term "society") of events: "the Heart is a clock / around which clusters / or which draws to itself / all which is the same / as itself in anything."[6]

For his inner circle of poet friends, Olson's leadership in the negotiations between poetry and science was a crucial part of his achievement. Robert Creeley attributed much of what was most valuable in Olson's poetics to his recognition of the need for a new stance toward the world, a recognition that "had come primarily from scientific thinking, as it might be called."[7] Robert Duncan paints Olson's interest in science as part of a wider American history of development that analyzes "the Westward movement of scouting, exploration, traffic with the natives, first settlements, raids and massacres, exploitations, scientific observations as a great psychic happening, a drive into the mythopoeic."[8] Robin Blaser claimed that Olson's main project was "the translation of science into poetry"[9]— a secular twist on Wordsworth's endorsement of the role of the poet in negotiations with science as lending "his divine spirit to aid the transfiguration" of science into "a form of flesh and blood."[10] Blaser's "translation" implies epistemic equality, whereas Wordsworth's "transfiguration" implies that poetry transcends science. Blaser makes these large claims because he believes Olson has shown that the sciences open new doors onto the universe and provide new tools for studying what is then revealed: "what I have noticed in the poetry and poetics of the most important poets is that they are arguing, weaving, and composing a cosmology and an epistemology."[11]

Almost all Olson's commentators and critics have acknowledged the scale of Olson's epistemic ambitions. He has been fortunate in the scholars he has attracted: George Butterick, Ralph Maud, Benjamin Friedlander, and other meticulous scholars have ensured that we have accurate texts, reliable glosses, and plenty of contextual data (we can't say this for many American poets of that era). George Butterick's introduction to the more than eight hundred pages of glosses devoted just to Olson's *Maximus Poems* describes them as "a poem that seeks to be as wide as the world."[12] At the outset of *Charles Olson's Reading*, Ralph Maud justifies his exhaustive bibliographic study by saying that he and his contemporaries "felt we were in the presence of the man, for our time, almost complete in knowledge."[13] Maud's book is subtitled "A Biography" because it is assumed that to study Olson's acquisition of knowledge through reading is to provide a biography of the poet. The first generation of Olson critics did a good job of combing through Olson's published writings and constructing sympathetic syntheses of his theories of poetry, history, and the cosmos. Paul Christensen argues that "the intention of all this breadth of reference, while not excusing its worst excesses, at least explains what it is attempting to do: to show a mind that has ventured out beyond its own routines of thought and experience and sought to immerse itself in areas of experience, of the remote past and of other cultures, that Western civilization has deliberately or unwittingly ignored."[14] Sherman Paul provides a totalizing exegesis of Olson's thought, offering such gestures at the scale of Olson's epistemological and ontological ambitions as this: "The epic is no longer a poem that contains history, it is his-story, the story of creation, of actual willful man."[15]

Where Olson's earlier readers were generally able to overcome their doubts about his epistemological ambitions, the succeeding generation, beginning perhaps with Robert von Hallberg, try to locate Olson not as mentor and guru for a generation of poets and readers with whom they identify but as a representative figure in the modern literary tradition in whom they discern eccentricity, intellectual weakness, and even incoherence.[16] Taken together, these judgments amount to a considerable weight of negative critical opinion, much though not all of which is concerned with Olson's failings as a researcher. Von Hallberg calls Olson "an untrustworthy teacher,"[17] though he still admires the poet. Andrew Ross can find little to praise at all, discerning a psychotic side in his imagery and ambitions.[18] Tom Clark depicts many deflationary moments in his iconoclastic biography,[19] Susan Howe observes a masculine blindness,[20] and Charles Bernstein sees too much of "shadows and haze."[21] Susan Howe honors the importance of Olson's example for helping her and others find a way of writing historical poetry because it gave her "a vocabulary for going forward into a past which is part of the living present," helping her to be a "radical poet," but she also regretfully points out that his sometimes-cruel misogyny reads as an attempt to exclude women writers and readers from his poetry's community of intellect and therefore creates an aporia that severely limits its ambitions to re-create the polity it aspires to.[22]

Charles Bernstein borrows the title of his review of the complete *Maximus Poems*, "Undone Business," from Olson's poem "Maximus, to Himself," as if to say that Olson ought to have heeded his own judgments more. Not only is Olson blind to the inequities consequent on gender and ethnicity in America, but he also overloads his poems with information that is not adequately transmuted into poetic form (Bernstein compares Charles Reznikoff favorably to Olson in this regard). The late Olson's "groping for occult explanations in the archaic stems from a refusal to accept the limits of knowledge," so that instead of respecting the methods of empirical inquiry and the painstakingly constructed domains of reliable knowledge (not least by the sciences), he projects his own preoccupations onto unreliable constructs.[23] In a similar vein, Libbie Rifkin devastatingly attributes Olson's "self-given license to travel across institutional domains" to his "amateurism."[24] Heriberto Yepez argues, from a standpoint critical of American treatment of its southern neighbors, that by promoting a "pantopic" and "postmodern" stance, Olson "constructs the most comfortable, most USAmerican of ideas: the myth that exonerates the Oxident of all its crimes," especially the crimes attributable to imperialism.[25]

To this day it is difficult to bring Olson's achievement into focus. Critical opinion is divided. The early hero worship gave way to iconoclastic critiques as a later generation of readers began questioning their elders' apparently uncritical admiration for him. I don't have anything like a full answer of my own about how to value Olson, only what I hope will be seen as measured appreciation of a small number of his poems accompanied by a narrative of his changing interest in scientific method. I think there is substantial justification for many of the criticisms I have so briefly summarized, especially Olson's insensitivity to gender, "race," and imperialism and his "amateur" unwillingness to credit and engage intellectually with many of his sources of information and ideas, particularly when he appears to have been embarrassed by his reliance on them or critical of their flaws. But I also think that his poetic practices are more understandable than is suggested by the charge of an amateur's refusal to accept the limits of what can be known if we give more attention to his early ambitions for a scientific poetics of history that had affinities to other social-scientific researches of the time. But seeing his activities as understandable is not necessarily to condone them all, and I think my narrative of his development does not exonerate him from the justifiable criticisms. In this way I suppose I too am part of the later, critical generation.[26]

My claim is that Olson's continuing significance as a modern American poet crucially depends on his struggle to create a poetics responsive to the ascendancy of the sciences. Whether awed or critical of his vast knowledge, most readers have not thought of Olson as a poet-scientist. Many other roles have been attributed to him: scholar and shaman, researcher and visionary, bureaucrat and busker, social theorist and mythographer, professor and amateur local historian, working-class boy and Harvard-trained politician, modernist poet and postmodernist anti-poet. In the eyes of his teachers, friends, family, and readers, Charles Olson was all these and more. He has not been thought of as what Willard Van Orman Quine drily calls a "lay physicist" nor as a social scientist, scientist of mythology, or philosopher of science, despite clear evidence for the significance to him of all these activities.[27] Thomas Merrill believes that "Olson vigorously contends that Art is a much superior epistemological tool than Science."[28] The common view can be represented by Michael André Bernstein's dismissal of the significance of Olson's interest in the sciences: "a consideration of topology, Riemannian geometry, or quantum physics, contributes surprisingly little to an understanding of *The Maximus Poems*." Bernstein dismissively characterizes Olson's syncretic statement that "the structures of the real are flexible, quanta do dissolve into vibrations, all does flow, and yet is there, to be made permanent, if the means are equal" as a "curious mixture of Max Planck, Heraclitus, and Alfred North Whitehead."[29] The adjective "curious" not only marks his concern that Olson finds too much joy in science but also hints that this rapture clouds his judgment so much that he makes incoherent connections between science, philosophy, and poetry.

One partial exception to the dismissal of Olson's enthusiasm for science can be found in von Hallberg's argument that Olson attempts "to construct aesthetic theories that would answer the challenge posed by twentieth century sciences and technologies," an argument based on linking Olson's search for a persuasive ontology to both Objectivism and Whitehead.[30] In support of his interpretation, von Hallberg cites a sentence by Whitehead that Olson annotated with the words "usable definition": "An 'object' is a transcendent element characterizing that *definiteness* to which our 'experience' has to conform."[31] The question though is what we understand by answering the challenge of science: embrace or confrontation. Joy surely points to the former; Whitehead's answer to the challenge of science is more the latter. Von Hallberg tacitly attributes Whitehead's answer to Olson also, but as we shall see, Whitehead's response to the challenge is to turn physics into metaphysics, practical epistemology into philosophical theory. Olson's response to science, I shall argue, changed over time and was not always the same as Whitehead's.

As his remark about "double-trouble" indicates, Olson himself was aware of the difficulties of being a poet who redeploys new scientific methodologies for poetry. He could be quite unsure how to characterize his own practice. In November 1952 he wrote his entry for *Twentieth Century Authors*, a brief essay that has become known as "The Present Is Prologue." Here he takes the theme of how to name his poetic vocation further, calling himself an "archeologist of morning" rather than a poet: "I find it awkward to call myself a poet or a writer. If there are no walls there are no names. This is the morning, after the dispersion, and the work of the morning is methodology."[32] This dislike of the self-designation "poet" could be taken as modesty, but the insistence on interdisciplinarity ("no walls") and on the importance of methodology, as well as the fanciful self-ascription of the profession of "archeologist of morning," suggests that the difficulty was primarily one of accommodating his underlying "joy in science." "Archeologist of morning" was a clever self-designation. Archeology was widely treated as an honorary member of the natural sciences, one of the few social sciences to be regularly represented in the general-science journals such as *Scientific American* and *Science*.

These self-descriptions are instances of the many occasions where Olson clearly adopts both the twentieth-century scientific world picture and its dominant epistemic image of scientific research as a venture into an unknown imagined as outside the territory of scientific knowledge. A poet of the present day needs to know as much as possible about the actual world of space, time, matter, and organic existence currently being researched by the sciences because this is where we live. In his Black Mountain "Lectures in the New Sciences of Man," Olson makes clear how important it is for students of literature to incorporate similar scientific ambitions in their poetics. In the lecture entitled "Beginning of 3rd Inst," he starts his discussion of "the crucial 'human' science" with a sketch of what he calls "the phenomenon of man," a sketch that draws on references to genetics and physics. Then he pulls back and reflects on his methodology: "You will note I am at some pains to state all this rather with the vocabularies of biology & physics than with any of those of the several modern psychologies—or, for that matter, so far as I can avoid it, with any of the languages of my own speciality, what can be simply put down as the science of image."[33]

Why does Olson try to state his method in the language of physics and biology, and why is he wary of psychology? In this and the next two chapters I set out to explore the midcentury moment when a rapidly expanding scientific methodology claimed knowledge of every aspect of human culture, making it appear necessary to turn to the natural sciences for better methodologies. I shall concentrate on physics and its central concept of the field because in practice it was this science and this epistemic metaphor that influenced thinkers across many disciplines, including poetry, at midcentury. And I shall start with the publication of Olson's "Projective Verse" at the very midpoint of the century in 1950.

THE MIDCENTURY POET

A reviewer of John Ciardi's *Mid-century American Poets* observes approvingly that the midcentury poet is ready to "construct an image around a bomb-sight and employ test tubes, tabloid headlines and X-rays as the natural props of poetry."[34] Poetry as news of war and peace was not news. What this reviewer, Rolf Fjelde, thought was new about midcentury poetry was the expectation that a poet would be something of a scientist, perhaps Quine's "lay physicist." Revealingly, Fjelde did not see any need to elaborate on what cultural changes had drawn these midcentury poets to science. He apparently shared the widespread view endorsed by Wallace Stevens in his 1948 essay "Imagination as Value" that American poets were living in "a civilization based on science."[35] And he could find in the anthology itself considerable direct evidence for a scientific turn. Richard Eberhart writes as if to be an American is indeed to be a lay physicist, saying in a statement of his poetics that through their reliance on the creative power of imagination "science and poetry join, as in Einstein's Unified Field Theory."[36] Karl Shapiro's "Elegy for a Dead Soldier" envisions the American future now lost to the GI as "[t]he quantum of all art and science curled / In the horn of plenty."[37]

Fjelde's optimism about the rapprochement between science and poetry appeared to be well justified at the time. As an editor of *Poetry New York*, where he published the review, he and his fellow editors had chosen to showcase a striking manifesto for poetry entitled "Projective Verse," by the then-unknown poet Charles Olson.[38] This midcentury poet may not have been included in Ciardi's anthology, but he shared an interest in science, arguing that poems should arrange their elements to create a field of forces held together in a "high energy-construct," which sounded analogous to the principles behind those other high-energy structures of the time, the synchrocyclotron and the atomic bomb.[39] In addition to Olson's concept of the nuclear poem, Fjelde could look to plenty of other proximal examples of poetry engaging with science. The same issue of the magazine also contained an extract from book 4 of William Carlos Williams's *Paterson* comparing money to uranium and speculating that humanity has a choice whether to employ financial radioactivity as a destructive energy (a "hurricane") or to use its "Beta rays" constructively to "cure the cancer / —the cancer, usury."[40] Never mind that Williams had misremembered his physics, confusing the fluorescing electrons once called beta rays by their discoverers Ernest Rutherford and Paul Villard at the beginning of the twentieth century, shortly before Williams went to university, with gamma rays, the electromagnetic radiation actually used in nuclear medicine. The orientation of his metaphorics was clear. Poetry should look to the new physics for inspiration. Another contributor, Muriel Rukeyser, had written a biography of the American physicist and chemist Josiah Willard Gibbs, whom Daniel Kevles describes as "one of the major theoretical physicists of the nineteenth century," in which she contended that "the world of the poet . . . is the scientist's world."[41] She had also published just the year before a book-length defense of poetry, *The Life of Poetry*, that called for a poetry modern enough to work with "the methods of science" rather than just report the "discoveries of science."[42]

Olson, Williams, and Rukeyser were part of a wider trend. Wallace Stevens, Marianne Moore, and others were also actively exploring the affinities and dissonances between poetry and the sciences. In a 1948 lecture, Williams spoke for many of them when he dismissed the poetry of a "ruined industrial background of waste and destruction" associated with Auden's generation because it is "becoming old-fashioned with the new physics taking its place."[43] By alluding to the latent powers of destruction in nuclear energy, Williams also made an implicit connection between the new physics and its social impact. Even those who disliked Williams recognized what he was doing. In his *Essay on Rime* Karl Shapiro expresses fascination tinged with skepticism about Williams's interest in science: "I for one / Have stared long hours at his discoveries / That seem at times the germs of serious science, / At times the baubles of the kaleidoscope."[44] Shapiro may have wondered whether these traces of "serious science" were more like infections or toys than discoveries that would contribute to "the horn of plenty" that America aspired to, but even he recognized that America's future depended on science as well as the arts and tacitly acknowledged that it was poets such as Williams who have had the most to say about science and poetry. And a significant number of poets were far more optimistic about the possibilities for a union of poetry and science, believing in the words of the science journalist Fred Dudley that it is "through poem, novel, drama, essay, we experiment with science until at last we know what to do with it—and are of course confronted with still newer science."[45]

Fjelde is exaggerating when he depicts *all* midcentury American poets as scientifically adept. In the prose statements about poetics that Ciardi insisted his poets contribute to the anthology alongside their poems, several of them make plain that they are resistant to the blandishments of the scientific attitude. Yet in doing so they underscore just how important it could be for a poet to define a position in relation to the sciences. Richard Wilbur insists that "a poem ought not to be fissionable," alluding both to the debate about the superbomb, or hydrogen bomb as we now know it, and to the antiformalism of modernist fragmentation.[46] And Randall Jarrell justifies his demurral at being asked by the editor to identify the audience he anticipates for his poetry by likening the poet's situation to the scientist's: "No one would say that a mathematician or scientist is chiefly or directly responsible to his readers; it is a mistake to say that a poet is."[47] Jarrell also contributes an essay on Robert Lowell as a placeholder for Lowell's own missing poetics statement and describes what is distinctive about his friend by saying, as if this were now unusual enough to deserve comment, that "Mr. Lowell has a completely unscientific, but thoroughly historical mind."[48] Lowell himself later remembered things differently. In the transcript of a talk he gave around 1960 looking back at his early development as a poet, he says that he was then aligned with those for whom "poetry was a form of knowledge, at least as valid as scientific knowledge."[49]

Fjelde was therefore being accurate when he suggested that the midcentury poets were very aware of the sciences, since even those who had no interest in science felt it necessary to justify this. Indeed, back in 1948 the leading Chicago critic R. S. Crane talked of "a morbid obsession" with "the problem of justifying and preserving poetry in an age of science" that "has resulted in an extraordinary florescence of modern apologies for poesy."[50] The question that divided both critics and poets was whether this florescence signaled a burgeoning of intellectual curiosity comparable to the work of the new natural sciences or a loss of poetic vitality and autonomy that risked leaving poetry as no more than a dried flower to be put under the scientific microscope.

Crane was prescient. Leading literary critics continued to rush into print to defend poetry against science, and 1950 saw the publication of two substantial studies of the dangerous influence of science on poetry: Hyatt Howe Waggoner's *The Heel of Elohim: Science and Values in Modern American Poetry* and Douglas

Bush's *Science and English Poetry: A Historical Sketch, 1590–1950.*[51] Waggoner depicts poets engaged in "defensive reactions against science," struggling with the idea that "man is not free, that he is the product of his genes and his conditioning and the vagaries of his id."[52] When Bush, a Harvard critic who had recently produced a volume in the magisterial Oxford History of English Literature, *The Early Seventeenth Century* (1945), issues his harsh judgment that "all modern poetry has been conditioned by science, even those areas that seem farthest removed from it," he may have thought he was doing little more than endorsing Cleanth Brooks's judgment made ten years earlier that "all poetry since the middle of the seventeenth century has been characterized by the impingement of science upon the poet's world."[53] Yet his conceptual addition of "conditioning" makes the new prosody resonate with the barking of Pavlov's dogs. The appearance of two books defending poetry against science must have seemed strong confirmation of Crane's acerbic observation.

Were postwar American poets really *conditioned* by science? Bush certainly still thought they were under the influence of science when he added a new preface to a 1967 reprint of *Science and English Poetry*, saying that the physical sciences are now "the governing and directing authority to which all else must bow," though the absence of the concept of conditioning is a reminder that the idea had lost favor by then.[54] Such fears that poetry was ceding epistemic and imaginative authority to the sciences were shared by a poet famous for his refusal to defer to governing authority. Allen Ginsberg states firmly in Donald Allen's landmark anthology, *The New American Poetry* (1960), that true poetic imagination arises from "unconditioned spirit." Although Ginsberg does not specify who or what might try to condition the poet, we can infer what would be doing the conditioning from his assertion that his poems are "Angelical Ravings" that are "trans-conceptual & non-verbal" and therefore have "nothing to do with dull materialist vagaries about who should shoot who." A "conditioned spirit" would not only conform to the modern American norms of sexual, economic, and aesthetic behavior, but also accept the scientific materialism that appeared to underwrite the political logic of the atomic bomb and the Cold War. A conditioned spirit would not be listening along with Ginsberg to "the music of the Spheres," sharing an experience of the prescientific cosmos.[55]

THE GENESIS OF "PROJECTIVE VERSE"

The January 9, 1950, issue of *Life* magazine led with a report on the recent American Association for the Advancement of Science (AAAS) conference in New York under the headline "U.S. Science Holds Its Biggest Powwow." An accompanying photograph shows a bearded Norbert Weiner standing on a platform in front of rows of impassive men in suits and ties, discussing the possibility of remedying deafness with the use of "new vibrators attached to fingertips."[56] In fact, technological developments in audiology were far from the most striking "news from the extended frontiers of knowledge" reported by the magazine. The Kinsey report was the subject of a special symposium (allegedly, the first special symposium organized by the AAAS since the appearance of *The Origin of Species*). Even more exciting than new insights into human sexuality, according to *Life*, was the news of a major breakthrough by Albert Einstein. A separate piece by Lincoln Barnett, "The Meaning of Einstein's New Theory," explains in detail the significance of Einstein's achievement: "Of all the news that emerged last week from the AAAS convention it is likely that the announcement that Dr. Albert Einstein had at long last completed the mathematical formulation of his Unified Field Theory will be remembered longest as a landmark on the unfolding frontiers of human knowledge."[57] The frontier metaphor had been standard since Vannevar Bush famously called scientific research "the endless frontier" in a widely circulated policy paper on the need, once wartime financial support for the hard sciences ended, to develop new mechanisms, such as a National Science Foundation espoused by Harlow Shapley, for peacetime government investment in research.[58] Although the AAAS scientists may be wearing suits, they are the new frontiersmen for whom field theory is even more interesting than sex.

The magazine made sure that its readers appreciated the significance of these developments by giving prominence to the views of the president of the association, Elvin Starker, who challenged the widespread perception that society was not yet ready for the responsibilities of the atomic age: "Science cannot stop while ethics catches up . . . and nobody should expect scientists to do all the thinking for the country."[59] Catching up with the scientists, finding a poetics capable of historicizing frontiers, and thinking about fields for the country are exactly what Olson would set out to do a month later as he began drafting the essay whose placement in Donald Allen's *New American Poetry* would confirm the essay's status as a manifesto for several generations of postwar American poets. By expanding the idea that poetry, like physics, can also work with fields, "Projective Verse" would implicitly claim that poetry can engage with similar concepts and materials to those used by nuclear scientists. Therefore, poems could be sites of inquiry just as vital to the present moment as physics. Poets too might be frontier explorers.

The timing of the publication of *Life*'s report of the announcement of Einstein's unified field theory at the AAAS conference and Olson's proposal for a field theory of poetry in "Projective Verse" is more than a coincidence. Indirect evidence suggests that Olson read Barnett's essay and that it provided the impetus to write out ideas that he had been gathering for some time. On the day after the issue of *Life* appeared, Olson wrote about Einstein to his friend (and later his lover) Frances Boldereff: "Before I forget it: somewhere last week, reading, I ran into—I think it was a propos Einstein—a terrific confirmation of that polar experience of yrs which I got into THE BABE, of the smallest and largest, and how that tension is what has blown the old concepts sky-high."[60] Olson is referring to his poem "The Babe" and its lines "Who it is looks, can look / into the smallest thing, and says / it was made for us both, we are committed / to be just, who says, to the very largest, / none is over us / when we have earned it."[61]

A similar questioning of the implications of the extremes of scale possible in the universe revealed by the natural sciences occurs in Barnett's *Life* magazine article. There he explains that Einstein's long-term aim has been to reconcile the seemingly incompatible theories of relativity, a cosmological perspective that pictures the universe "as a total and homogenous cosmic field composed of three dimensions of space and one dimension of time, these dimensions being as inseparable as the three spatial dimensions of Euclidean geometry," and the infinitesimal world of the particles and forces of quantum physics, because at present "they do not, as it were, speak the same language." The hope is that "if Einstein's new field equations achieve their full purpose, this gulf between the great and the small will be bridged, and man will be able to describe all the multifarious phenomena of nature in terms of a single harmonious edifice of cosmic laws."[62] Implicit in Olson's allusion to the article is a similar aspiration to find a poetic means to "speak the same language."

Of course, Olson remembered where he had read the "terrific confirmation," since the magazine had been published only the day before he wrote to Boldereff, but he probably felt coy about mentioning that the deep insight he was offering to his friend was borrowed from a drugstore magazine. No doubt he was also affected by the grandeur of the language in which the science was described. Barnett was a powerful science writer, and as we saw earlier, Olson had read and retained an earlier essay by Barnett, a 1949 profile of J. Robert Oppenheimer, because quotations from it found their way into a surviving draft memorandum probably written early in 1952 to the faculty of Black Mountain College.[63] In those early weeks of 1950 Olson must have been mulling over his ideas for "Projective Verse" because in a letter to Frances Boldereff written a month after news of Einstein's breakthrough, Olson mentions that he is "going to try and set down the limits & advantages of, this day," a new type of poetry he calls "PROJEC-TIVE / (projectile / (prospective / VS. / non-projective, or, closed / VERSE."[64]

A week later he sent Boldereff a draft of the essay. Despite the title "Projective Verse" its topic is not an existing body of poetry but the method by which such poetry can be written, the process of "composition by field." The phrase is repeated sufficiently often to make very clear the centrality of this idea. Using numbered subheadings, Olson argues that the poetic field has three key features. First, this field can be understood, similarly to the field in physics, as a distribution of energy. "A poem is energy transferred from where the poet got it (he will have some several causations), by way of the poem itself to, all the way over to, the reader. Okay. Then the poem itself must, at all points, be a high-energy construct and, at all points, an energy-discharge."[65] Second, just as physical fields arise from a scientific law, so too the field poem also depends on a "law," although this is a law not of matter but of poetic structure: "FORM IS NEVER MORE THAN AN EXTENSION OF CONTENT." And third, similarly to a field in physics, this field comprises a space of singular points that are in movement: "ONE PER-CEPTION MUST IMMEDIATELY AND DIRECTLY LEAD TO A FURTHER PERCEPTION. It means exactly what it says, is a matter of, at all points (even, I should say, of our management of daily reality as of the daily work) get on with it, keep moving, keep in, speed, the nerves, their speed, the perceptions, their, the acts, the split second acts, the whole business, keep it moving as fast as you can, citizen."[66] This unexpected interpellation of the aspiring poet not as scientist but as citizen is a reminder that Olson does not think of the physics of energy fields as the preserve of a scientific elite. The American poet, like other Americans, is, or should be, a *scientific* American by virtue of being a citizen. In fact, as we shall see later, other citizens were availing themselves of the new concepts. Physics was not the only science making use of fields, as Olson himself knew, and nor was he the only poet to be talking about the poem as a transfer of energy, an idea that also interested Muriel Rukeyser.

In advocating a field poetics, Olson might well have backed the wrong horse, for it soon turned out that Einstein had not solved the conceptual problem of constructing a unified physical science. The following year Einstein said, with considerable pathos, that "the unified field theory has been put into retirement."[67] To be fair, it was not Einstein himself who had claimed a breakthrough but his publishers, Princeton University Press, who were at the AAAS conference promoting a new edition of Einstein's *The Meaning of Relativity* based on the addition of a new chapter, a revised version of his 1948 paper "A Generalized Theory of Gravitation."[68] For several years Einstein had been ruefully telling friends that he thought his search for a unified field was never going to succeed. Unfortunately, Princeton University Press's ploy to push the new edition gave such an exciting impression of a great breakthrough that a front-page *New York Times* article immediately reported their announcement under the inflationary headline "New Einstein Theory Gives a Master Key to Universe." The article explains that this theory "attempts to interrelate all known physical phenomena into one all-embracing intellectual concept, thus providing one major master key to all the multiple phenomena and forces in which the material universe manifests itself to man."[69] Three "alls" in one sentence describing a theory of everything ought to have alerted even the most credulous that something was amiss. The publishers even displayed a copy of the manuscript at the conference as if to prove their claim, and the *New York Times* printed a page of equations instead of a photograph. All this stir confirms just how ready the public were to believe that nuclear physics was capable of providing a "master key" to the universe in the form of a grand synthesis of all material knowledge in one foundational theory. And for a careful reader such as Olson the message appeared to be that the universe and everything in it could be represented as a *field*.

"Projective Verse" might have been sunk by extensive metaphorical references to a defunct theory, but Olson was lucky. Einstein's unified field key to the universe might not open the door, but other fields were proving much more useful at opening up new possibilities for physics. While unified field theory moved to the margins of the discipline, *quantum* field theory was becoming central to nuclear physics and would remain so for several decades, long enough to ensure that "Projective Verse" retained its appearance of timeliness. The final paragraph of the *New York Times* article on Einstein's alleged discovery mentions his ambition to unlock the secrets of quantum physics, which he notoriously mistrusted, with his own field theory. But it would be quantum physics in the form of a newly completed quantum electrodynamics that would unlock at least one new door into the unknown. QED, as it was known, remains one of the most important achievements of modern physics.

The theory was first broached in the late 1920s, but mathematical difficulties with the treatment of infinities limited its use. Then in the late 1940s several young physicists, notably Julian Schwinger, Richard Feynman, Freeman Dyson, and Sin-Itiro Tomanaga, reconfigured the theory in a series of major papers in theoretical physics. Their new theory had its origins in awkward discrepancies between mathematical predictions of the behavior of the electron in an electromagnetic field and experimental measurements of actual electrons. How did the electron interact with its own field? As Feynman's biographer puts it, previously the mathematics showed that theoretically the particle could gain infinite extra mass/energy under certain conditions, while experiment showed the opposite, that the effect was negligible. More sophisticated experiments revealed only that the effect was measurable, at once "theoretically infinite and experimentally real"—that is to say, there was a gain but not an infinite one.[70] The young physicists combined mathematical research with new modes of visualizing the processes to create a successful model of the electron based on justifiable mathematical elimination of unwanted infinities in the equations by a process known as "renormalization" (another of the many latent metaphors whose resonance the literary critic is tempted to unpack). The *New York Times* reporter William Laurence reported twice on Schwinger's triumphal presentation of his new calculations of the energy of the electron, introducing him as the person "whom American physicists regard as the heir-apparent to the mantle of Einstein."[71] Schwinger had come to the "rescue" of Paul Dirac's theory of the electron, which needed saving from new experimental results seemingly incompatible with the otherwise very successful theory. "With this threat to the basic theory of the electron removed, Dr. Schwinger told the physicists, the road has been opened to attack another formidable citadel of the universe, the realm of the mysterious meson, which holds the key to the very foundations of the physical universe."[72] This time the key did unlock the door to the cosmos. Thirty years later, QED field theory itself could still be described by Richard Feynman as "the jewel of our physics—our proudest possession."[73]

"Projective Verse" is so familiar to readers of postwar American poetry that the initial impact of its novel style mixed with already widely circulating mid-century ideas about poetry and science is no longer easy to recognize. Today what were once seen as its virtues are overshadowed by its now more evident idiosyncrasies and conceptual weaknesses. In the pedagogic contexts where it is mostly to be found, its prescriptiveness fits almost too well the desire for a set of guidelines for students of poetry on how to compose verse in the modernist lineage. For more experienced readers of poetry, its confident, sometimes-strident mentoring tone and its seductive hints of secrets or vast cultural conflicts in which poetry participates can easily look like an orator's tricks. Disembedded from its original historical moment and from its place of publication in *Poetry New York*, the specificity of its metaphors based on high-energy physics blurs into worn generalizations about poetic vitality.

The first appearance in public of "Projective Verse" was not auspicious. It must have seemed unlikely that readers would discern either its relevance or its targets among the decidedly mixed company offered by the inexperienced editors of *Poetry New York*. Behind the magazine's back, Olson was dismissive. He wrote to Cid Corman, who would become Olson's first literary impresario, that he hoped Corman would not repeat the mistakes of *Poetry New York*: "And why should I weep to see you get out only another of such MAGS as Hudson, or PNY [i.e., *Poetry New York*], or NINE?"[74] These magazines and the poets who published in them cared too much for the "*literary* inheritance" or "cultural tradition," a preoccupation with the literary scene that led them to create just the sort of work presented in this edition of *Poetry New York*, which Olson cruelly parodies as "the 'O, I am here, and O, I am human, and O, isn't it, weary-or-howlyrically [*sic*] lovely' (Barbara Gibbs) or 'o not pretty yet but will be' (Rukeyser or the lazy leftists) or 'it stinks, because, tho I don't say, I stink, which is what humans always have done, look at Diogenes' or any same, which you may document, even on such a high level (so much a source as) T. S. (GI) Eliot?"[75] Apart from a poem by William Carlos Williams, the verse was barely modernist, and as Alan Golding says: "Corman rightly pointed out, 'almost all the verse is NON-projective.'"[76] This relative weakness of most of the poetry in the magazine actually helps make Olson's point about the need for a new poetics in a way that publication in a journal with much stronger material might not have done. The only other obvious modernist text is an extract from Paul Gauguin's journal, a self-consuming meditation on the ridiculousness of his own theories in which he identifies with Alfred Jarry's scurrilous, mad puppet Ubu. Juxtaposition with Gauguin's self-mockery may well have irritated Olson more than the undergraduate quality of most of the verse. Only Fjelde's concluding review of Ciardi's anthology, which I mentioned earlier, gives a sense of the intellectual context with which Olson is concerned.

Perhaps this unpromising context in *Poetry New York* actually helped trigger the slow process of canonizing Olson's essay. Accompanied by weak poetry from a supposedly upcoming generation of poets, his call for a new poetics might have seemed necessary and timely, while Fjelde's attribution of a new scientific outlook to the midcentury poets reinforced Olson's tacit endorsement of scientific literacy. The setting of Olson's essay could be said to dramatize the experience that Olson alludes to in "Human Universe" when he talks of a person being "astonished he can triumph over his own incoherence."[77] But this struggle and its urgency are also a puzzle. Why does projective composition urgently matter to the world? When Olson tries to cash out the benefits of the new techniques in the second half of the essay, he has surprisingly little to offer. This may be why his attempt to explain why bettering the condition of poetry in 1950 is of great cultural urgency has been much less discussed than the account of compositional method and, to judge from some responses to his essay, was readily forgotten. Certainly, William Carlos Williams appears to do so when he reprints only the first half of "Projective Verse" in his *Autobiography*. It was publication in Williams's autobiography that first gave Olson's ideas wide circulation, though not everyone noticed they were his. The philosopher Henry Bugbee credited the impetus to begin his influential experiment with a philosophical journal in August 1952 to what he assumed to be Williams's suggestions (Olson is not mentioned) that in the poem "one perception must move instantly on another" and that "form is never more than an extension of content." These were helpful because Bugbee believed that "philosophy is in the end an approximation to the poem."[78]

Although Olson's essay insists that the postwar world demands a new poetics, it is notable for what it doesn't talk about: contemporary history and politics; specific poets and poems other than a couple of uncontroversial classics; poetic movements such as Surrealism that it might have been expected to notice; and specifically identified contemporary areas of knowledge. Most remarkably, it has nothing to say about the whole issue of what should constitute poetic content. Olson appears to see a threat to the future of poetry that justifies his claim that new methods are needed, yet for some reason he is unable or unwilling to be specific about the threat. Perhaps the difficulty of articulating the causes of this urgency arises from the challenge Olson encounters in conceptualizing what it means for poetry to be "pushing knowledge forward at the edge of the unknown" in an age of nuclear physics. What was it about nuclear physics that so appealed to a poet wanting epistemic authority for poetry? Whatever the roots of his sense of urgency, his strategic decision to be nonspecific, which may have looked like a weakness at the time the essay was written, helped it travel forward in time to a decade when the baggage of references to now-eclipsed poetic luminaries of the 1940s could well have hindered its wide reception.

THE MIDCENTURY ATOM

The midcentury atom was highly mathematical. This created challenges even for the experts, many of whom shared an anxiety that the mathematics might be creating too much distance from actual processes. In 1949, for instance, Freeman Dyson writes that "one must assume provisionally that the mathematical formalism corresponds to something existing in nature."[79] But what? For many nonscientists encountering the writings of the nuclear physicists, the mathematical formalism was so opaque that it corresponded to nothing. The following equation shows what electrons looked like to readers of a physics journal:

$$K^{(1)}(3, 4; 1, 2) = -ie^2 \iint K_{+a}(3, 5)K_{+b}(4, 6)\gamma_{a\mu}\gamma_{b\mu}$$
$$\times \delta_+(S_{56}^2)K_{+a}(5, 1)K_{+b}(6, 2)d\tau_5 d\tau_6$$

This particular formulation of what was known as the Dirac equation, which Richard Feynman describes as "our fundamental equation for electrodynamics," occurs in an article by Feynman in which he attempts to provide a "considerable simplification" of the mathematics of quantum interactions so that the matrix equations can be "understood directly from a physical point of view."[80] The article also contains some of his famous diagrams, drawings that look like a set of obliquely angled arrows connected by springs and, to the nonexpert, are little more comprehensible than the equations. To anyone not trained in advanced mathematics, the symbols in the equations for the behavior of electrons would have appeared as unintelligible and as confined to an elite world of specialists as did the Chinese ideograms of Ezra Pound's later Cantos to nonspecialist, non-Chinese-speaking readers. How then did nuclear physics entice so many scientists from other disciplines, along with intellectuals in the arts, to emulate its concepts and methods?

The usual answer is that nuclear physics owed its cultural influence to the creation of the atomic bomb, an argument made at the time by many scientists and other intellectuals. John Dunning opens a survey essay, "Atomic Structure and Energy," by saying, "The atomic bomb dramatized the opening of this so-called Atomic Age" (the prevalence of the theatrical metaphor probably encouraged Olson to give Oppenheimer the not especially exalted profession of "stage manager").[81] The atom's energy had been not only released but politicized. Poets were no different from other intellectuals; everyone wanted to know more about this science, whose constantly changing names (atomic physics, nuclear physics, quantum physics, particle physics, and high-energy physics, for instance) confusingly indicated the rapidity with which this knowledge was developing and altering its empirical and theoretical orientations. They wanted to know how it could be "that energy and mass are different aspects of the same basic cosmic stuff, and that the two can be converted one into the other."[82] This was how Selig Hecht, about whom Olson once said that he "wrote a prose which taught me a good deal," explained the explosive power of uranium in an introduction to the science of the atomic bomb.[83] This and many similar attempts to explain in nontechnical language the theories and discoveries of the new physics opened the exciting possibility that a discourse about the structure of materials and the material world could be converted into a discourse about this "cosmic stuff" *energy*, which was a much more familiar notion than atomic structure.

The midcentury atom was also secret, which no doubt made knowledge of nuclear physics all the more alluring. In January 1950, after a series of secret committee meetings in which Oppenheimer and Conant argued against further development of nuclear weapons, President Truman announced on the radio that the United States would go ahead with the hydrogen bomb, and a fierce struggle between physicists and government over control of information about nuclear science broke into the open. Oppenheimer made new enemies by going on television and arguing that "it is a grave danger for us that these decisions are taken on the basis of facts held secret."[84] The April issues of both *Scientific American* and *Bulletin of the Atomic Scientists* had to be withdrawn and reprinted to censor paragraphs in articles Hans Bethe wrote about the hydrogen bomb, articles in which the Atomic Energy Commission feared he had given away too much information about the importance of using heavy isotopes of hydrogen in the manufacture of the atomic bomb.[85]

Many publications of the time never saw the light of day. Consider the cover page for a typical research paper from 1949 on nuclear physics commissioned by the Atomic Energy Commission and authored by Herman Feshbach, Julian Schwinger, and John Harr, only now available online.[86] It is stamped with the words "This document is PUBLICLY RELEASABLE" and the date "3/4/05," indicating that it was not released to the public for more than fifty years after its composition. The cover bears a further index of the pressures of secrecy in an odd caveat written at the time it was issued: despite being one of its authors, "Professor Schwinger has not yet had the opportunity of reading this manuscript." Looking back on the past fifty years, Peter Galison estimates that "the classified universe" of secret documents may be larger than the unclassified one, and a significant part of this hidden archive is derived from nuclear research: "nuclear weapons knowledge is born secret."[87]

The bomb's importance might be enough to explain why nuclear physics should generate curiosity and critique among researchers and intellectuals in many disciplines. But is this enough to explain why the very epistemology of nuclear physics should have been so widely emulated? Why should physics, particularly nuclear physics, appear the epitome of the methodology of rigorous inquiry even in quite different realms such as the study of mind or society, let alone the creation of poetry, given its uncompromising mathematical abstraction, reliance on heavy machinery for empirical research, and focus on inanimate matter? Actual physics experimentation required messy engineering: the endless problems that 1940s physicists had with the fiddly, temperamental cloud chamber typify this.[88] All the high-energy machinery would seem very far from the needs of psychology or genetics, let alone poetry.

To explain the hold over the midcentury imagination exercised by nuclear physics we need to reflect on the totalizing power of its theory and on a paradox of this totalization: its deliberate acting out of epistemic uncertainty. Hans Bethe's populist portrait of the atom is a typical, if striking, example of the universalizing claims of nuclear physics in the 1940s: "The atom is the hero of the day. It is the most democratic hero you can choose, because it is everywhere and in everything."[89] Democratic? Bethe is not just aligning physics with the defense of democracy that defined the politics of the Allies but also alluding to the universality, vast explanatory power, and cosmological reach of atomic physics, which treats everything in the material world as if it were constructed from different combinations of the same very few humble constituents.

From atoms to galaxies, or in Olson's words from the "smallest thing . . . to the very largest," the methods of inquiry used in quantum physics can yield an explanation of everything that exists.[90] Dunning's friendly formulation that "you and I and all the world around us are made of some 92 basic types of atoms which we call elements," like Bethe's democratic atom, makes it possible to elide the many supervening levels of organization between an atom and the ego and to infer that selfhood, language, belief, and the body are all naturally constructed, independently of human agency or cognition, from the constituents that the discourse of the reductionist science has clearly in its purview: energy fields and particles.[91] A poet might reasonably conclude that a poetry curious to reveal more about what is not well understood should be investigating not only the politics of nuclear energy but also the energy and particles that make poetry and its world possible. Perhaps even poetry could be analyzed into fields and fundamental units.

Nuclear physics was enjoying a peacetime success that appeared to confirm the power of its pan-theoretical ambitions. At this time about twenty particles had been discovered, the more recent ones—two mesons, the pion, and the muon—by studying cloud chamber tracks of cosmic rays. In 1948 *Time* magazine announced the artificial creation of mesons at the Rad Lab in Berkeley and expressed the hope that this would "lead in the direction of a vastly better source of atomic energy than the fission of uranium."[92] Great things were definitely expected of nuclear research. It was predicted that airplanes would be powered by atomic engines, and nuclear radiation would cure cancer. In August 1950 *Science News-Letter* confidently anticipated "a mathematical concept or law that will apply to the universe as a whole as well as to the microcosmos of the atom will be worked out."[93] Behind much of this exaggerated optimism was the impressive success of quantum electrodynamics, a field theory of the electron. Success yes, but also making ever more plain just how elusive certainty was.

The midcentury atom was not only a government secret but itself secretive. Physicists repeatedly talked of the difficulty of understanding the phenomenal world. In a 1948 article in *Physical Review* (and therefore aimed at the specialists) Julian Schwinger worries about the risk of distortion in attempts to conceptualize recent experimental findings that clash with prediction: "Attempts to avoid the divergence difficulties of quantum electrodynamics by mutilation of the theory have been uniformly unsuccessful. . . . The elementary phenomena in which divergences occur, in consequence of virtual transitions involving particles of unlimited energy, are the polarization of the vacuum and the self-energy of the electron, effects which essentially express the interaction of the electromagnetic and matter fields with their own vacuum fluctuations."[94] The success of the theory that eliminated the unwanted infinities, quantum electrodynamics, was encapsulated in its clever acronym QED, although for these physicists there was also a kind of protective irony: no one was quite sure what the *quod* was, nor how best its *demonstrandum* could be achieved, and they were open about the problems of the *erat*. In his lectures on QED, Feynman tells his audience "not to turn away because you don't understand it. You see, my physics students don't understand it either. That is because *I* don't understand it. Nobody does."[95] Altogether, the elusiveness of the realities behind the mathematics, the unobservability of particles, and the imperfections of theory all meant that the midcentury atom was an epistemological puzzle. It was not surprising that Freeman Dyson sounded so tentative when he said that "one must assume provisionally that the mathematical formalism corresponds to something existing in nature."[96]

"*Something* existing in nature": but what? Indeed, what did this midcentury atom actually look like? No one was quite sure. A physicist writing for a nonspecialist audience in the *Science News-Letter* on the future of nuclear physics admits to its lay readership what Schwinger had said in the jargon of physics: that "just visualizing the atom is a strain upon the imagination." As a result, he continues, "no puzzle is more troublesome just now than to try to figure out the action within this terrifically concentrated atomic core, made up of protons and neutrons." [97] In 1950 as a student at MIT, Murray Gell-Mann, the future architect of quark theory, was pondering this problem of representation as he tried to finish his thesis and struggled with the question of who was right: those who thought the nucleus was more like a planetary system of neutrons and protons orbiting in shells or those who thought it was more like a drop of water (some physicists were even fudging things by combining both images in the model of "single-particle levels in a suitably proportioned potential *well*" [my italics]).[98] How could physicists communicate a picture of this elusive, infinitesimal world to each other, let alone to the wider public?

The cloud chamber came to their rescue. Evans Hayward, a researcher at the Rad Lab, explains that this detection apparatus is "a machine that makes visible, and therefore enables one to photograph, the paths of charged particles: electrons, protons, mesons, etc." (this is a very eloquent "etc.").[99] Cloud chamber photographs were the closest the nonspecialists could get to seeing the midcentury atom, which appeared as white particle tracks accompanied by many peripheral white dots and dashes on a deep-black background, all of it looking like nothing so much as chalk scratches on a heavily used blackboard. The residual unexplained marks acted as visual reminders that much about these particle interactions was yet to be understood. These photographs could indeed be read by nonphysicists as unusually vivid images of the intersection of the known and the unknown. The philosopher of science Andrew Pickering wittily calls such a zone "the mangle," deliberately using the image of a homely laundry device to represent a space of discovery that is "temporally emergent," constantly altering and unpredictable in its resistance to the capture of material agency (in the form of particles, cloud chambers, electromagnetic fields and so on) by the scientific apparatus.[100]

Physicists made a great virtue of this epistemic uncertainty. Everywhere in the writing of the physicists of this time is an excitement about the unknown, and they talk lovingly about it, as if they had full rights over it, as if this was *their* unknown. In a radio talk in the same series in which Bethe talked about the democratic atom, Isidor Rabi, another of the Los Alamos scientists, describes the nucleus as the site where "vast energy is locked" and "mysterious forces of attraction" hold protons and neutrons fast together against powerful forces of repulsion. Talking about the "mysteriousness" of these forces is deliberate; Rabi admits that he is trying to point out the "limitless unknown" of nuclear physics to call attention to the opportunities in the field for "young men and women to make great discoveries" (he is unusual in appealing to both genders).[101] Theoretical physicists were inclined to think that it was theory rather than experiment that was taking nuclear physics into the unknown, not least because they had already theorized uncertainty. A telling reminder of just how important this concept had become at midcentury occurs in a graduation address given at Brown University in 1947 by the theoretical physicist Richard Tolman, former chief science adviser to General Leslie Groves on the Manhattan Project. Having just a few minutes to talk about the "new and improved ideas provided by quantum mechanics," he mentions at once Werner Heisenberg's uncertainty principle: "we can now see that a complete knowledge of the state of a physical system at any given instant is not sufficient to permit an exact prediction of that system's future behavior" due to "the uncontrolled disturbances introduced by the very act of observation."[102]

By midcentury, release from the extreme conditions under which the sciences had operated during the war had unleashed research energies in many new directions as well as instigating a sharp-elbowed struggle for continuing government investment, which necessitated new rhetorical advocacies for the sciences, destabilized disciplinary boundaries to an unusual degree, and in some cases rendered existing scientific and epistemic authority questionable. We can observe a modern version of a process that Mary Poovey describes in her study of literature and economics in the eighteenth and nineteenth centuries, when "writers tried to differentiate among kinds of writing—so that they could rank them, acquire social authority for some but not others, produce disciplinary norms, and claim for themselves institutional positions."[103] By the mid-twentieth century, the legitimation of disciplinary knowledge was increasingly the responsibility of tightly regulated networks of review and publication that delegated representative status to their writers, making the acquisition of social and epistemic authority in domains over which natural and social scientists were extending their research increasingly hard for those, like poets, who were outside the institutional networks.

One way to enter the fray was to join those exploring currently authoritative epistemic metaphors and open up the field concept, a concept with a good scientific pedigree that has continued to prove extremely "fertile" in repeatedly suggesting further lines of investigation.[104] In her classic study *Force and Fields* (1961), Mary Hesse traces the long history of the concept in physics back to the attempt to understand the action of forces at a distance on which Newton's physics depended.[105] She carefully categorizes the concept of a field as a "model" (which we could also call a "construct"), a provisional description of the actual world in metaphoric terms. The actual implications of its constitutive metaphors will always need to be checked empirically against experimental findings. This history of the field concept is the backstory to its modern uses in relativity theory and quantum physics. According to the philosopher of physics Martin Krieger, the field concept is the third step in a process of abstraction by which physicists make the universe available for experimentation. First, the physicist has to design "walls" that will achieve the isolation of experimental systems from the noisy turbulent mixture of materials that make up our everyday world. Experimenters need quiet. The walls they build form boundaries that can also act as "interfaces, functions, skins, and dynamical processes." Many different sorts of structure, even regular crystal lattices, can function as boundaries that crucially limit the degrees of freedom (effectively, the range of variables) of the universe to make them manageable. Walls, however, are only one part of the research strategy; physicists "are concerned with what is enclosed by those walls, the rooms or particles, and with what is excluded, the outdoors or the field."[106] These particles can be any kind of object, as long as they are localized, nominalized, and individuated.

Physicists then introduce a third key epistemic tool, the concept of field, "something like a flowing fluid or a magnetic field" that results from a source such as a particle, "a field that provisions that empty space with properties or degrees of freedom, such as the electric field at every point." Physicists keep their metaphor under tight control. A field is generally smooth, is connected to particles, balances the inflow and outflow of energy, and enframes the idea of a *path* through the field as a means of understanding the instabilities and peculiarities of specific particles: "Fields also provide for local interactions, by, in effect, transporting the effect of one particle over to another particle with which it is to interact."[107] Although Krieger is specifically discussing particle physics, the features of the field concept that he identifies as most important tend to be the most salient for other disciplines.

The concept of the field not only was conceptually fertile but had also recently proven immensely successful in answering fundamental questions about the material universe. Other disciplines were irresistibly drawn to try opening the field for themselves. For the most part, however, they reflected on what they were doing. For although the reiteration of this metaphor across multiple disciplines can be tracked back to authoritative usage by prominent, attested theories in nuclear physics, each discipline attributed to the field concept its own endogenous conceptual architecture, whose wider epistemic status was a matter of considerable negotiation and dissent, both within disciplines and even more so between them. Underwriting all this epistemological competition, to borrow a term from Sarah Winter, was the assumption that quantum physics was a theory of everything.[108]

EPISTEMOLOGICAL COMPETITION

Norbert Wiener was a natural scientist who aspired to unite natural and social sciences and believed that information itself could act as "the negative of the quantity usually described as entropy in similar situations."[109] He was one of those scientists from outside nuclear physics who was impelled by its swaggering epistemic authority to copy its moves for another arena of human knowledge. His Humboldtian account of the universe in *Cybernetics* (1948) merges a theory of feedback in machines and animals with the new information theory of communication and, like *What Is Life?*, was widely read by poets. Robert von Hallberg aptly describes the appeal of cybernetics for poets as a "dramatically expansionist approach to knowledge, for its advocates propose to explain and manage vast fields of inquiry."[110] Wiener is as at home talking about Leibniz, Locke, and Hume as he is talking about computers, steam engines, catalog systems, and brains and their pathologies such as ataxia or Parkinson's disease. He repeatedly reminds his readers that information theory and its statistical foundations have strong affinities with quantum theory because the tendency to informational error seems "to have something in common with the contrasting problems of the measure of position and momentum to be found in the Heisenberg quantum mechanics, as described according to his Principle of Uncertainty."[111] A poet reading his final chapter might well be reminded of the Ezra Pound of the recently published *Pisan Cantos*. The opening paragraph of the chapter moves rapidly from references to classical Greece, the Soviet Union, Hobbes, Leibniz, and blood to references to the cells of living organisms, before it winds up with a technical description of the internal structure of the Portuguese man-of-war jellyfish. Both Rukeyser and Olson were influenced by Wiener's expansive vision of a new science: Rukeyser discussed *Cybernetics* in *The Life of Poetry* and Olson actually cited from it in his breakthrough poem "The Kingfishers" (1950).

Nuclear physics had such hegemonic cultural authority at the time that even thinkers whose domains might seem very far indeed from that of quantum physics nevertheless drew on its metaphors and epistemic status. One of the reasons for the power of Quine's famous critique of the tenets of logical positivism, "Two Dogmas of Empiricism," is its knowingness about the degree to which recent philosophy has leaned on the authority of physics.[112] He argues against what he provocatively calls the dogmas that, first, there is a firm boundary between synthetic and analytic propositions (roughly the distinction between statements that depend on references to the state of the world and statements that depend solely on their logical form) and, second, that true propositions are ultimately grounded on nonconceptual direct experience (a version of "reductionism," or what Wilfrid Sellars would call "the myth of the given").[113] Quine repeatedly alludes to various aspects of scientific authority and, by implication, to its use and abuse by intellectual borrowers. He alleges that Rudolf Carnap's philosophy of language reduces all the tangles of ordinary language to "constructions" based on "logical particles" whose truth can be analyzed into "state-descriptions" or "exhaustive assignments of truth-values to the atomic, or noncompound, statements of the language." Quine argues that it is impossible to treat truths and knowledge as divisible in this manner. It is only when we consider the whole picture that we can evaluate its truth: "The totality of our so-called knowledge or beliefs, from the most casual matters of geography and history to the profoundest laws of atomic physics or even of pure mathematics and logic, is a man-made fabric which impinges on experience only along the edges."[114]

Struggling to explain concisely and vividly how to picture this fabric of knowledge, Quine rescues himself with a field metaphor: "Or, to change the figure, total science is like a field of force whose boundary conditions are experience."[115] Despite having rapped Carnap on the knuckles for misusing atomic metaphors, he too finds that he needs to draw on the epistemic authority of those "profoundest laws." Then in the final section Quine goes much further and concedes that, as far as conceptual schemes are concerned, physical objects, energies, forces, and even mathematical classes are not epistemologically different from the gods:

> As an empiricist I continue to think of the conceptual scheme of science as a tool, ultimately, for predicting future experience in the light of past experience. Physical objects are conceptually imported into the situation as convenient intermediaries—not by definition in terms of experience, but simply as irreducible posits comparable, epistemologically, to the gods of Homer. For my part I do, qua lay physicist, believe in physical objects and not in Homer's gods; and I consider it a scientific error to believe otherwise. . . . Physical objects, small and large, are not the only posits. Forces are another example; and indeed we are told nowadays that the boundary between energy and matter is obsolete. Moreover, the abstract entities which are the substance of mathematics—ultimately classes and classes of

classes and so on up—are another posit in the same spirit. Epistemologically these are myths on the same footing with physical objects and gods, neither better nor worse except for differences in the degree to which they expedite our dealings with sense experiences.[116]

Although Quine surely does not intend his readers to draw this inference, poets might well have concluded that he would support the stance of Robert Duncan, who believed that scientific concepts of forces and fields should be treated on a par with figures from theosophy, mythology, and the poetic imagination.[117]

Literary critics were also ready to lay claim to a little of the epistemic authority of the new physics. In a nervous gesture in the penultimate paragraph of *Theory of Literature* (1949), René Wellek and Austin Warren try to dispel the impression that they are advocating an effete European, idealist model of literature rather than a robust and up-to-date American one by insisting that their literary theory aligns well with scientific developments. "If we reject some of the preconceptions of nineteenth-century scientism—its atomism, its excessive determinism, its skeptical relativism—we are thereby in agreement with well-nigh all of the physical and social sciences, for with them today, revolutionary concepts such as patterns, fields, and *Gestalt* have superseded the old concepts of atomism, and with them determinism is no longer a generally accepted dogma." Even if they were not aware that the dominance of the field concept in particle physics had been vindicated by the success of quantum electrodynamics at the end of the 1940s, both philosophers and critics assumed that field theory carried the imprimatur of Albert Einstein himself. In their final paragraph Wellek and Warren optimistically claim that the whole "nineteenth-century epistemology" that reduced the humanities "to the status of pseudo-sciences" is no longer valid. Their unexplored yet prescient implication is that literary theory itself might in future lay claim to scientific status of some kind.[118] Around 1950, it seemed as if every field of inquiry, even literature, had to have its own field theory. Poets were no different.

Genetics, cybernetics, psychology, and even philosophy might vire epistemic authority from quantum physics, but they were, after all, further investing in their already-existing status as valid forms of *research*. Their reputable cognitive methodologies provided frameworks of inquiry that led to recognized knowledge. Poetry started much further back. Poetry was, as many thinkers did not stint to argue, seemingly incapable of inquiry and could produce only versions of Richards's "pseudo-statements," those fake propositions with no truth-value. If poets were to claim some territory on Vannevar Bush's endless frontier of knowledge, standing shoulder to shoulder with physicists, cyberneticists, social scientists, philosophers, and literary theorists, they would need methodological credit. Could they too base their claims on an endorsement of the physics paradigm of reducing complexity to fundamental particles and their interactions? If so, how could they also plausibly lay claim to the idea that poetry might be a mode of inquiry? The next chapter shows that Muriel Rukeyser and Charles Olson found most helpful the work of social scientists who had worked through the implications of borrowing concepts from physics.

* 4 *

CONCEPTUAL SCHEMES

*The Midcentury Poetics of
Muriel Rukeyser and Charles Olson*

THE FIELD CONCEPT

"Can science save us?" This melodramatic address to popular hopes for a better future after the war was both the title and theme of a 1947 defense of sociology by George Lundberg, a former president of the American Sociological Association and the author of the college textbook *Sociology*. His widely read book *Can Science Save Us?* not only advocates the adoption by the social sciences of methods based on those used in the natural sciences but also indulges an imperialist tendency that appears to threaten to displace the humanities for good. Salvation depends on being as scientific as physics, while customizing its methods to fit social research: "Can science save us? Yes, but we must not expect physical science to solve social problems. . . . We cannot expect atomic fission to reveal the nature of the social atom and the manner of its control. If we want results in improved human relations we must direct our research to the solution of these problems."[1]

Lundberg is a determined player in the epistemological competition among research disciplines. He may say that "we cannot expect atomic fission to reveal the nature of the social atom and the manner of its control," but his very manner of saying it, as well as his unequivocal answer to the question posed by his title makes clear his conviction that sociology can emulate the scientific methods of the natural sciences. Social scientists are, for instance, developing "laws of social behavior comparable to the physical laws."[2] Many sociologists, psychologists, anthropologists, and archeologists were then arguing hard for the right to call themselves scientists and went on to campaign in the 1950s to have their fields of science recognized by the National Science Foundation. Lundberg's book helped keep up the pressure and was republished in a new edition in 1961. At the same time, he and other social scientists were busy annexing for study by these new scientific methods the very domains of human experience that poets and artists had thought they had rights to explore.

According to Lundberg, science can "save us," but not everyone can be a rescuer. Lundberg is blunt. "Throughout history, and to a considerable extent at present, literature has been regarded also as a sort of social science." Now things are different, although "unfortunately, it is not yet clear to many writers that, in the event that the portrayals of literature do not check with scientific fact, science must take precedence as a guide to all practical achievements." Social science has replaced literature as a source of knowledge. Old-fashioned thinkers may believe "that the poet, the philosopher, the novelist, and the classical scholar, rather than the social scientist, are still the authorities on human relations," but they are wrong: "literature and the other arts are not substitutes for social science." Poets should give way gracefully to their successors, because even if "social science seems to encroach upon traditional, vested areas in the academic world," rational thinkers will soon grasp that it is inevitable that the arts should relinquish their vested interests, since these are based on emotion, not reason. If poetry does any future cultural work, it will be to write copy for the scientists. Poets risk being self-regarding fantasists unless they make the effort to become reliable cultural brokers for science. Fortunately, many have already tacitly recognized the superiority of the sciences to the arts: "It should not be forgotten either that poets have found in science and the scientific attitude liberation for the spirit as well as the body of man."[3]

Although similar beliefs were widely held, Lundberg was not a major figure outside his own field. Far more influential in the drive for social studies to become more scientific was the German émigré social psychologist Kurt Lewin, who arrived in America in 1933 and quickly established himself. He too was excited at the prospect that the new social sciences could expand into the human domain that was once firmly associated with the arts and philosophy, but unlike Lundberg, he had a sophisticated grasp of methodological issues. His new mode of group, or "field," psychology laid exclusive claim to discovery across a large area of what Charles Olson called "the close world which the human is," territory that he and poets such as Muriel Rukeyser believed poetry had at least equal rights to investigate.[4] Lewin's expansionist ambitions on behalf of his field psychology would, however, unexpectedly yield fertile territory to the arts and especially poetry through his far-seeing, detailed elaboration of the epistemic metaphor of the field. Variously geometric, topological, spatial, temporal, interactive, and, above all, dynamic, his "field" would prove as helpful to late-modernist poetry as to the study of human groups. Less well known today outside psychology, Lewin was a dominant figure in his own time. An obituary ranks him with Freud: "Freud will be revered for his first unraveling of the complexities of the individual history, and Lewin for his first envisioning of the dynamic laws according to which individuals behave as they do to their contemporaneous environments."[5] A history of communication theory goes further still, setting Lewin alongside Darwin, Marx, Freud, and contemporary social theorists Wilbur Schramm, Paul Lazarsfeld, Carl Hovman, and Norbert Wiener.[6]

Lewin, as Rosmarie Waldrop first observed, is likely to be one source of Olson's concept of "composition by field."[7] The sheer scale of the claims Lewin made for field psychology could well have impressed Olson. The aim of any new scientific theory in social psychology, says Lewin, should be to integrate a wide range of phenomena and "treat cultural, historical, sociological, psychological, and physical facts on a common ground," ground that can be both metaphorical and literal.[8] The researcher, says Lewin,

> finds himself in the midst of a rich and vast land full of strange happenings: there are men killing themselves; a child playing; a child forming his lips trying to say his first word; a person who having fallen in love and being caught in an unhappy situation is not willing or not able to find a way out; there is the mystical state called hypnosis, where the will of one person seems to govern another person; there is the reaching out for higher, and more difficult goals; loyalty to a group; dreaming; planning; exploring the world; and so on without end. It is an immense continent full of fascination and power and full of stretches of land where no one ever has set foot. Psychology is out to conquer this continent, to find out where its treasures are hidden, to investigate its danger spots, to master its vast forces, and to utilize its energies.[9]

No one has ever set foot there? Poets and other artists would surely have begged to differ. This social scientist is determined that the psychologists will be out on a frontier of knowledge that sounds remarkably like terrain that until recently would have been regarded as the very field in which artists and poets were already indigenous. And this is just how the founder of communication studies in America, Wilbur Schramm, remembers Lewin: "He was the Columbus, the Francis Drake, the Captain Cook of social psychology. I have never seen him when the excitement of exploration did not come through in everything he did and said. He was always in the process of 'conquering the endless continent.'"[10]

When Olson referred to the field of the poem, was he just borrowing a metaphor that was in the air thanks to Lewin and his followers, or did Olson know that the field had these specific social science credentials? There is circumstantial evidence that he did, and that in using the concept of field he understood its underlying epistemological commitments. While at Harvard Olson met and befriended Henry (Harry) Alexander Murray, a social psychologist who was running the Harvard Psychological Clinic and shared with Olson a passionate interest in Melville.[11] This "particularly important friend" of Olson's, as Tom Clark describes him, had other important friends, notably Carl Jung, whose direct and indirect influence on Olson would be marked.[12] Kurt Lewin became a good friend of Murray's at the time Olson was at Harvard. In 1939 and 1940 Murray invited Lewin to give seminars at the Harvard Psychological Clinic, and Lewin's field theory clearly made a big impact. Lewin's influence on Murray is evident in Murray's textbook *Personality in Nature, Society, and Culture* (1948), cowritten with Clyde Kluckhohn, in which they talk about personality using terms from physics and geometry: "Following Lewin and Erikson we are calling the action tendencies *vectors*, each of which is a physical or psychological *direction* of activity." And they recommend the use of the field concept too: "However, the formulation can be put more neatly in terms of field. There is (1) the organism moving through a field which is (2) structured both by culture and by the physical and social world in a relatively uniform manner, but which is (3) subject to endless variation within the general patterning due to the organism's constitutionally-determined peculiarities of reaction and to the occurrences of special situations."[13]

Several motifs in Kurt Lewin's 1939 landmark essay "Field Theory and Experiment in Social Psychology," where Lewin set out the basic principles of his theory for American researchers, have verbal and conceptual resonances in Olson's early poetics: the importance of "constructs," the notion of a "universe of discourse," and the use of a post-Euclidean geometry. A new social psychology must, according to Lewin, "use a framework of 'constructs'" that "do not express 'phenotypical' similarities, but so-called 'dynamical' properties," and "represent certain types of interdependence."[14] Such a framework of constructs is, according to Lewin, necessary "for any science which wishes to answer questions of causation." It was this emphasis on the need to construct theories to manage empirical data that led to Lewin's warm reception at Harvard, where Lawrence Henderson had already introduced the idea that to be properly scientific any social theory must frame a "conceptual scheme" for itself.[15] When Olson says the new type of poem should be a "high-energy construct," he could be admitting poetry to the club.

Olson was also receptive to a key feature of Lewin's essay: its advocacy of geometry as a research instrument for social studies. Because it is crucial that the "conceptual properties of the constructs" are well defined, it is necessary to find "a geometry which is able to represent the psychological and social field adequately."[16] Olson's preoccupation with projective geometry (he read closely H. M. S. Coxeter on geometry) and with non-Euclidean geometries is similar to Lewin's. This statement by Lewin could almost have been written by Olson: "Since Einstein it has been known that Euclidean geometry, which previously was the only geometry applied in physics, is not best fitted for representing the empirical physical space. For psychology, a recently developed nonquantitative geometry, called 'topology,' can be used satisfactorily in dealing with problems of structure and position in a psychological field."[17] Compare Olson telling his Black Mountain students that "history is the practice of space in time" or the opening of his lecture "The Topological": "the world hasn't been the most interesting image of order since 1904, when Einstein showed the beauty of the Kosmos."[18] Instead of the worldly chaos of mundane affairs, writers should look to the new relativistic vision of the universe.

We cannot be certain that Olson read Lewin's 1939 essay. What we can say is that there are many echoes of the thinking of Lewin and other social scientists to be found in Olson's early writings. His bold opening claim in "Human Universe"—"There are laws, that is to say, the human universe is as discoverable as that other"—is an echo of the kind of claim made by social scientists such as Lewin, who prescribed the future for the social sciences in similar terms: "the systematization of facts by 'classification' should gradually be replaced by an order based on 'construction,' 'derivation,' and 'axiomatization' of laws."[19] The problem for Olson and poetry was that the social scientists were not issuing any invitations to poets to join them. How could poets negotiate with these developments without becoming no more than Lundberg's sources of hypotheses for subsequent scientific research or publicists "communicating and dramatizing scientific truth"? What would a poetics as scientifically valid as Lewin's social psychology look like?

I shall argue that Rukeyser was also asking similar questions in her book-length proposal for a new poetry, *The Life of Poetry* (1949). By comparing her reasoning with Olson's we can better understand the demands that he placed on poetry to respond to midcentury physics. Rukeyser was in a position to help poets grasp why Lewin was relevant because she had written an ambitious biography of the nineteenth-century American theoretical physicist Josiah Willard Gibbs in order to work out for herself just how poetry and science might be connected.

Amid all the stir about the sciences at midcentury, Muriel Rukeyser and Charles Olson stand out as the two poets most aware of the stakes and of the implications of working with scientific knowledge. Both use metaphors from physics to do epistemic work: Rukeyser argues that by being as informed as possible about current natural-scientific theories and methods, the poet can create poems that through observation, reasoning, and the full use of the poem as a potential experimental system might contribute to the knowledges valued in contemporary American society.[20] Olson would push back at the encroaching sciences, especially what he saw as the carpet-bagging social sciences, claiming for poetry a superiority as research if it adopted a fallibilist type of inquiry founded in a synthesis of materialist and idealist thought. Neither proposal was well understood at the time. Although both *The Life of Poetry* and "Projective Verse" would later become widely influential, their genesis in the epistemological competitions with other sciences for a share of the epistemic authority of nuclear physics would be largely forgotten.

THE LIFE OF POETRY

"It is a great thing to come to the unbegun places of our living and to say: Now we will find the words."[21] Rukeyser was already a well-known poet when she drew together her notes for adult-education classes into a polemical book-length study of the possibilities of contemporary poetry. Indeed, for a brief time in the 1940s she was on the edge of Olson's world of New Dealers and poets, having had a similar though less senior position in a government office full of idealists, the Graphics Workshop of the Office of War Information. Later she was for a short time part of Robert Duncan's wider San Francisco intellectual circle and sympathetically reviewed his first book of poetry. In practice, she does not always quite find the best words for the emergent culture. As a sympathetic reviewer of her biography *Willard Gibbs* noted, her style can be "oracular" as she drifts into Rooseveltian speechmaking, even echoing the "four freedoms" speech of 1941, and as difficulties she has explaining the relevance of Gibbs and his ideas of phase space remain unresolved because of her haste in writing.

Although she had already achieved prominence as a Yale Younger Poet, been the recipient of a National Institute of Arts and Letters Award, and published the biography of Gibbs, her poetry had also stirred up fierce controversy. Her poem sequence "The Book of the Dead" (1938), which treated poetry as a space for political journalism, using documentary techniques to expose the exploitation of the silica miners in West Virginia, was considered left wing. Then just four years later she was accused of being too right wing. Her poem *Wake Island* (1942), about the heroic defeat of a small American garrison on this tiny Pacific island, was, in the words of James Brock, a "public poem meant to encourage Americans to regard Wake Island within the larger context of the war against fascism" but led to the accusation that she was a poet whose "poetic equipment was available, on short notice, for any patriotic emergency."[22] Socialists might think she was shifting to the right; the FBI meanwhile was compiling a large file on a person they saw as a potentially dangerous Jewish homosexual Communist and subjecting her to extended interrogations in the early 1940s and scrutiny by informants that continued for many years. When she worked at the Office of War Information in New York, the FBI, encouraged by a hostile Congress, investigated the OWI for communist infiltrators, and the *New York Times* put her in the spotlight: "Poetess in OWI Here Probed by U.S. as Red."[23] Going to the opposite coast after she resigned was no help either. The agency still spied on her every move: "Confidential informant T-8, of known reliability, advised during June, 1945 that MURIEL RUKEYSER was at that time employed as an instructor in poetry at the Communist Political Association dominated California Labor School . . . [and] obtained a renewal of her library card at the University of California Library."[24] When she wrote *The Life of Poetry*, she did so against this background of successful experimentation as a poet, a sometimes painful awareness of the public cultures of poetry, and a growing need to work out her position in a postwar world increasingly dominated by the sciences and widespread suspicion of progressive social theories.

"The poems of the next moment are at hand," announces Rukeyser in *The Life of Poetry*. Midcentury is a moment when political hopes for "one world," a secular vision latent in both the unity of science and the "unity of imagination," will finally be realized.[25] *The Life of Poetry* is a passionate defense of poetry's potential to intimate, in Adorno's phrase, "the possibility of the non-existing."[26] "It is a great thing," says Rukeyser, "to come to the unbegun places of our living and to say: Now we will find the words."[27] By the time late in her book that Rukeyser begins a detailed discussion of science and poetry, she has not only explored modern poetry's affinities with the contemporary arts of music, song, film, and radio and discussed poetry's historical burdens in America but also crucially acknowledged that scientists now "claim their right of experiment and inquiry" over most aspects of modern life.[28] Poets should not just shrug and walk away from science. She is scornful of the idea that the most poetry can aspire to in its dealings with the sciences is to "provide a fairly reliable index of the extent of popularization of major scientific advances," doubting whether a valuably scientific poetry could ever be created solely from what she calls the "answers," the "discoveries," or "the by-products, the half-understood findings of science."[29] If poets want their poetry to be a valid "kind of knowledge" for their time, the equal of science and not its primitive antecedent, she advises them to turn instead to the "questions" or "methods" of science.[30] The tricky question is how.

After approvingly mentioning several modernist precursors who have paved the way for midcentury poets, Rukeyser concludes her chapter on science and poetry by telling her contemporaries that she expects them to learn from physics, the dominant science of their time, that poets as well as physicists are working with energy: "But to go on, to recognize the energies that are transferred between people when a poem is given and taken, to know the relationships in modern life that can make the next step, to see the tendencies in science which can indicate it, that is for the new poets. In the exchange, the human energy that is transferred is to be considered."[31] She likes this idea so much that she begins the next chapter by repeating it: "Exchange is creation. In poetry, the exchange is of energy. Human energy is transferred, and from the poem it reaches the reader. Human energy, which is consciousness, the capacity to produce change in existing conditions."[32] Olson uses similar terms a year later to describe a poem working at full tilt: "a poem is energy transferred from where the poet got it (he will have some several causations), by way of the poem itself to, all the way over to, the reader."[33]

Rukeyser does not let physics envy blind her to the realities of human labor, the relationships and social interactions, behind this "energy." Where Olson strips energy of humanity, she insists on it. Although he does say that poetic energy is "an energy which is peculiar to verse alone and which will be, obviously, also different from the energy which the reader, because he is a third term, will take away," he nudges his readers toward thinking of the domain of poetic language as capable of being conceptualized by a model constructed purely from the abstractions of physics.[34] Doing so enables his model to encompass an analogy between the poem as a "high-energy construct and, at all points, an energy-discharge" and the high-energy discharges being studied by nuclear physicists, as well as to link it with the concept of energy as spatialized across a "field" of forces (interactions) between subatomic particles and the waves or radiation that they generate.[35] *Energy, transfer, causation*: these abstractions borrowed from physics elide questions that mattered to Rukeyser, questions about the normative force of the *doxa* that lay behind these working abstractions. Such *doxa* have normative force because, in the words of the philosopher Robert Pippin, they are in effect "social interactions within communities over time, collectively self-constituted norms."[36] Rukeyser wants her poets to create poems whose energies are responsive to the emotional and intellectual labor or energy that make possible this collective achievement.

What are we to make of this convergence between Olson and Rukeyser? A quick look at some of the most salient points will begin to suggest that their significance lies as much in a shared problematic as in any question of precedence or borrowing. *The Life of Poetry* surveys many midcentury cultural practices in an effort to hasten the poetic realization of such unbegun potentials in radio, in the blues and popular song, and in the use of "quick, rhythmic juxtapositions" in cinema, a rapid montage technique similar to Olson's rule that "one perception must immediately and directly lead to a further perception."[37] He makes no reference to cinema as poetry's role model—he praises the typewriter's capacity to register the rhythm of juxtapositions, because its controlled spacing can "indicate exactly the breath, the pauses, the suspensions even of syllables, the juxtapositions even of parts of phrases."[38] *The Life of Poetry* anticipates several other themes in "Projective Verse." Just as Olson's projective verse assumes that "form is never more than an extension of content," Rukeyser's futural poem will recognize that "form and content, relation and function, reach and merge," because when a poem grasps "imaginative truth," then "it finds its form, for the truth of a poem is its form and its content, its music and its meaning are the same."[39] Rukeyser is a little diffuse in her expression of the idea that form and content are completely interdependent, perhaps because of her commitment to a scientific thoroughness. Olson goes for the sound bite, attributing to Robert Creeley the pithy admonition that in poetry, "Form is never more than an extension of content."[40] Characteristically, Olson also gives to the idea of a relation between form and content a geometric image of *extension*. In her discussion of Walt Whitman, Rukeyser foreshadows yet another Olson preoccupation, the handling of the poetic breath: "Out of his own body, and its relation to itself and the sea, he drew his basic rhythms . . . of the relation of our breathing to our heartbeat," rhythms that came "not out of English prosody" but "from the rhythms of pulse and lung." Poets will come to recognize that the layout of the poem can be a kinesthetic notation, that "punctuation is biological . . . the physical indication of the body-rhythms which the reader is to acknowledge."[41] Olson will later make a similar point—the rhythms of the poet's body are to be made into poetic punctuation.

The evidence of chronology, plus verbal and conceptual echoes, suggests that it is possible that Olson had some idea of Rukeyser's book in his mind as he began to write "Projective Verse," whether through reading the text, from a review, or from verbal reports of it. Interpreting the implications of this connection is not quite as straightforward as it might seem. Olson was writing a terse manifesto for the midcentury, a polemic designed to shock its poet readers into action, not a survey of the state of American poetry at midcentury or even a teaching guide on how to write and read poetry. If Olson did borrow from Rukeyser, it could have seemed quite reasonable to the almost-unpublished, and certainly unknown, poet to take current ideas about the future of poetry where he could find them and use them as departure points for his own process of thinking through how to express a poetic call to action. And the Rukeyser demonized by both Right and Left would have been a difficult figure to be seen to be indebted to. Nevertheless, it does seem all too characteristic of the treatment of most modern women poets that Rukeyser's possible contribution to the poetics of the New American Poetry has gone largely unexplored, a neglect reinforced by Olson's own tendency not to acknowledge intellectual sources about which he was ambivalent.[42]

If it was only a matter of Olson's looking round for a convenient roundup of current ideas about poetry with which he could work to formulate his own condensed poetic manifesto, then the story would end with questions about originality and the ethics of unacknowledged creative borrowings. But verbal and metaphorical echoes are not the most interesting aspect of the convergence. There is a bigger story to be told about this convergence, a story about the influence of ideas about how to be scientific that arose in the development of the social sciences between the 1930s and 1950s. Olson's interest in the field concept that was central to Kurt Lewin's field psychology has a counterpart in Rukeyser's interest in the concept of phase space that played a central part in her biography of Willard Gibbs. During the 1940s both Rukeyser and Olson had become captivated by the possibility that poets, like social scientists, could introduce scientific method into their work by emulating the way physicists and other natural scientists construct models or idealized images of the material world. Rukeyser based her genealogy of scientific modeling on the history of Gibbs and his influence.

WILLARD GIBBS AND SOCIAL SCIENCE

In her diaries, the poet Marya Zaturenska records a visit in late 1941 from her friend Muriel Rukeyser, who "spoke with excitement about her book on Gibbs, making it sound more like an epic poem than a biography."[43] Zaturenska and her husband, poet and biographer Horace Gregory, had introduced Rukeyser to their neighbor, the chemist Theodore Shedlovsky, whom Rukeyser credits with having brought Gibbs's work to her attention.[44] After studying at MIT, Shedlovsky spent his working life at the Rockefeller Institute for Medical Research (now Rockefeller University), where he did pioneering research on electrolytes; he is now probably best known as the author of a textbook on electrochemistry. A biographical memoir tellingly describes him as having "an uncanny instinct for bringing together people who had problems with people who had ideas and suggestions."[45]

The problem with which Shedlovsky could help Rukeyser was her sense that there were no histories of America that gave the sciences the prominent role they had actually played in its cultural life. Their absence signaled a lack of understanding of the arts and poetry too. "The reception of work in science in this country has always been a reliable indication of the American attitude toward creative effort."[46] The aim of *Willard Gibbs*, as Catherine Gander says, was to trace the "continuing legacies" of Gibbs, Whitman, and Melville "to the American mind."[47] This is an inventive style of biography; Louise Kertesz calls it "history in film flashes."[48] Revealingly, Rukeyser borrows strategies from Van Wyck Brooks's account of the transcendentalists, *The Flowering of New England* (1936), occasionally elbowing the reader as if to say that Van Wyck Brooks has told only half the story.[49] Brooks may compare Webster's dictionary to the Declaration of Independence—historians of science, according to Rukeyser, have compared Gibbs to Webster.[50]

Brooks proves an imperfect model for Rukeyser, however, because he lures her into attempting to make Gibbs into a far more pivotal figure in the development of the sciences and the arts in America than the facts justify. She does this by associating him with the major cultural narratives of the time and by endowing him with a putatively rich inner life that can be recounted as an education of the sentiments. Both strategies falter. After a brief introduction, Rukeyser sets the scene for her narrative of the life of Gibbs with an extended, suspenseful account of the *Amistad* mutiny, a choice that might seem strikingly irrelevant. She chooses to begin this way because Gibbs's father, a professor of theology at Yale, whose full name confusingly was also Josiah Willard Gibbs (hence his physicist son's use of Willard), employed his skills in comparative linguistics to help translate the language spoken by the Mendi Africans who had taken possession of the *Amistad* and, after their capture, been incarcerated in a prison in New Haven. Once the *Amistad* story is told, Rukeyser soon digresses again, this time into a biography of the poet James Gates Percival, whose "headlong drive toward death . . . was checked by two forces only: poetry and science." Percival's story enables her to create an ideal figure against whom she can later measure Gibbs.[51] Percival, she tells us, was "the poet who ushered in the expression of Gibbs."[52] After citing lines from Percival's poem "Prometheus," she makes this extraordinary claim: "Here is the free spirit, the scientist and poet, the first poet in America who dared to deal with these materials, the man who worked for the progressive sciences, knowing himself in poetry, but feeling that mathematical truth could only withstand modification, grasping in this poem at the science that Shelley had only tentatively touched—writing lines that mount and mount, suggesting the past of Newton, and the future of Clerk Maxwell and Einstein."[53] This panoramic cultural biography is largely projection. It is Rukeyser, not Percival or the young Willard Gibbs, who finds herself in poetry and, like many radicals of the 1930s, feels that mathematical truth and the new sciences could be directed to progressive social purposes if only visionary reformers with a poetic imagination were to come forward. Similar attempts to link Emerson's "Poetry and Imagination" with Gibbs's student days also feel forced. Nor is it helpful to liken, as she does, the Gibbs family with the Adams or James families.

Rukeyser does better when she concentrates on the science. She makes a reasonable job of providing a narrative history of science into which she can place Gibbs, of outlining his research achievements, and of showing how later scientists, even such seemingly distant figures as Fritz Haber, who discovered how to fix nitrogen from the air into fertilizer and explosives, were directly assisted by Gibbs's theories. She is also alert to historical research aligning scientific concepts with political ones and repeatedly attempts to make connections between scientific and social developments in the nineteenth century. Sometimes these narratives are well grounded; sometimes they veer into wishful rhetoric: "A West opened up before mathematical physics, a whole unexplored forest ready to be known."[54] Where she prevaricates is in straining the evidence to create a protagonist with a rich imaginative life. At times the biographer almost sighs with relief as she diverts to talk about some far more richly documented life such as that of William or Henry James, whose letters are quoted at length. As she admits more than once, Gibbs is a challenge to the biographer because his "life is shadowy"; there just is not enough documentary material about his inner life, a situation exacerbated by the refusal of his estate to let her see family papers on the grounds that this was not an authorized biography.[55] Sentences like those already quoted about James Percival and about physics as a new frontier, as well as many sentences like the following—"Young Willard Gibbs rode out into the hills, torn by the tearing of his world" (as the young man's father is dying)[56]—are probably why Rukeyser's friend Zaturenska eventually confided to her diary that she had changed her mind about the biography, annotating the entry she wrote when she first saw the book's "brilliant flashes" and "real poetry" with the harsh private judgment: "it's a dishonest book, I wanted to believe in it."[57] Such sentences are symptoms of an attempt to claim for Gibbs a representative cultural status that the biography itself, when it concentrates on his work as a theoretical physicist, gives little support for.

Why should Rukeyser risk seeming to celebrate a history she does not wholly believe and engage, even in a limited manner, in what her friend felt was biographical dishonesty? Indeed, why choose such a seemingly unpromising protagonist for a heroic account of the growth of American science and culture? What did a young poet passionately committed to radical politics and poetry itself, a literary descendant of Percival rather than Gibbs, see in a nineteenth-century theoretical physicist still relatively little known outside the profession even today? If this is a book that she wrote because, as she puts it, "I needed to read it," what drove that need?[58] One likely reason was the growing importance of Gibbs in the 1930s, not as a theoretical physicist at all, but as the exemplar of a way of understanding scientific method. Social scientists attempting to place social theory on a more scientific footing believed that Gibbs could give them the answer. In the first sign of her interest in the physicist, her poem "Gibbs," written well before she started the biography, Rukeyser attributes to the scientist a commitment that is really the product of historical hindsight: "He binds / himself to know the public life of systems."[59] This commitment to a more scientific understanding of society and its politics, which she herself so brilliantly demonstrated in her series of poems on the silicosis scandal in West Virginia, "The Book of the Dead" (1938), is what drives her interest in Gibbs. She thinks of him as making possible vital new modern American theories of the public life of *social*, as well as physical, systems, theories that point to a progressive American future. To understand how she comes to think of Gibbs in this way, we will need to trace his subsequent influence on the history of science and, especially and at first sight surprisingly, on the social sciences in the decades up to midcentury.

Although this interest in Gibbs as the source of a scientific method that could be transposed to the study of society and psychology permeates the whole biography, Rukeyser is never explicit about it. Even a partial rationale for concentrating on Gibbs's theories begins to surface only late in the story, when she briefly mentions the famous economist Irving Fisher because he took courses with Gibbs and then applied his vector theories to economics. Milton Friedman described Fisher as "the greatest economist the United States has ever produced."[60] Others are harsher in their judgment. In his critique of neoclassical economics for its lack of understanding of the consequences of borrowing conceptual metaphors of energy and field, Philip Mirowski argues that Fisher was a major contributor to a long history of poorly "reprocessed physics" in economics up to the present day.[61] Mirowski himself, however, does not simply dismiss such reprocessing as physics envy, concluding instead that what is needed is not to expel these metaphors but to give more attention to the epistemological stakes. Why have they been valuable and do they remain so?

Such questions were raised at Harvard in the 1930s, though not so much in the discipline of economics. After Fisher's momentary appearance, Rukeyser introduces an influential sociologist at Harvard, Lawrence J. Henderson, saying simply that "he has written a study of Pareto in the light of Gibbs."[62] This is an understatement. Henderson is the main reason for Rukeyser's whole interest in Gibbs. Henderson is largely forgotten today because his influence lay not so much in major publications or his own research into public health but in his mentoring of others. Henderson is the main protagonist in the previously mentioned compelling study *Working Knowledge* by the historian Joel Isaac of how Harvard social science fundamentally shaped the work of Thomas Kuhn in *The Structure of Scientific Revolutions*. Isaac attributes to Henderson a central role not just in the background of Kuhn's intellectual development but in the development of the whole of social theory at Harvard (and hence in American sociology), which eventually led to the theory of scientific paradigms. Henderson was already a significant biochemist before he became a specialist in public health and a sociologist, a background that led him to see the possibility of combining his long-standing chemist's interest in the use that Gibbs made of phase space together with his sociologist's interest in Vilfredo Pareto's systematization of the concepts necessary to model the workings of a society. Pareto is now best known for his work on the circulation of elites, but it was this systematization of such processes as the "residues of combination" (i.e., the way that social groups imagine causal relationships on the basis of their familiar interactions—e.g., immigration and underemployment) that mattered most to the researchers at Harvard. The result of thinking of Paretian analysis in terms that echoed Gibbs could have been a rigid, unusable physics of society, but instead, Henderson's odd marriage of statistical mechanics and Paretian social modeling gave birth to principles for the constitution of any scientific method in a new domain: the use of a provisional structure of concepts, a conceptual scheme or working model, as a valuable scientific instrument. Isaac explains that Henderson treated the various abstract components of Pareto's social theory as if they "functioned in the same way as notions of temperature, pressure, and concentration in the physicochemical theory outlined by Gibbs: they were attributes of a system, and thus furnished the core concepts of a general social theory."[63]

Henderson was fully aware of what he was doing: "There seems to be a psychological, if not a logical, advantage when conceptual schemes consist of *things* which have *properties* (or attributes) and *relations*" (his italics).[64] In her biography, Rukeyser puts this idea in terms familiar to the literary world: for Henderson, "the social system is a fiction, not the result of the application of facts."[65] She is extrapolating from his own somewhat less generalized comment that "Gibbs's system is plainly a fiction"—only by extension could the systematic model of a society be called a fiction.[66] More interestingly still, she calls Pareto's conceptual scheme metaphoric: "Pareto's work is not an application of science; it is, however, a metaphor of a scientific scheme. . . . His social system was one of molecules set in motion by tastes, and subject to checks."[67] Perhaps by introducing the term "metaphor" she hopes to leave the door open to exchange between social theory and poetry.

Isaac offers a compelling account of how Henderson persuaded many of his peers and students at Harvard that the route to scientific rigor lay in the development of what he was probably the first to name a "conceptual scheme." In the 1930s and early 1940s "Harvard hedged on the human sciences."[68] The university was so reluctant to give full institutional stability to the fields of psychology, anthropology, political science, and sociology that it compelled many researchers within these fields to fend for themselves within the interstices of the university and its satellites. As a result, a whole range of informal and formal activities, from reading groups and intellectual societies to small institutes and interdisciplinary programs, grew up around the edges of the large formal departments. These rough-and-ready groupings of researchers made up what Isaac inventively calls the "interstitial academy," a more informal structure than that of departments and faculties. Typical of those around the edges of the institutionalized departments who were engaged in the creation of these practical epistemologies were the members of a discussion group called "The Levellers," which included sociologist Talcott Parsons, anthropologist Clyde Kluckhohn, and psychologist Henry Murray.[69] The group had its roots in attendance at Lawrence Henderson's seminar on Pareto and scientific method.[70] Olson's fascination with the creation of temporary interdisciplinary "institutes" at Black Mountain probably has roots as much in his own experience of this interstitial world as in his fascination with Oppenheimer's Princeton Institute. Participants in the interstitial academy found themselves closely "attending to the foundations of knowledge" in order to justify the very right of their new, unwalled activities to exist.[71] These largely self-fashioned social and human scientists became very attentive to the methodologies they used, picking carefully over the points where epistemology, actual research practices, and pedagogy intersected with each other in the study of human affairs. They were constructing what Isaac aptly calls "working knowledge" or do-it-yourself epistemologies.

Conceptual schemes came as part of a package. Isaac identifies three other components to the practical epistemologies that developed at Harvard: scientific philosophy (broadly speaking, logical empiricism and its later variant, logical positivism), Percy Bridgman's operationism, and the case study method. Operationism need not detain us here, since it plays little or no part in the development of Rukeyser or Olson. Case studies were important, however. Rukeyser's "Book of the Dead" and Olson's *Call Me Ishmael*, the final version of his unfinished PhD and Guggenheim projects, are case studies written along Harvard lines even if not written in a Harvard style. The case study, according to Isaac, was also integral to Henderson's methodology: his "sociology and theory of science presupposed an understanding of scientific philosophy in which the study of *cases* was central" (his italics).[72] The case study method emerged piecemeal from a long history of its use in teaching at Harvard, first in legal theory, then medicine, and, in the 1930s, history and anthropology. As the name suggests, this was a methodology of studying actual cases or specific histories, retaining all their idiosyncrasies while at the same time eliciting patterns that might reveal glimpses of more general laws of human behavior. Isaac cites several instances of research produced by the interstitial Harvard social thinkers, including Clyde Kluckhohn and Dorothea Leighton's *The Navaho* (1946). In an introductory note to his essay "A Navaho Personal Document with a Brief Paretian Analysis," Kluckhohn underlines the close tie between conceptual schemes and case studies: "Between 1935 and 1941 the distinguished physiologist, L. J. Henderson, conducted a course at Harvard University which was known as 'Concrete Sociology.' Physicians, lawyers, business executives, governmental administrators, and representatives of the various social sciences presented and discussed a 'case' with which they were personally familiar. The conceptual scheme in terms of which all the concrete materials were considered was that of Pareto."[73] Kluckhohn's approach is also captured in a letter to his colleague Talcott Parsons outlining their differences: Kluckhohn insists that "it is proper and indeed useful to behave experimentally with reference to conceptual schemes."[74] Experimenting with conceptual schemes: these are provisional schemes, not the totalities that later philosophers such as Donald Davidson would inveigh against.

It is important to notice that even in the 1930s, the case study method did not always sit comfortably with the use of conceptual schemes. Biographers and critics of Olson often mention his receptivity to the historian Frederick Merk's 1937–38 course on westward expansion.[75] A student who took the same course a couple of years later recalls that Merk made "significant excursions into related topics: national politics, of course; economic and business history; the history of agriculture and extractive technology; the history of American foreign policy."[76] This was not only interdisciplinarity but the case study at work. Merk was not, however, entirely sympathetic to the newfangled ideas of his social-scientist colleagues. In a 1936 review of a sociological study of the frontier, Merk is skeptical of the value of searching for "laws of human behavior" in frontier societies around the world using existing historical studies rather than firsthand research.[77] He objects not so much to the idea of using a conceptual scheme as to overreliance on its explanatory power and to failures to test such schemes in the field. This same issue would become increasingly important to both Rukeyser and Olson in the 1950s.

Conceptual schemes such as Gibbs's model of phase space were, even if not so named, already widely used in the natural sciences. According to Henderson himself, it is the "well-chosen simplifications and abstractions that make possible a systematic treatment of complex phenomena" within Gibbs's schema.[78] Henderson takes a pragmatic approach. This sort of conceptual scheme is only a provisional model of a selected set of phenomena in the actual world, and it certainly does not provide a template for all of science. Such schemas, though useful, are most needed by emergent sciences where there is not yet much agreement even about what is to be studied. Disagreements among researchers in the well-established natural sciences occur "most often at the frontiers of knowledge," which are clearly recognized as such, and the divergences can be "ordinarily settled by observation, experiment, or some other method that all accept."[79] Readers of Olson are likely to be attuned to metaphors of the frontier, but it is worth underlining the connection that Henderson makes here between conceptual schemes, new domains of inquiry, and disputed frontiers of knowledge. Henderson recognizes that the human sciences are nearly as fortunate as their cousins in the natural sciences. Writing in the mid-1930s for sociologists and psychologists in search of a more rigorously scientific methodology, Henderson argues that the social and human sciences need conceptual schemes if they are to take further steps toward full scientific status, and he encourages his contemporaries in the human sciences at Harvard to explore their potential. By implication, as the human sciences become better established, conceptual schemes might become less vital.

Today we are likely to recall Donald Davidson's devastating critique of the notion of a conceptual scheme as a recidivist dualism and therefore find it hard to believe that the notion was ever taken seriously. How could philosophers think that conceptual scheme and phenomenological reality were distinct, so that when free of the ideological or conceptual mediation of a conceptual scheme, we could experience reality with sparkling freshness? How could those earlier thinkers imagine that a conceptual scheme could take over one's mind to the point where it determined every aspect of experience? Conceptual schemes, we are likely to suspect, belong with 1950s fantasies of total conditioning or with that weird preoccupation of 1950s and 1960s science fiction, zombie possession. As it happens, Davidson's main targets are two Harvard philosophers: Thomas Kuhn, who believed that successive scientific "conceptual categories" could be "incommensurable," and W. V. O. Quine, who as a self-declared empiricist thought of "the conceptual scheme as a tool."[80] Davidson takes the zombie view of conceptual schemes and assumes that for both thinkers these schema were equivalent to the total possession of the mind by an inescapable world picture. Against this interpretation, Isaac argues persuasively that when Henderson's idea of a "conceptual scheme" became foundational in the social sciences at Harvard in the late 1930s and 1940s, the *Truman Show* version of the conceptual scheme that Davidson mocks was not at all what the human sciences had in mind. They understood the typical conceptual scheme in a far less totalizing manner as a voluntarily adopted and temporary model, constructed from a limited system of useful abstractions that could provisionally simulate the behavior of a chosen subset of processes in the natural or social world.

A conceptual scheme such as the theory of electromagnetism, or statistical mechanics, was rather like a pair of surgeon's spectacles, a model through which to magnify some aspect of the world. The guide to social science pedagogy that Talcott Parsons wrote (in conjunction with Harvard colleagues J. F. Dunlop, M. P. Gilmore, and Clyde Kluckhohn), which I previously cited in the introduction, makes this modest scope evident:

> If we did not select, if we did not abstract, the writing of history would take as long as the making of history. And so it is with all the social sciences and indeed with science generally. Since the scientist cannot deal with events in all their uniqueness, the best he can do is to construct a conceptual model which reflects with a minimum of distortion certain important relationships which prevail between the phenomena.[81]

Selectivity is the precondition for rigor in social-scientific method. This essay's origin in the classroom typifies the intersection of epistemology, research, and pedagogy that, Isaac argues, made Harvard social and human sciences so distinctive.

Rukeyser's biography of Gibbs was inspired by Henderson's idea that Gibbs provided the intellectual foundations for a generalized scientific method based on the conceptual scheme. In writing about Gibbs she believed she was writing not just about an important American physicist but about the possibility of a new kind of progressive social science in which the poet and poetry could play a part. Whether or not she was aware when she began her biography that Talcott Parsons, after working through the possibilities of Henderson's ideas, had recently published one of the first great American works of social theory, *The Structure of Social Action* (1937), she was aware that many thinkers believed that it might be possible to bring scientific methods to bear on the social issues about which she cared. Her poem about the miners who contracted silicosis was an attempt to do this before she had learned about conceptual schemes. In her biography, Rukeyser chides Woodrow Wilson for misusing scientific theories, so that "a key formula becomes a basis for hifalutin fortune-telling."[82] Wilson criticizes attempts to describe government as a Newtonian machine only to repeat the mistake by likening it to a Darwinian world. This, says Rukeyser, "is the error of a rigid analogy, of using the discoveries of science instead of the methods themselves in dealing with other material."[83] Methods were likely to be provisional conceptual schemes. She liked her summary of the proper relation between natural science and other disciplines so much that, as we saw earlier, she made it a key point in *The Life of Poetry*, saying that for those who try to bring insights from the natural sciences into other domains, "The trap is the use of the discoveries of science instead of the methods of science."[84] Both she and Olson were determined to try to avoid that trap. By comparing their attempts to do so in *The Life of Poetry* and "Projective Verse," we should be able to see more clearly just what it meant to attempt to bring poetry into dialogue with the sciences at midcentury.

RUKEYSER'S POETICS OF THE SYSTEM

How then did Rukeyser bring to poetry the ideas about conceptual schemes that she learned from reading Henderson and Gibbs? In *The Life of Poetry* she identifies conceptual schemes (though she does not call them such) as the latest scientific development: "This gathering-together of elements so that they move together according to a newly visible system is becoming evident in all our sciences" (as we saw was the case in Henderson's system-based social theory) because it is able "to deal with any unity which depends on many elements, all interdependent."[85] Such conceptual schemes could obviously help poetry develop strategies for the gathering and coordination of rapidly moving elements needed to manage the rapid juxtapositions of aesthetic elements already widely used in cinema and other arts, a high-speed montage that she, like Olson, believed a truly modern poetry needed to develop for itself. She argues that an unfolding process of disjunctive transitions can be understood in the geometric language of Willard Gibbs's version of "phase space" set out in his *Statistical Mechanics*:

> The poems which depend on several emotions, each carrying its images, move like a cluster traveling from one set of positions to another: the group ABCDE, say, moving to A′B′C′D′E′; a constellation.
>
> This gathering-together of elements so that they move together according to a newly visible system is becoming evident in all our sciences, and

it is natural that it should be present in our writing. Wherever it exists, it gives us a clue as to a possible kind of imagination with which to meet the world.[86]

Gibbs's version of the phase space model depicts the molecules in an enclosed gas in just such terms of translocation across a mathematically constructed space that represents the changing phases, or states (cool, hot, excited, and so forth), of the gas molecules. Phase space is a sophisticated extrapolation of the principle of the pie chart. It helps one to visualize relationships that are otherwise too abstract or too widely distributed across a period of time for ordinary cognition to manage to grasp their patterns easily. According to Rukeyser, the phase space model gives poets a means "to deal with any unity which depends on many elements, all inter-dependent." Despite the awkward struggle to justify her analogy between a poem's movement and the phase space mathematics, her proposal has some promising features: it emphasizes the interdependent relations of *all* the elements of a poem (which may include linguistic, affective, cognitive, visual, or aural elements of a poem—a list that is not exhaustive because there may be others yet "unbegun"), and it also recognizes that the resultant system is not static; it is a complex series of unfolding transformations that occur as a reader engages with the poem.

In making much of phase space Rukeyser is not disappearing into an obscure corner of the history of science. Unfamiliar as it may be to cultural theorists, phase space is the adjustable wrench of mathematical modeling. David Nolte makes strong claims for its significance: "Listen to a gathering of scientists in a hallway or a coffee house, and you are certain to hear someone mention phase space. . . . Though it was used originally to describe specific types of dynamical systems, today 'phase space' has become synonymous with the idea of a large parameter set: Whether they are stock prices in economics, the dust motes in Saturn's rings, or high-energy particles in an accelerator, the degrees of freedom are loosely called the phase space of the respective systems. . . . In his popular book *Chaos* on the history and science of chaos theory, James Gleick calls phase space 'one of the most powerful inventions of modern science.'"[87] A phase space is a clever mathematical device for transforming something very hard to imagine into something more readily comprehended and visualized, and therefore more easily manipulated. The different potential states of a system—such as a gas enclosed in a container—are treated as if they are points in a geometric space for the analysis of which mathematicians have already created many helpful formulae. So useful has been the concept of modeling the changing conditions within a closed system by means of phase space that it has been further generalized to represent potential choices in all sorts of other situations. The Black-Scholes equation for modeling the behavior of derivatives, which has been blamed for the financial crash of 2008, is also a descendant of the phase space concept.

Rukeyser believes that once poets adopt a poetics of the system, the art of poetry will be able to hold its head up alongside physics as a "*kind* of knowledge" (her italics).[88] Her main justification for this claim that poetry can produce knowledge rests on her interpretation of Gibbs's own explanation of how his phase space modeling creates new knowledge by revealing new truths about materials:

> Truth is, according to Gibbs, not a stream that flows from a source, but an agreement of components. In a poem, these components are, not the words or images, but the relations between words and images. Originality is important before the accord is reached; it is the most vivid of the means in a poem, and the daring of the images allows the reader to put off his emotional burden of association with the single words, allows him to come fresh to memory and to discovery. But when the whole poem has taken effect—even its first effect—then the originality is absorbed into a sense of order, and order then becomes the important factor. All of these words were known, as the results leading to a scientific discovery may have been known. But they were not arranged before the poet seized them and discovered their pattern. This arrangement turns them into a new poem, a new science.[89]

Here Rukeyser offers the important insight that the constituents of the poem that generate a system are not the molecules of language and imagery but the set of interrelations across the whole poem that are generated by the provisional systems brought into being by the placement of words and by their interactive meanings. Poetic knowledge, in other words, does not occur at the level of direct propositions about actual states of affairs but is a second-order activity. Discovery will occur not through naming of new phenomena or statements about how things are but through the revealing of new patterns of connection.

The puzzling final sentence—the new poem is a new science—points to an underlying problem with her argument. This unconvincing claim, that a new poem potentially opens up a whole new science, results from the distortion of lifting much of this passage (as she did at several other places in *The Life of Poetry*) directly from her biography of Gibbs, without making necessary adjustments to accommodate the new context. In her poetics essay it is the poet who makes a discovery worthy of being called scientific. In the biography it is the scientist who carries out the work of revealing a new pattern by taking observational data and then using it to shape a new model of such phenomena. Here is the biography:

> Truth is, according to him, not a stream that flows from a source, but an agreement of components, an accord that actually makes the whole "simpler than its parts," as he was so fond of saying. It is truth flowing through the world, depending on an accord in great complexity. Not originality, but order, becomes the important factor; the point of view and the arrangement may be different, he says of this. "These results, given to the public one by one in the order of their discovery," were not arranged before he seized them and discovered their pattern. This arrangement turns them into a tool, a new science.[90]

From the original context it is clear that Gibbs is concerned to counter any mistaken assumption that his models can be validated by their direct correspondence to what is happening in an actual gas. In the preface to his major work *Statistical Mechanics*, he disavows allegiance to any correspondence theory of truth for his new science, saying that "there can be no mistake in regard to the agreement of the hypotheses with the facts of nature, for nothing is assumed in that respect." What he aspires to is "agreement between the premises and the conclusions"— that is, to a mathematically rigorous, coherent modeling of the mechanics all the way from premise to result. [91] The truth of his results does not issue straight from any source material, such as a specific cylinder of gas, under observation. His models are intended to make it easier for researchers to study the complex interactions at work in accessible, idealized models of how gases change as external conditions of pressure and temperature alter. With these calculations, the researcher can then estimate what an actual gas might do. Gibbs concentrates on the internal coherence and mathematical rigor of his systematic models. I am not confident that Rukeyser quite grasps this insistence on formalism in his analysis, nor his resistance to correspondence theories of scientific truth. She feels able to rework this passage for *The Life of Poetry* by cutting the sentence in which Gibbs reintroduces the metaphor of flow in order to retain his contrast between a dynamic, changing structural pattern and the idea of a flow from a liquid source like a spring or a pipe.

When Rukeyser translates Gibbs's claim that truth can be found in "the agreement of components" into the idea that a poem achieves new truths, new discoveries, through the creation of a new arrangement of its components, discoveries worthy of standing alongside those of science, she is conflating two arguments. One is that Gibbs believes that for any research practice to be deemed scientific it must adopt what Henderson calls a conceptual scheme, but what Rukeyser confusingly calls a system. The other is that Gibbs's own research practice uses a specific type of conceptual scheme that takes the form of a system (the type of system known as phase space).[92] The result of this understandable conflation is that she starts to think of conceptual schemes solely as modes of system building and loses sight of the possibility that conceptual schemes might be constructed around other sorts of bricks and mortar, other sorts of metaphors and images such as frameworks, fields, paradigms, or even specific myths.

By saying that a poem is a system, Rukeyser is therefore comparing the poem to a mathematical model that can absorb empirical data and discover new patterns within it. Unfortunately, this epistemic metaphor of phase space, even when generalized into the idea of system, does not do all the work that Rukeyser hopes for. Its weaknesses are threefold: as we have seen, it blurs the distinction between method and specific schema; its formalist approach to truth also leaves it unable to reveal areas of uncertainty; and finally, the internalism of the concept of a system makes it hard to conceive how the poem's patterns could then be related to the rest of the textual universe. The issue of indeterminacy is particularly significant. In his *Cybernetics*, Norbert Wiener criticizes Gibbs for his inability to represent uncertainty and the indeterminacies of knowledge. "In the complete Gibbsian theory it is still true that with a perfect determination of the multiple time series of the whole universe, the knowledge of all positions and momenta at any one moment would determine the future. . . . The great contribution of Heisenberg to physics was the replacement of the still quasi-Newtonian world of Gibbs by one in which the time series can in no way be reduced to an assembly of determinate threads of development in time."[93] Gibbs's Laplacean vision of a possible total description for every particle of the material universe remains latent in Rukeyser's ideas of the closed energy system.[94] The limitations of Rukeyser's poetics are also apparent when we ask how the discovery of a pattern throws light on the rest of the textual universe. Aren't all poems patterned, a pattern being a manifestation of intentionality, discursive regularities, or underlying laws, and don't they all implicitly claim that their patterns have some elucidating relation to the world? Why should the poet be credited with some special, quasi-scientific achievement just for arranging words in a pattern? How can this be compared to the methodological rigor, and empirical fallibilism, of the kinds of cognitive inquiry on which the sciences rely? Without more discussion of what the discovery of patterns is useful for—as when the geometric phase space patterns help the researcher understand more about the physics and chemistry of gases—the claim remains too formalist to be of much help.

Valuable as it is to think of the poem as a system, or hypothetical model, this is an epistemic metaphor that doesn't distinguish enough between the principle and specific epistemic metaphors, doesn't make room for uncertainty and disagreement among the elements of the poem, and doesn't make visible the epistemological competition with other knowledges. From this standpoint we can see why, if poetry were to keep up with the epistemological competition, a *system* theory of poetry might not be as useful as a *field* poetics. Olson would of course try to supply this field poetics, even citing Wiener himself in "The Kingfishers." In a retrospective interview Rukeyser looks back at her interest in Gibbs and appears to hint that she has come to realize that her earlier idea of a poetics of system was not wholly adequate for the creation of poems in which many elements are coordinated as they change and, as she puts it, "move together." She says: "The reason I came to do Gibbs was that I needed a language of transformation. I needed a language of a changing phase for the poem. And I needed a language that was not static, that did not see life as a series of points, but more as a language of water."[95] The field concept, as we have seen, is a concept of flow, based on everyday experience of the "language of water." Its ability to model a flow of continuous gradations of energy transformations is what led many physicists to adopt it rather than to model transformations of energy states in terms of systems. And as I shall argue in a moment, Olson's adoption of the field has proven almost as fertile for poetics as it has for the sciences.

Finally, it is important to grasp that one reason that Rukeyser was vulnerable to the conflation of conceptual scheme with a specific model of system derived from Gibbs's physics was that she believed in science. She was committed to acceptance of the validity of scientific knowledge. When she goes into detail about the elective affinities between science and poetry, as she does with Gibbs, it is evident that she accepts the primacy of the epistemic authority of the new natural sciences. Unlike Heidegger and Whitehead, Olson's favored thinkers, she does not believe that science is simply a subset of a wider, more inclusive knowledge-in-the-making nor, in the words of the philosopher Jay Brassier writing in another context, does she patronize "scientific assertions about the world as impoverished abstractions" that have been imposed on a "more fundamental sub-representational or pre-theoretical relation to phenomena."[96] She does not imagine that poetry might have some privileged access to the wellsprings of knowledge or, like Olson, believe that poetry itself might one day play a starring role in a new kind of all-encompassing knowledge of everything, whether belonging to the subjective or to the objective world of experience. At the same time, she does not think that scientists should have sole rights of inquiry into the physical or social worlds, and she wants poets to seize the opportunity offered by new ideas of inquiry developed by physicists like Gibbs.

OLSON AT HARVARD

Rukeyser and Olson both grasp the difficulty of ascribing to poetry the power to produce new knowledge, yet neither wants to relegate poetry to a support role far behind the frontiers of knowledge. Like Lewin, they want to be there in the thick of the action. Rukeyser's strategy is to start with the claim that poetry is a kind of knowledge and then argue that it can hold its own in a scientific culture because scientific method is simply an extension of a universal human tendency to be inquiring beings. Poets too can share in the enterprise of discovery, and at best their resulting poems will offer knowledge, perhaps different from scientific knowledge, yet still a form of knowledge. She then has to explain why such poems are *valid* kinds of knowledge, and her attempts to justify this bold claim rely on analogies with the way scientists since Gibbs construct conceptual frameworks logically independent of empirical sense data. Using these frameworks, scientists can anticipate potential discoveries by manipulating the model and then carrying out exploratory tests on carefully prepared ingredients from the material world. Similarly, new models of poem may create new knowledge.

Olson's strategy differs from Rukeyser's in a crucial way. He does not concede epistemological primacy to the natural sciences. He tacitly proposes that poets have their own mode of inquiry, one that is more universal than scientific inquiry and can therefore encompass all that science does and much more. Myths, dreams, particulars, Gloucester streets, sailors, and ships can all be included. Throughout his career he would remain committed to the use of a conceptual scheme, though he gradually moved from the idea of a provisional self-made schema to the ready-made version of Alfred North Whitehead's process philosophy. Olson's poetic strategy has its roots in his encounter with Harvard social science.

As an active young doctoral scholar who was himself a part of the interstitial academy, Olson would have encountered the lively debates about how the scientific study of human life could be built from a pragmatic mix of scientific philosophy, case studies, and conceptual schemes, and he would have seen how new methodologies were emerging from the close coexistence of actual practices of research, teaching, and careful reflection on theories of knowledge. In the late 1930s he enrolled in the emergent History of American Civilization doctoral program, a course comprising English, history, and sociology. Leo Marx recalls that it "began life in a scandalously 'untheorized' condition. It was introduced without fanfare, almost casually, as a strictly local experiment in interdisciplinary teaching and research. If a theory was implicit in this modest curricular innovation, it was a rationale for interdisciplinarity. In official announcements of the new project the mantra was *interdisciplinary*" (his italics).[97] Marx is right to emphasize interdisciplinarity, though his judgment that the program was undertheorized makes more sense from the standpoint of the theory-saturated intellectual scene of the decades from the 1970s onward rather than the late 1930s, because interdisciplinarity was then itself a developing form of theory.

The first, if somewhat delayed, fruit of Olson's immersion in the practical epistemology of Harvard social science was *Call Me Ishmael: A Study of Melville* (1947). After a brief anecdotal account of marine cannibalism, Olson immediately introduces the first elements of a conceptual scheme: "I take SPACE to be the central fact to man born in America, from Folsom cave to now." Space leads to other conceptual categories, such as nature, to technologies of transport, and to conceptual affinities between the vast interior of America and the oceans. Olson attempts to use these categories to construct an ambitious conceptual scheme. Sounding like a nuclear physicist adding up the growing number of fundamental subatomic forces, Olson adds together the tensions between key forces, "the strongest literary force" represented by Shakespearean tragedy and "the strongest social force" instantiated by American democracy. Olson also reflects on the value of the uncertainty principle for his methodology: Heisenberg's theory, "a first act of physics," is taken to be a guide to a further refinement of the conceptual scheme. Whaling in American history is an "OBJECT in MOTION," which means that the researcher will have to "learn the speed at the cost of exact knowledge of the energy and the energy at the loss of exact knowledge of the speed." Throughout his speculative essay, Olson is attentive to the interrelations that Melville's novel finds active between science and poetry: "The body of the book supports the bulk of the matter on the sperm whale—'scientific or poetic.'"[98]

Olson goes as far as to depict Melville as a social theorist avant la lettre, a human scientist working with a conceptual scheme: "Melville raised his times up when he got them into Moby-Dick and they held firm in his schema." The novelist treated the whaling industry as if he were a social scientist working in the Harvard tradition, seeing the industry as "a problem in the resolution of forces solved with all forces taken account of." Olson's Melville intuitively grasped the need for the scientific method of the participant observer and the conceptual scheme: "Melville did his job. He calculated, and cast Ahab. BIG, first of all. ENERGY, next. PURPOSE: lordship over nature. SPEED: of the brain. DIRECTION: vengeance. COST: the people, the Crew."[99] Melville even sounds a little like Kurt Lewin, since according to Olson, he analyzed whaling as "a problem in the resolution of forces solved with all forces taken account of." If some of this language of physics sounds stretched in the context of a novel about whaling, this strain is as much an artifact of the use of a conceptual scheme constructed with partially unsuitable elements of physics and topology as due to Olson's attempt to fit scientific schema to poetic novelist. These strains are signs of what was to come in Olson's work. The history of the conceptual scheme from Henderson to Davidson follows a trajectory in which the sort of strains that appear in Olson's discourse lead to greater and greater abstraction of the process. But, first, it is time to look more closely at the handling of scientific borrowings in "Projective Verse" and how the essay negotiates challenges similar to those faced in *The Life of Poetry*.

THE CONCEPTUAL INFRASTRUCTURE
OF OLSON'S FIELD

Olson's "Projective Verse" is a fraction of the length of Rukeyser's book, yet it condenses many of the same ideas about breath, form, scoring, speed, energy, and science into its terse, vocative prose, as if his essay were the accompanying practical instructions to her extended defense of the unbegun poetry of the later twentieth century. Despite these affinities, his essay distinguishes itself sharply from Rukeyser's book in two ways.[100] First, unlike Rukeyser, Olson never mentions science explicitly. He does not say, as Rukeyser and Williams had done, that a new poetry must engage with the new physics, yet the essay's repeated use of scientific discourse leaves readers with the impression that the new poet should resemble one of Fjelde's midcentury scientifically literate poets, especially when the essay says that the poet should aim to create a poem that is a "high-energy construct and, at all points, an energy-discharge." Not mentioning science enables Olson to avoid having to locate his poetic practice in relation to other natural and social sciences and to negotiate the issue of their authority. Such questions can be sidelined.

Second, whereas Rukeyser's book is a defense of poetry that spends much of its time tracing cultural tributaries to the poetry of the future, Olson's essay is a manifesto looking almost entirely to the future, consciously employing the public-relations strategies of earlier Futurist, Dadaist, and Surrealist manifestos. Even its title alludes to one of the three features that Martin Puchner argues can be found in all manifestos: "theatrical posing, unauthorized speech, projective positioning."[101] By calling the essay "*Projective* Verse," Olson knowingly invokes the manifesto's foundational power to call a new movement into being through its projection of a new aesthetic practice, while giving a name to this new, or "unbegun," poetry. With considerable chutzpah, given his marginality at the time, Olson confers names on both the new poetry ("projective") and its poetics ("composition by field"). The names stuck. Projective verse became widely used to designate the type of poetry written by Olson and some of his acolytes, though the concept of the "projective" was rarely used to analyze their practice. Field composition became an even more pervasive name for the practice of Olson and the New American poets, as they were often called, usually in the phrase "open field poetry," and it was interpreted as a conceptually rich descriptor of this textual practice. Almost every critic who has written about these poets beats a path through the field.

By introducing these two terms, "projective" and "field," and not directly using the catchall concept of "science" with its implicit exclusion of the arts, Olson is able to shake off preconceptions about the possible interrelations between poetry and science. At midcentury, both terms were actively referent to theories and practices in domains that included physics, social science, philosophy, and aesthetics, and each had different degrees of connection to the scientific ideal. Moreover, informal usage associated the terms "project" and "field" with, respectively, two key features of any organized research, the process of inquiry and the hoped-for outcome, better knowledge. Although Olson's ideas of "projective" verse and "field" composition might appear distinct, the structural tensions internal to the assemblage of semantic domains comprising each idea are quite similar, as we shall see. Both ideas have a similar network of connections to contemporary thought, and it is the degree of success of this conceptual architecture in the epistemological competition of the time that makes them significant.

Olson's sources for the idea of the "projective" included the following: the use of the word to describe a research enterprise; the mathematical theory of projection that he had read in H. S. M. Coxeter's books on geometry and polytopes;[102] A. N. Whitehead's geometric metaphysics; the psychoanalytical idea of the unconscious imposition of a remembered person or relationship onto a person in the present; and Martin Heidegger's concepts of *entwerfen* and *Projektion*, particularly as somewhat misleadingly explained by Werner Brock in his introduction to the first main translation of Heidegger's *Existence and Being* (1949) in terms of project and projection.

All but the connection to geometry, which we have encountered in relation to Kurt Lewin, requires a little unpacking to reveal its contribution to the complex structure of the overall concept of the "projective" in Olson's poetics. Whitehead's metaphysics is a speculative geometrical transformation of Einsteinian physics in which subjectivity, history, society, and imagination are all translated into geometrical terms. Whitehead links subjectivity, energy, and geometry: "Our perception of this geometrical order of the Universe brings with it the denial of the restriction of inheritance to mere personal order. For personal order means one-dimensional serial order. And space is many-dimensional. . . . There is thus an analogy between the transference of energy from particular occasion to particular occasion in physical nature and the transference of affective tone, with its emotional energy, from one occasion to another in any human personality. . . . But the human body is indubitably a complex of occasions which are part of spatial nature." He also describes projective geometry as a "science of cross-classification" whose interest lies in its being a field of mathematics that studies numerical relations that are not directly based on measurement.[103]

Heidegger's philosophy is not usually thought of as similarly heavy on lines and rectangles. *Existence and Being*, a collection of four newly translated essays by Heidegger, then hardly known to readers without German, first introduced the philosopher to an Anglophone audience. Brock's book-length introduction to the essays manages to make Heidegger sound far more scientific than he was. For Brock the new physics exemplified the potential of the projections of *Dasein* to shape sciences. In his extended prefatory outline of *Being and Time*, Brock claims that a science is made possible when objects and beings are considered as *vorhanden* ("present-at-hand" or, in Brock's faulty interpretation, "existent"), and he gives as "the classical example" of such an orientation toward the world "the genesis of mathematical physics, which is decisively guided by the mathematical 'project' of Nature itself" because "only in the light of such a 'project' of Nature can 'facts' be discovered and 'experiments' be planned." Earlier, Brock has explained that Heidegger's thought distinguishes between "thrownness" (*Geworfenheit*) and "project" (*Entwurf*): "The 'project' of understanding is always essentially concerned with 'potentialities,' in all possible respects." Brock all but calls for a projective methodology as he parses the word *Dasein* in the following sentences. He says that "such 'projecting' has nothing to do with a well thought-out 'plan'" because "Dasein has always 'projected' itself already, and continues to 'project,' as long as it is."[104] Brock makes it sound as if the new physics is the leading edge of *Dasein*'s potential, manifested in its projective acts.

Later translators have challenged this interpretation of Heidegger. John Macquarrie and Edward Robinson, the translators of *Being and Time*, caution English readers that "project" and "projection" do not quite translate the sense of "throwing off," the pro-jection, implied by *Entwurf* and *entwerfen*.[105] And Miles Groth, in his study of the translations of Heidegger, shows that Brock doesn't sufficiently distinguish between Heidegger's use of *vorhanden*, which Brock misleadingly interprets as equivalent to "existent" rather than the now generally accepted meaning of "present-at-hand," and Heidegger's term *zuhanden* for what is "immediately on hand."[106] The strong support that Brock's Heidegger appeared to offer for the idea that a poetry that wants to share in the "model character" of the new sciences should enact that "primary project"—that is, should write from the "projective" character of *Dasein* itself—was based on misreading.

This is what the conceptual architecture of Olson's idea of the projective looked like around 1950. A projective verse would be rooted in the deepest potentialities of being, it would operate in both the psychic and material worlds, and it would productively emulate the work of scientific research at those margins of knowledge where the actualities are least controlled by existing rigorous science.[107] Despite this rich mix, almost no critics have managed to find conceptual resources for poetic analysis in the idea of a projective poetry. I would speculate that the ultimate weakness of "projective" as a conceptual matrix for the new poetics, unlike the concept of the poetic field, stems from the largely negative associations with the psychological idea of projection, since these suggest that to be projective is to be self-deceived, and this negative association with imagination runs counter to the scientific referents that emphasize feats of inventiveness. Olson overlooked the pejorative psychoanalytic connotations because the concept could be tied both to his long-term guide to the philosophy of science, Whitehead, and to the newly translated Heidegger.

The metaphor of field was considerably more robust, and historically has proven far more generative and coherent as a conceptual matrix both for the analysis of poetry and for literary theory. As we saw earlier, not only was the concept well established in physics, but it had been remade in the theory of quantum electrodynamics (QED). Already well established in economics and, in the work of Kurt Lewin and the group psychologists, in psychology, it was also firmly entrenched in aesthetics and literary studies. I. A. Richards, who in addition to helping create academic literary studies did so much to lay the groundwork for the study of poetry and science, drew on his own knowledge of behavioral psychology and philosophies of art to talk extensively of "fields" of "attention," "stimulation," and "phenomena," as well as fields of "investigation" and "facts."[108] Similar discourse was widely used in aesthetics in the 1940s. When William Carlos Williams gave his lecture "The Poem as a Field of Action" at the University of Washington in 1948, it was the convergence of these literary concepts with the increasing scientific prestige of the term in physics that encouraged him "to propose sweeping changes from top to bottom of the poetic structure," although he has no specific application of the idea to the construction of poems.[109] It is the idea of the poem as a field or space of action that excites him.[110] But it was the use of the term in nuclear physics, and especially its association with the great achievement of QED, that gave the concept its authority and salience around midcentury, while its "fertility" continued to suggest further lines of investigation.

Olson realizes that the concept of field can do more than signal poetry's advance toward the frontiers of knowledge. It can help address two seemingly intransigent problems for the analysis of poetic composition without encountering the limitations that Rukeyser faces: the difficulty of figuring out how each actual component and type of component contributes to the overall meaning of a poem and the difficulty of accounting for the relations between these components and knowledges outside the poem. To address the first difficulty it appears as if we need to identify every species of component individually and then produce a determining theory of exactly how that species operates (rhythm requires prosody, words require linguistics, images require an elaborate theory of symbols, and so forth). However hard we try, too many components—the white spaces, the force of diacritical marks, the pauses for breath, and so on—lack such functional explanation or are interpretively undecidable. To address the other difficulty we seem to need some theory about the relation between poetic propositions and propositions in other fields, such as the sciences. Olson's concept of the field absorbs these uncertainties or degrees of freedom. Now we start with the idea that every constituent, every particle of the poem, whatever it is, contributes to the overall field of the poem, and that this field in turn interacts with others, whether they are poems or sciences.[111] The field is what matters. No wonder Olson frequently invoked Heisenberg's uncertainty principle and was so drawn to nuclear physics as a discipline defined by epistemic frontiers. His idea of field composition provides a means of thinking through the many implications of treating poetry as a mode of inquiry that might be as valid as those of other sciences.[112]

Olson realizes that instead of talking about new patterns or systems, it would be helpful to take the field concept from physics in order to explain how poems can be valid forms of research. Adopting the field concept enables him to circumvent the problem that bedevils all attempts to establish a theory of literary knowledge. Where, the skeptic asks, pointing to a text, is the knowledge in this novel or poem? The question hides two disabling presuppositions for the defender of literary knowledge: knowledge is a state of understanding that, once achieved, is permanent, and it doesn't matter at what moment in history the question is asked; and knowledge is the defining feature of science, what all the experimentation, evaluation of results, modeling, mathematics, technology, and so forth actually amount to. Olson's idea of composition by field resists any attempt to locate knowledge in the poem because any such knowledge would only be a part of the field. Instead, Olson's concept points strongly toward the idea that the poem can be treated as a mode of inquiry, or knowledge in progress. But he pays a high price for this concept of field. Unlike Rukeyser, he ultimately needs to claim that poetic inquiry is a higher or more all-encompassing form of inquiry than scientific research. For this very reason much of his later work is devoted to the writings of Whitehead. As we saw earlier, the resulting claim that the poetic field, or any metaphysical system, can be so much bigger than science that it can swallow it whole remains, to the say the least, controversial.

✳ 5 ✳

STORIES, GEOMETRIES, AND ANGELS

Muriel Rukeyser, Charles Olson, and Robert Duncan in the 1950s

"THE UNIVERSE IS MADE OF STORIES, NOT OF ATOMS"

How well did the conceptual schemas of field and system serve Olson and Rukeyser in the 1950s, and what influence did their ideas about poetry and science have on other poets? Rukeyser got off to a flying start with a project whose prospectus sounds remarkably like a source book that might have accompanied *The Maximus Poems*. In her new archival study of Rukeyser, Catherine Gander reveals that around 1950 Rukeyser began planning to edit, with a physicist, a major anthology of writings on science and literature, to be titled *In the Beginning*.[1] In her notes for the project, Rukeyser explains that her aim is to "present the materials of man's endless search to explore the nature of creation . . . since the earliest time of records."[2] She sounds a little like the "archeologist of morning" Olson called himself when she sketches her plan to investigate "the mythologies and the sciences, the philosophical and theological findings, the opening of consciousness by aesthetic and philosophical means, and the new images of creative power." Gander reports that Rukeyser "notes her wish to unearth a 'buried' human history, beginning with 'the primitives,' who, 'in coming toward the unknown,' represent 'a closeness with living scientists and children and poets.'"[3]

Rukeyser's proposed coeditor, Professor Philip Morrison at Cornell, had been a student of Oppenheimer's and later worked at Los Alamos, an experience that led him to join with others to found the Federation of Atomic Scientists and write for the *Bulletin of Atomic Scientists*. He recalled what he saw at the Trinity testing of the prototype atomic bomb: "There was just one enormous, flat, rust-red scar, and no green or gray because there were no roofs or vegetation left. I was pretty sure then that nothing I was going to see later would give me as much of a jolt. The rest would be just a matter of details."[4] In a 1949 essay on the dangers of nuclear warfare, he set out in grim detail the picture of just what a bomb would do to New York: "the lingering death of the radiation casualties, or the horrible flash burns, of the human wretchedness and misery that every atomic bomb will leave near its ground zero."[5] Morrison's writings suggest that he and Rukeyser shared sensibilities and values, so it is a little surprising that the project did not go forward in some form or other. Gander speculates that the project may have been too ambitious to be a viable proposition for the publisher. Or maybe Rukeyser wanted to concentrate on her own writing and teaching, or perhaps she was becoming less certain that her vision of a new poetics based on the idea of phase space and its systems could work. Maybe physics was not worth envying.

Some commentators on Rukeyser look to a later poem as a guide to her postwar attitude to science. In one of her best-known poems, "The Speed of Darkness," the title poem in a volume that gained her widespread recognition from feminist writers, she appears to announce an estrangement from physics: "The universe is made of stories, / not of atoms."[6] The reductionism of physics that made the concept of phase space possible by treating the universe as no more than the result of particles and forces now looks oversimplified. Stories, or temporal narratives, are needed too. Yet a decade later, in the preface to her *Collected Poems* (1978), Rukeyser appears to remain committed to a close relationship with the sciences. She explains that she is delighted to be published by McGraw-Hill because it "is a small experimental press as far as poetry is concerned, but a great vast publishing house of science; and I care very much about that meeting-place, of science and poetry."[7] Although most of her later poetry did not have any obvious scientific content, she did continue to be interested in the biographies of others who worked with both science and politics, notably Wendell Willkie in her hybrid poetic biography *One Life* (1957) and the Elizabethan explorer and poet Thomas Harriot, about whom she published a biography in 1971.

As early as 1949, we can glimpse the shape of the problem that Rukeyser would come to have with science in her own poetry. In the same year as *The Life of Poetry*, Rukeyser also published "Orpheus," a lengthy narrative reenactment of the murder, scattering of the limbs, and gradual resurrection of Orpheus. Violence invades the allegorical body of Orpheus, whose transformations represent the fundamental principle of change as he is reduced to matter in its chaotic state, "all forms and no form."[8] An account of the composition of "Orpheus" provides the poetics statement that Rukeyser wrote for *Mid-century American Poets* (although the poem itself was not included in Ciardi's anthology). She starts with the large claim that "the laws of exchange of consciousness are only suspected," an exchange that readers of *The Life of Poetry* will know she thinks of as an energy transfer. She supports her claim with a quotation from Einstein in which he speculates that science will one day discover a "much stricter and more closely binding law than we recognize today" for comprehending such patterns of causation. But although she mentions reading the nineteenth-century work by Geddes and Thompson, *The Evolution of Sex*, and more recent work on memory and "dislocated nerve centers," as well as the Orphic hymns, and listening to the Gluck opera, a process that results in "pages of notes" in which there were "whole lines, bits of drawing, telephone messages in the margin," it is not easy to connect her science research to the poem she wrote.[9] The story hides the atoms, and allegorical form hides personal history.

Lorrie Goldensohn speculates that Rukeyser is allegorizing her own otherwise-impossible-to-articulate experience of birth accompanied by an unauthorized and unwanted hysterectomy.[10] In addition to that history of personal distress, the allegory obviously alludes to the war and to questions about the connection between poetry and human violence. Any interest in science is at first sight hard to see in Rukeyser's intense mythic and sometimes-surreal narrative of bodily dissolution and reincarnation. Given that the poem is about transformation, we could see it as a highly poetic version of her phase space theory of the poetic system, though the insistence on formlessness does not encourage this interpretation. There are a few traces of scientific knowledge, however, traces that hint at an unsuccessful struggle in the poem to bring together the Orpheus myth with scientific thinking. At one moment when Orpheus is just a mass of wounds, completely dead, the poem avoids the vitalist language we might expect to be used to describe him and instead uses the language of physics—"No space / is here, no chance nor geometry"—as if to be alive is to embody space, geometry, and probability.[11] When Orpheus is reborn as a god of poetry, one of the first things he does is to remember history and bind together its constituents, including "the firewind and the cloud chamber" (his images of the atomic bomb and the nuclear research apparatus), which he is able to do because "he knows the nature of power, / the nature of music and the nature of love."[12] Here, as in the poem's conclusion, we seem perilously close to wishful thinking. It is hard to imagine this figure from classical legend standing in a laboratory with the Manhattan Project physicists.

An extended review of Rukeyser's *Selected Poems* (1951)—shortlisted for a National Book Award—by Isidore Salomon in the April 1952 issue of *Poetry* reveals that despite the excitement about science expressed by some male writers, the literary world was unlikely to be receptive to a woman writing poetry about science. The reviewer argues that her work always risked slackness of structure and expression, sometimes even giving way to a "devil-may-care tone," but fortunately the recent work shows that she has seen the error of her ways: "No, Miss Rukeyser has not sold her birthright for a mess of facts." "Orpheus" is proof that she has instead finally turned "towards pure poetry." The reviewer is clear what the myth of Orpheus means: "Orpheus murdered and dismembered by women" is then "reborn the god of poetry and song."[13] Given that a relatively sympathetic, if critical, reviewer, an English teacher and poet living in New York, can dismissively interpret the poem as a parable of women's threat to male creativity and disparage documentary or scientific realism in poetry, it is evident that Rukeyser would have had difficulties even publishing such work.

Rukeyser's major poetic work of the 1950s, *One Life*, was a poetic biography of Wendell Willkie combining prose and poetry, a work in which, as she had said of her early poem "The Book of the Dead," "poetry can extend the document." Michael Davidson glosses this aim of extending the document in the early poem as a means of providing "a voice for individuals who live at the margins of a national scene" and rectifying the muffling effect of national forums, which can neutralize "critical perspective on the conditions that prompt one to speak."[14] Her attempt to extend the document in her poetic life of Willkie has met with almost universal dismay. Poorly reviewed at the time, and still largely neglected by critics, the book has suffered the ignominy of having its poems omitted from the most recent *Collected Poems*. Even a critic as sympathetic as Louise Kertesz concedes that *One Life* "fails repeatedly," though she tries to soften this judgment by adding that "the risk and the successes, one feels, are worth it."[15] The failure is not due to a lack of ambition. Rukeyser is as interested as Olson is in his Maximus poems by questions about the scale of human achievement, and she also draws on documents as well as dreams for evidence of the placement of recent human experience in a much longer history. Willkie's slow conversion to Roosevelt's belief in the value of government management of such public goods as energy fascinates Rukeyser because she thinks of it as a story of how modern American politics actually operates. Gander mentions an unpublished preface for *The Speed of Darkness* (1968) in which Rukeyser explains that she had always been interested in two sorts of material for poetry: "the unverifiable fact" as opposed to document or verifiable fact that offers historical "evidence of the world."[16] *One Life* uses very similar techniques to those used in the poems about silicosis in *U.S. 1*, though on a much larger scale.[17] Historical narrative, fictionalized inner thoughts, dialogues based on the historical record, and poems mingle freely.

One Life ends not with Willkie's heart attack in 1944 but a year later with the dramatization of fractious congressional hearings to confirm David Lilienthal as chair of the new Atomic Energy Commission. Lilienthal was originally Willkie's opponent at the time of the setting up of the Tennessee Valley Authority in the 1930s, but by 1945–46 the commission represented an ideal of governance in which Willkie had come to believe, and Lilienthal represented a promising political stance for the future. Rukeyser's dramatization is oddly reminiscent of a musical. Senators unexpectedly break into poetry that Rukeyser constructs from transcripts with an ironic eye for rhetorical patterning, as if to say that even politicians who despise the arts rely ungratefully on the art of poetry. At times an unidentified choric voice meditates on the metaphorical and metaphysical implications of nuclear energy:

I have tried to show the atom as a source
A source of energy.
I have touched on another question:
Might energy become a source of atoms?
If this relationship is real,
The universe passes along a way of cycles.
A process of matter dissolving in the stars,
Turned into radiation, passing through forms
Again to matter; again, perhaps, to birth.[18]

This could be the author speaking or a physicist, perhaps Arthur Haas, who is cited in the previous paragraph, or a choric amalgam of poets, physicists, and politicians. It feels like a weakness in the poem that this uncertainty does not enrich what is said. To go from atom to forms to birth is a reminder that physics relies on metaphors from everyday experience, as well as a restatement of a recurrent theme, summarized in the final poem in the book, "In Praise of Process," as "one law moving and given: the form of the love of growth" that takes place in "the human light of meanings."[19] Rukeyser chooses to conclude her book, in other words, not with Willkie's global vision (set out in his book *One World*) but with debate about the future of American political management of atomic energy, which Senator Vandenberg calls "this world-wrenching mystery."[20] In the final poem in *One Life* she writes as much about her understanding of the significance of this mystery as about the mystery of a representative human life such as Willkie's.

The failure of *One Life* to make a lasting impact on readers of poetry must have had repercussions for Rukeyser's sense of poetic possibility.[21] The poems in *The Speed of Darkness* repeatedly hint at a long-standing crisis in writing poetry, no more so than in the opening poem, "The Poem as Mask: Orpheus," whose resounding phrase "No more masks!" was adopted as the title of one of the most widely read anthologies of feminist poetry.[22] "The Poem as Mask: Orpheus" is a retrospective disavowal of *Orpheus*: "When I wrote of the women in their dances and wildness, it was a mask . . . when I wrote of the god . . . it was myself, split open, unable to speak, in exile from myself." Readers may remember that the phrase "split open" was used by Louise Bernikow for her historical anthology of feminist poetry, *The World Split Open* (1974), for which Rukeyser wrote a preface.[23] To have two phrases from one poem become titles of key feminist anthologies is more than impressive; it underlines how significant the idea of overcoming silence had become.[24] "Orpheus" was an attempt to write about tearing oneself apart in response to personal tragedy, war, the atomic bomb, and the desire to speak. Now that she wrote openly of birth, of the love of other women, of women's silences, she could write about the birth to which knowledge of the atom might lead and speak to a whole new generation of readers. Her attempts to bring science and poetry together had led far beyond atoms and their systematic phase spaces to a universe made of stories of birth, the body, and history.

THE MESSAGE OF OLSON'S "THE KINGFISHERS"

In the early years of the decade Olson continued to celebrate the marriage of poetry and science in the conceptual scheme exemplified by the field, but toward the middle of the decade his growing investment in Alfred North Whitehead's process philosophy led him to set aside the work of building new constructs and schemas, leaving Whitehead to do that work for him. Instead, he would concentrate on history and the poet's encounter with it. These developments were foreshadowed in his first major poem, "The Kingfishers," which conforms well to the desiderata of "Projective Verse": its collage structure does move quickly from one perception to the next; its stylistic features such as its terseness, its syntactical instability, its shifting line lengths and indents, and its absence of punctuation, all suggest a field of barely controllable high energies; and its themes, the enigmas of change and continuity, as well as the dark mystery around the origins of violence, point to what "Projective Verse" calls "the secrets objects share." The poem draws attention to the incommensurability of different knowledges that have been brought to bear on kingfishers: anthropology, archeology, biology, cybernetics, historiography, and information theory are all referenced.

For many readers, the most obvious sign of science in the poem is likely to be the intrusion of an indigestible chunk of information theory in part 4. The poem splices classical philosophy with this very contemporary form of physics in a passage alluding to Ammonius's speech in Plutarch's essay "The Letter E at Delphi" where the philosopher Ammonius meditates on how it is that we think of ourselves as having a singular identity over time. The poem looks to science for the answer:

We can be precise. The factors are
in the animal and / or the machine the factors are
communication and / or control, both involve
the message. And what is the message? The message is
a discrete or continuous sequence of measurable
       events distributed in time[25]

The final, unattributed sentence in this passage is taken verbatim from Norbert Weiner's 1948 book *Cybernetics; Or, Control and Communication in the Animal and the Machine*.[26] This large claim that the puzzle articulated by the ancients has now been solved is curiously empty, however, as if we had been informed that the solution to Ammonius's dilemma is a series of words without being told what they are. Our uncertainty is exacerbated by a literary critical joke. The poem helpfully answers a familiar pedagogical question that troubled generations of literature students: what is the message of the poem? The message of this poem is that a "message is a discrete or continuous sequence of measurable events distributed in time." The message of the poem is that a message is a message, and words are words. But the irony of this circularity also rebounds onto cybernetics, which instead of adding to knowledge is left dressed in the emperor's new clothes. Olson has staged a confrontation between science and the literary in which the result is a Mexican standoff.

Cybernetics is far from the only scientific allusion in the poem. Biological details about the kingfisher's life cycle and the legends that have grown up around it, taken from the eleventh edition of the *Encyclopaedia Britannica*, are another salient example of science in the poem.[27] Olson rearranges the words a little, keeping some of the distinctive lexical items (e.g., *"rejectamenta"*) and the awkwardness of some phrasing ("bored by itself" could mean that the bird is lonely and bored on its own, though it is intended to describe the tunneling into the bank), all with the effect of a tacit irony at the expense of the encyclopedist with no ear for the resonances of language. These details about the kingfisher represent what is ordinarily thought of as scientifically derived knowledge based on the foundation of expert research. This normative manner of presenting scientific knowledge sets aside individual differences between actual birds in favor of a species ideal. In his *Principles of Literary Criticism*, I. A. Richards dismissed such sources as the *Britannica* as irrelevant to poetry, saying, "the kind of information which we can acquire indefinitely by steady perusal of Whitaker or of an Encyclopedia is of negligible value."[28] Olson's liberal inclusion of this encyclopedia material is a tacit questioning of assumptions about what constitutes knowledge, scientific or otherwise.

In addition to these two different forms of knowledge, cybernetics and biology, the poem also offers data and theories from anthropology and archeology, which were then both becoming increasingly scientific disciplines. Olson's joy with science is therefore very evident in the thematic materials of "The Kingfishers." I want to argue, however, that its structure is where Olson's interest in scientific method is most active. "The Kingfishers" enacts the process of inquiry that Olson thought of as fundamental to scientific method. As many readers have noticed, the continuing power of this strange, even clumsy, poem resides more in its awkwardness, its naive surprise, and its incompleteness than in any wisdom, eloquence, or demonstration of a new "projective" poetics. Indeed, "The Kingfishers" can make other, more polished late-modernist poems seem by comparison merely slick. Instead of giving readers the conclusions of a thought process, the poem enacts the stumbles and false starts that occur as the poet becomes curious about the significance of the Mesoamerican trade in kingfisher feathers.

It enacts the process of inquiry in two ways. First, the poem plunges us from the start into a Dantean landscape of obscurity where an unidentified man, who might be the poet/narrator or might be a friend who recounted this experience to the narrator, wakes disoriented in an entanglement of perceptions, memories, and dreams. For a while it is not even clear in the pronominal whirl who is who. This image of the barest form of inquiry as a man awakening in a dark place could have seemed tired given its roots in Dante and the Cartesian philosophical tradition if it was not that by starting in this manner Olson is also able to hint that poets need to wake not just from their dogmatic slumbers but from their dogmatic adherence to older ideas of what constitutes the subject of poetry. Gradually, epistemological confusion gives way to a cryptic question about the trade in kingfisher feathers and then to further questions about human violence and its legacy. These questions are a sign of the second strategy for enacting inquiry in the poem. In a 1951 essay on sociological methodology, Paul Lazarsfeld and Allen Barton "recommended that the researcher involved in creating concepts should start 'with fairly concrete categories.'"[29] The researcher could simply adopt existing social concepts, but "in most exploratory research . . . the investigator will have to develop his own categories."[30] Developing his own categories and concepts for research is what Olson aims to do in this poem, and the process becomes more and more explicit as the poem proceeds.

Given the overexposure this poem has had in the past fifty years, its scrupulous depiction of faltering inquiry can be easily overlooked. As we get to know the poem better, its stumbles toward knowledge tend to fade from our minds, and we are likely to interpret the poem as being equivalent to its paraphrasable conclusions about change and violence in human history. The intensity of the poem is, however, found less in the shock of the horrors uncovered and more in identification with its sometimes very disorienting momentary states of epistemological and ethical uncertainty. It is not the explicit allusions to contemporary and ancient sciences that make this a "high-energy construct" and, thus, a *scientific* poem. It is the structuring as inquiry that emulates scientific methodology of the time, a structuring that is carried over into *The Maximus Poems*.

The Maximus Poems can be read as Olson's equivalent of Wordsworth's *Prologue*, a prepoetic attempt to explain the shaping of the poet, the sources of his poetic authority, and a preparation for the poetry and poets that will follow. Its mixture of local history, personal memory, polemic about the current state of America, dreams, and judgments on literary culture is held together by the dramatized persona of a researching pedagogue, Professor Maximus, finding out facts, changing his mind, reaching conclusions, and always emphasizing the act of communicating his judgments. Almost wherever one looks in the first volume of *The Maximus Poems* the poet is trying to enlist the reader into a shared research project, even when the nature and details of this inquiry are so unresolved as to leave little more than the abstract form of inquiry remaining. Trust inquiry, the poet tacitly pleads. We have the appeal to data: "There is evidence / a frame // of

Mr Thomson's / did // exist"; the methodological statement: "in *Maximus* local / relations are nominalized"; the historical argument: "the Continental Shelf // was Europe's / first West, it wasn't / Spain's / south: fish, / and furs, // and timber, / were wealth, / neither plants, / old agricultural / growing, from // Neolithic"; the excitement of discovery: "Here we have it—the goods—from this Harbour"; and allusion to anthropological discovery (in this case Java Man): "if you are drawn / if you do unite, / if you do be / pithecanthropus." The poet constantly uses an inclusionary rhetoric that encourages readers to believe they are sharing in an important project: "on founding: was it puritanism / or was it fish? // And how, now, to found, with the sacred & the profane—both of them— / wore out."[31] Even the frequently changing visual appearance of the text on the page presents itself as an occasion for suspenseful discovery. What especially interests the poet in the dream poems such as "The Twist" and "Letter 22," as he explores oneiric Gloucester, are glimpses of possible lawlike workings of this human universe. Most of the poems in this first Maximus sequence offer readers not findings so much as acts of finding out.

It might be objected that these examples are a reminder that Olson was primarily concerned with history, as if that were incompatible with scientific inquiry. I hope by now that I have established that social research was inventively working with scientific methods at the time. Many respectable thinkers also believed it was possible to have a genuinely scientific historiography.[32] The final chapter in the philosopher of science Ernest Nagel's life's work on the foundations of scientific method, *The Structure of Science* (1961), which builds on his publications of the previous decades, is devoted to "problems in the logic of historical inquiry." Historians, he says, "aim to assert warranted singular statements about the occurrence and the interrelations of specific actions and other particular occurrences." They do not, as the natural scientists often do, assume that "for an event to be explained it must be subsumed under a strictly universal law serving as a premise in the explanation." And sometimes history is "practiced as a fine art, comparable in some respects to poetry." Nevertheless, historical inquiry does rely on broad generalizations about social processes and does assume varying degrees of determinism in apportioning causes of historical events, and therefore Nagel concludes his largely promissory account of scientific historiography on an encouraging note for the would-be scientific historiographer: "However acute our awareness may be of the rich variety of human experience, and however great our concern over the danger of using the fruits of science to obstruct the development of human individuality, it is not likely that our best interests would be served by stopping objective inquiry into the various conditions determining the existence of human traits and actions, and thus shutting the door to the progressive liberation from illusion that comes from the knowledge achieved by such inquiry."[33] This invitation for history to become scientific is a reminder that Olson's stance was shared with his contemporaries in other human sciences, even though none of them were contemplating poems as the best vehicle for publishing research.

Olson's interest in myth was complex and perhaps contradictory, but here too he was interested in treating myth and science not as antitheses but as complementary modes of knowledge. His Black Mountain lecture "The Science of, Mythology" asks what would be the best methodology for this "science of man" and answers that it must recognize the fundamental determining fact of embodiment. The task is to "be as scientific as those who have sold us the idea matter alone can be measured" and "to bring back explorations" of our own organism because "who I am is an organism I also am beside my body, inside my body, if you like, with structure and event and population so like the physiology."[34] The lecture title's odd placement of the comma is intended to signal a double intention: this lecture is about both the possibility of a new science of mythology and the protoscientific features of premodern mythologies. The new methodology will have affinities with those of the botanist Edgar Anderson, the archeologist Christopher Hawkes, and the geographer Carl Sauer:

And with that, I turn scientist. I propose to try to do what any scientist has to do first: to determine and isolate, the field proper to the study, to take up the materials he is interested in working out technologies to further the knowledge of, whether it is one man's bugs, another's—like Anderson's— cultivated plants, or Hawkes' microliths and barrows, Sauer's fibres and roots fisher-folk of the Pleistocene shredded to make nets and lines by, and found, instead that they didn't need the nets, the juice of the plants poisoned the water and the fish turned up on the surface dead.[35]

This summary is about as close as Olson ever gets to an explicit outline of what makes his method scientific: science involves selecting a set of objects and events from the swirl of phenomena and then working out the technologies needed to study them. Although we might be inclined to think of test tubes and other machines as the technologies referred to, his examples are all instances of *conceptual* categories used as research tools by the scientists in their fieldwork. Olson's argument rests on two assumptions: that scientific method requires a conceptual modeling of a determinate field of materials and that mythology was once a similar process, equally trusted and equally real to its users. He is trying to persuade his listeners that the reflexivity necessarily involved in constructing a human science can best be understood in terms of the process of building a myth and then living within it.

WITHHOLDING: OLSON'S GROWING DOUBTS ABOUT CONCEPTUAL SCHEMES

Although Olson's 1953 Black Mountain lectures still advocate the virtues of understanding scientific methods as involving case studies within a conceptual scheme, by then his poetry is beginning to openly register the difficulties of doing it for oneself. One of the most striking examples occurs in an unpublished Maximus poem written around June 1953, "Maximus Letter #28," where he pronounces flatly: "There is no science / of human affairs."[36] "Maximus Letter #28" is part of a stretch of orphaned Maximus poems written between 1953 and 1957 that have only ever been published in *Olson: The Journal of the Olson Archives*, although the quality of some could justify inclusion in a collected poems. Olson rejected them because, according to his editor George Butterick, the poem sequence appeared to be getting out of hand as it extended outward from its initial location in Gloucester to encompass the entire westward settlement of America.[37]

This insistence that "[t]here is no science / of human affairs," that a *human* science is impossible, appears to contradict many other nearly contemporaneous statements, such as the opening claim in "Human Universe" (1951): "There are laws, that is to say, the human universe is as discoverable as that other."[38] Moreover, the stagey line break in the assertion "There is no science / of human affairs" hints that Olson might even be disenchanted with science itself, not just claims made on behalf of the *human* sciences. Yet as we have just seen, this same year he was proclaiming to a Black Mountain audience his "joy of science." How can he be making an impressively exuberant endorsement of scientific method, on the one hand, and dismissing scientific methods of social research, on the other? Is the canceled "Maximus Letter #28" a misstep whose assertions should be ignored since it was, after all, deleted from the epic sequence, or might this contradiction with Olson's other views about science be only apparent? I shall argue that this almost-forgotten poem is a sign of the strain of trying to base his poetics of inquiry on the development of his own workable conceptual scheme for the scientific study of history.

"Maximus Letter #28" contrasts two opposed ways of writing history, Henry Adams's meticulous respect for historical fact and Bouck White's imaginative re-creation of the life of the ruthless speculator Daniel Drew. The poem centers on Adams's disenchantment with his own record as a historian, a harsh self-appraisal made in the famous chapter of his autobiography, "The Dynamo and the Virgin." Adams begins his chapter by recalling the disorientation he felt at the Great Exposition of 1900 when he encountered the awesomely powerful new electric dynamos as well as the alarming new discoveries of the "anarchical" (a twenty-first-century equivalent would be "terrorist") X-rays and other mysterious radiations. Adams is particularly sensitive to the power of scientific explanation to construct a new universe of experience. Once upon a time the atom "figured only as a fiction of thought"; now everyone "had entered a supersensual world" revealed by the scientists who were handling these new forces.[39] Olson teasingly calls Adams a "piston" because of this fascination with new machines.

Olson's main attention is directed to the sixth paragraph of Adams's chapter. In section 4 of his poem, Olson quotes verbatim from Adams's scathing judgment on the purpose behind his own career as a historian:

> "for no other purpose than
> to satisfy himself whether
> by the severest process of stating,
> with the least possible comment,
> such facts as seemed sure,
> in such order as seemed rigorously
> consequent, he
> could fix for a familiar moment
> a necessary sequence of
> human movement."

Which he could, and did,
as any of us can,
for ourselves (as my man did,
who died) but never
as of what we are not actors in, agents of (why Jefferson

did let Burr go
so long (why history

hoisted the two of them, born
of Presidents, and thus
unable to practice
themselves, spectators
of both professions, weighed down
by statesmanship, the instinct
in them to fact gone
neither to act or image, the

alternatives. And both
one[40]

Adams eventually came to believe that a viable historical method requires the historian to be a participant observer, not the bystander to big historical events that Adams felt he had been. Olson adds his own further judgment that viable historiography also requires for its full realization either consequent action or facture as art (the image). By presenting Adams's sentence in free-verse form, Olson further intimates that Adams could have found an aesthetic form for his history that would have been truer to his ambitions.

Olson's poem has not finished with Adams's text. Section 5 paraphrases the following sentence from Adams, "The result had satisfied him as little as at Harvard College," before again quoting verbatim, this time from Adam's stark sense of failure for having been unable to persuade other historians of the validity of his interpretations: "Where he saw sequence, other men saw something quite different, and no one saw the same unit of measure."[41] Olson thinks that Adams is in part the architect of his own difficulties. Section 5 chides Adams for thinking that any such standardized unit for measuring human actions could be possible:

> "And no man saw
> the same unit of
> measure," missing
>> that just that is
>> the measure,
>> that no man does,
>> that just that is why
>> we are not
>> as flowers are (are not
>> birds[42]

To be human is to have one's own unique measure or makeup, and there can be no equivalent to the taxonomies of birds or flowers when it comes to taking the measure of the human condition. Not only are people too individual to be subject to some such scientific system of categorization, but history requires acknowledgment of effects that could never be measured by a human science, such as the effects of the greed evident in the stories of men like Daniel Drew. This consequence can be encompassed only by an aesthetic method capable of representing lived experience.

"The result had satisfied him as little as at Harvard College." Adams's disillusion would have had special resonance for Olson, for whom Harvard did once seem to provide a satisfactory science of human affairs, even though it failed him in other ways, not least in his ambitions to be a literary scholar. After all, Olson's Harvard had learned the lessons that Adams wanted to teach his readers, that the rigorous study of history must have concepts and measurements as well defined as those that made possible the dynamo. By 1953, however, like other social thinkers of the time, Olson was beginning to be dissatisfied with the Harvard version of practical epistemology. The conceptual scheme derived from physics began to look more and more designed to measure phenomena less sticky with human complexity than social life. He was therefore prey to doubts about the efficacy of systematizing theories of society, doubts about the validity of any generalization about human diversity, and doubts about the putative accuracy of social measurement, however well grounded in the kinds of conceptual schemes offered by Lewin and Murray. Over the next few years he would turn to another Harvard thinker, Alfred North Whitehead, because his comprehensive metaphysical system appeared to be well grounded in modern science and able to solve once and for all the problem of finding a viable conceptual scheme that would assist poetic inquiries into the human condition.

Evidence of Olson's struggle to rethink the principles behind the use of conceptual schemes also appears in another displaced poem of this time, "Maximus to Gloucester, Letter 27 [Withheld]," which contains lines frequently cited as evidence of Olson's radical poetics: "An American / is a complex of occasions, / themselves a geometry / of spatial nature." As is well known to readers of Olson, these are unacknowledged quotations. They and other phrases are taken from two adjacent short sections of Whitehead's *Adventures of Ideas* (1933) — "Space and Time" and "The Human Body" — from within a longer chapter on the philosophical issue of the relation between "objects and subjects." Today this poem is usually read, not as an argument with Whitehead, but as a statement of Olson's own principles underlying the whole project of the Maximus sequence.[43] Shahar

Bram, for instance, asserts that "beyond his own personal belief in Whitehead's metaphysical worldview, Olson also requires the reader to believe in it."[44] Joshua Hoeynck claims that in "Letter 27" "the complex of occasions appears to give birth to the form of the poem by simple evolution."[45] Reading the poem as a statement of Olson's beliefs is made plausible because he starts the poem with a cameo of his own early memories. The concepts and descriptions borrowed from Whitehead are inflated with the authority of the voice of the poem, a voice whose opening identification with a memory of his progenitors has the authenticating effect of a signature. I shall argue, however, that Olson treats these memories as a test of the truth of Whitehead's ideas. A reader who knows the source of such oddly distinctive phrases in the poem as the claim that "An American / is a complex of occasions" might assume that rather than a test, this is an illustration of Olson's commitment to Whitehead's metaphysical system. I shall challenge this assumption, because this interpretation of the poem does not give sufficient attention to the signs that Olson is not yet ready to commit himself to Whitehead and wants to investigate the worth of Whitehead's propositions about history, as he was checking the value of other theories of history in poems such as "Letter #28."

We need to give weight to Olson's withholding of "Letter 27." The odd square brackets in the full title of this poem, "Maximus to Gloucester, Letter 27 [Withheld]," derive from its publishing history. Its initial expulsion from the first volume of the Maximus sequence, along with others written in 1953 such as "Maximus Letter #28," was followed in this special case by its much later readmission to the fold in the second volume of *The Maximus Poems*. Other Maximus poems from this phase of his career were not so lucky. Ralph Maud suggests that Olson's initial uncertainty about his reliance on Whitehead's authority was the likely reason for withholding "Letter 27" from inclusion in the first volume of *The Maximus Poems*: "Olson was not ready to declare how much of an influence Alfred North Whitehead had become." Whitehead's insistence that "personal history" must be taken into account alongside "the welter of events and of the forms they illustrate" allowed Olson, according to Maud, "to assert his uniqueness."[46] Once Whitehead became a regular guest authority in the second volume of *The Maximus Poems*, there was no reason to withhold this earlier and obviously significant poem, which has subsequently become one of Olson's more reprinted and discussed poems. This interpretation is fine as far as it goes but is only part of the story; it underplays Olson's shifting investment in scientific method back in the early 1950s.

All the signs are that by 1953 Olson was beginning to have doubts about the viability of conceptual schemes, local or cosmic. I want to suggest that for a while even Whitehead fell afoul of this suspicion. If we compare the withheld "Letter 27" to the discarded "Letter 28," we notice that the latter poem also cites approvingly an intellectual authority, Henry Adams, and yet subjects him to explicit and implicit skeptical questioning about what it treats as his imperfect conceptual schemes. We might reasonably wonder if at the time of writing the less obviously marked allusions to Whitehead in the withheld "Letter 27" were part of a similar strategy: making a distinct historical voice of intellectual authority represent a particular viewpoint that can then be enframed by the poem. I shall argue that when he first drafted this poem, Olson thought he could manage Whitehead's authority simply by using techniques of self-reflexivity and rhetorical involution similar to those he had used to manage Adams in the discarded "Letter 28." Olson imagined he would test out what he assumes to be Whitehead's conceptual scheme, rather than a full-blown metaphysics, by juxtaposing it with a brief case study from his own personal history and then working out how, if at all, the schema might need alteration. Conceptual schemes were always in need of adjustment according to the findings of new research, new case studies.

At the time of the withheld "Letter 27," Olson was only beginning to come to grips with Whitehead. The influence of Whitehead's process philosophy on Olson and his friends, notably Robert Duncan, would grow stronger as the decade wore on. They were far from alone in this enthusiasm. Several studies of Whitehead appeared around 1960, including Ivor Leclerc's *Whitehead's Metaphysics: An Introductory Exposition* (1958), Robert M. Palter's *Whitehead's Philosophy of Science* (1960), A. H. Johnson's *Whitehead's Theory of Reality* (1962), and Victor Lowe's *Understanding Whitehead* (1962), to name the most prominent. Even Hannah Arendt makes extensive use of Whitehead's *Science and the Modern World* in *The Human Condition* (1958). Then in the 1960s Whitehead's appeal began to wane in America as a new generation of philosophers derided philosophical strategies allegedly widely used by their predecessors, including "the myth of the given," naturalized epistemology, and even the very idea of conceptual schemes. Although these critiques were by no means specifically aimed at Whitehead, it could be argued that he had fallen into all these errors. Whitehead's theory of prehensions is effectively a theory of the given and therefore open to many of the criticisms that Willard Van Orman Quine, Wilfrid Sellars, and others would direct toward the dogmas of empiricism in the 1950s (one possible defense of Whitehead would be that to interpret prehensions this way is to fail to recognize that Whitehead is transmitting insights from the American pragmatist tradition, such as C. S. Peirce's theories of intuitive knowledge, of firstness, secondness, and thirdness).[47] Meanwhile, in Europe prominent philosophers were becoming suspicious of any whiff of metaphysical smoke. Whitehead's science was also creaky. He was writing about science and the history of science as it was understood in the 1920s, when the wider epistemological and ontological implications of the new quantum physics were not well understood.

Today, the postwar enthusiasm for Whitehead may appear more intelligible since the rediscovery of Whitehead by Gilles Deleuze, Isabelle Stengers, Steven Shaviro, and others has rekindled interest in alternative narratives of late modernity. Current philosophers who find a new relevance in Whitehead's bold willingness to try to solve the biggest questions about how subjectivity and materialism are connected generally want to "end the divorce of science from the affirmations of our aesthetic and ethical experience."[48] Stengers describes this as a philosophical approach that "demands, with utter discretion, that its readers accept the adventure of the questions that will separate them from every consensus," even if this adventure will "shake our judgmental routines."[49] Shaviro's narrative of Whitehead's ideas is particularly effective because it starts from what he calls a "philosophical fantasy" in which Whitehead replaces Heidegger as the most influential philosopher on literary theory and then carefully maps how Whitehead's system differs. Shaviro suggests that the great question that drives Whitehead's thinking is "How are novelty and change possible?"[50] Literary scholars have begun to find new resources too. Miriam Nichols's highly original ecological reading of Olson thinks through the affinities between Whitehead and Deleuze and takes seriously Whitehead's reorientation of the relations between subject and cosmos.[51] Hoeynck argues that both Olson and Duncan believed that "Whitehead's relational cosmology provides a framework for conceptualizing how humans and nonhumans fold into each other over the course of deep time and in immediate, mediated experience."[52] In fact, philosophers of biology have also been showing an interest in exploring whether process philosophy might help them formulate better models of the complexity of the unending interdependencies of genetic processes, and this has carried over into ecocriticism. Postwar interest in process philosophy is also more explicable if we look back, not at the monumental *Process and Reality*, but to the more accessible version of the philosophy that Olson was reading at the time of "Maximus to Gloucester, Letter 27 [Withheld]": *Adventures of Ideas*, whose very title hints somewhat misleadingly at the sort of excitements normally reserved for boyish stories.

Adventures of Ideas successfully aimed at a wider audience than professional philosophers and was very much part of the intellectual culture of the 1950s in a way that the more technical volume *Process and Reality* never was. The American Mentor paperback edition stayed in print for many years—its cover immodestly assures readers that inside was "a brilliant history of mankind's great thoughts." *Adventures of Ideas* offers a single schematic ontology of both the material universe and human consciousness, not a conceptual scheme in the sense that the social scientists thought of such frameworks. This metaphysical system is not provisional, not open to revision by experiment, and its constituents are not part-metaphorical in the way the fields, forces, and vectors in Lewin's schema were. To poets reading the book in the 1950s, the philosophy would have appeared to be so deeply grounded in the recent history of science that Whitehead's argument that the new physics of relativity represented a crisis for all conceptual thought would have the ring of truth. Whitehead very much looked like someone poets should endorse, someone who could, in his words, "end the divorce of science from the affirmations of our aesthetic and ethical experience."[53]

Olson incorporates phrases and vocabulary from Whitehead into several parts of "Letter 27." As well as the aphoristic assertion that "An American / is a complex of occasions, / themselves a geometry / of spatial nature," he also borrows from Whitehead in other places: "This, is no bare incoming / of novel abstract form, this // is no welter of the forms / of those events"; "It is the imposing / of all those antecedent predecessions, the precessions / of me"; and "There is no strict personal order / for my inheritance." Looking now at one of the passages from Whitehead that Olson was poring over as he wrote his poem, it is possible to see how its merger of the language of physics with discourses of the body and mind might strike Olson as commensurate with the practices of Lewin or Murray.

Our perception of this geometrical order of the Universe brings with it the denial of the restriction of inheritance to mere personal order. For personal order means one-dimensional serial order. And space is many-dimensional. Spatiality involves separation by reason of the diversity of intermediate occasions, and also it involves connection by reason of the immanence involved in the derivation of present from past. There is thus an analogy between the transference of energy from particular occasion to particular occasion in physical nature and the transference of affective tone, with its emotional energy, from one occasion to another in any human personality. The object-to-subject structure of human experience is reproduced in physical nature by this vector relation of particular to particular. It was the defect of the Greek analysis of generation that it conceived of it in terms of the bare incoming of novel abstract form. This ancient analysis failed to grasp the real operation of the antecedent particulars imposing themselves on the novel particular in the process of creation. Thus the geometry exemplified in fact was disjoined from their account of the generation of fact.[54]

Hadn't Olson argued in "Projective Verse" for a poetics that recognized the "analogy between the transference of energy from particular occasion to particular occasion in physical nature and the transference of affective tone, with its emotional energy, from one occasion to another in any human personality"? Whitehead's confidence might also have struck Olson as admirable. By this stage in his career, Olson had been following in the footsteps of Lewin and others for more than a decade, attempting to construct from scientific spare parts the sort of tentative schemas he hoped would be capable of generating new insights into American and human histories of migration and settlement. Now in Whitehead's writings he finds a possible answer. Maybe his attempt in "Projective Verse" to construct a schema out of the abstract forces, fields, and particles taken from nuclear physics could be justified in Whitehead's terms.

Olson cherry-picks images from Whitehead's explanation of the interconnections between time and being, and between novelty and the self. According to Whitehead, Plato's doctrine of fixed transcendental Forms failed to account for the problem of novelty, the puzzle of how anything new can come into existence.[55] Whitehead believes that the spatialization of temporality in the theory of relativity, the Einsteinian insight that time is a fourth dimension somewhat analogous to the three spatial ones, although flawed, is an opportunity for philosophers and scientists to think again about the interrelations of space and time. Determinism, because of its attendant rigid notions of causality, assumes that knowledge of antecedent conditions could theoretically enable complete prediction of future events. Whitehead does not believe this. Events and perceptions can never be fully determined by temporal precedents because time does not chain events together. He prefers to use his own term "occasion" in a special sense, rather than more familiar words such as "event," "perception," or even the fancier word "becoming," to predicate a discrete unit of unfolding existence, or fragment of becoming, which has the capacity for genuine novelty. Whitehead insists that a satisfactory modern cosmology must headline novelty, the "novel particular in the process of creation."[56] His objection to Greek philosophy as represented by

Plato is therefore not so much against the Forms themselves, as to the lack of any explanation of how they come into being.

Whitehead also has an answer to any doubts about the shakiness or improbability of treating history as physics. He simply argues that what seems to be an analogy is actually our reality. In doing so, he diverges from the thinking of the human scientists who adopted conceptual schemes as provisional models of inquiry for guiding their case studies. For them, the conceptual scheme was an elaborate, self-conscious analogy. Whitehead believes he can demonstrate with philosophical rigor that the world, society, and the self *really are* made up of fields and forces, space and time. In the section immediately following the passage just quoted, he admits to a certain apparent flimsiness in the association between energy in persons and energies in physics: "this analogy of physical nature to human experience is limited by the fact of the linear seriality of human occasions within any one personality," contrasted with the multidimensionality of "the occasions in physical Space-Time." Limitations are built into conceptual schemes. They need constant tinkering and adjustment to make them work with experiments and case studies. Whitehead has no truck with such jerry-built schemas. He wants something far more comprehensive and enduring, built on deep foundations, so at this point in his argument he asserts that in reality human experience *is* multidimensional and appeals to our ordinary experience of embodiment to counter doubts (in *Process and Reality* he had already offered extensive philosophical arguments). After all, even "common speech does not discriminate the human body from the human person."[57] Or put in philosophical terms: "the human body is indisputably a complex of occasions which are part of spatial nature." The body is a walking, talking piece of space-time. The unity of self, sometimes called the soul, can be thought of as coordinated with the unity of the physiological body, which is not only the vehicle of the soul but also the interface with the scientific cosmos. The body "inherits physical conditions from the physical environment according to the physical laws."[58]

Although Whitehead seems to be talking about the self and history in the manner of other human scientists, he is really offering something quite different, a metaphysical cosmology whose logical framework can never be tested by case study or experiment. Only philosophical reasoning can expose possible flaws. This metaconceptual scheme presents "Maximus to Gloucester, Letter 27 [withheld]" with a challenge: how to test its epistemic authority? Olson's methods of poetic inquiry had been formed around the use of improvised conceptual schemes of the kind modeled by Murray, Lewin, and other human scientists. Historical details such as just who sailed which ship, and where, mattered. How can you test poetically a schema as comprehensive as Whitehead's?

Olson has a clever idea. He sets up a structural contrast between his own earliest formative memories and the "complex of occasions" that would presumably converge on his own embodied self, occasions that existing epistemology identifies as history, geography, and physics. By this means he thinks he can test the explanatory power of Whitehead's human geometry. If Whitehead's system had been just an expanded version of the conceptual schemes of the Harvard social scientists, this test would have been analogous to the usual to-and-fro tussling between a case study's empirical findings and the logical structures of the conceptual scheme that attempted to explain the scientific laws or processes behind those findings. Whether such a test could ever hold a metaphysical system to account, however, is less clear, and doubts about this may also account for the poem's withholding.

He starts therefore with a beguiling memory of his parents. In doing so, he does what confessional poetry does and faces the reader head-on as the poet-speaker bodies forth his own existence and its origins, thereby establishing the bona fides of his poetic voice. This memory is his and his alone, a sign of authentic, firsthand experience. The speaker can be trusted by the reader to be fully behind what he says, fully committed to the substance of his assertions, and therefore, the reader can confidently ascribe to the speaker reasons for these commitments. By leaving the last line of these nostalgic memories unfinished—"under one of those frame hats women then"—omitting the expected word "wore," the poem enacts the idea of a recurrent return to this origin, as if this recollection of his mother's cheeks and garments could go on and on and, indeed, will in future be replayed and perhaps made fuller by the inclusion of missing words and new details. If Whitehead's ideas are pertinent, Olson's memory will have to manifest just this many-dimensional ordering, and his formation as a new person, a novel emergence of life, will have to be traceable, by Whitehead's means, to this geometrically visualized cosmos of space-time.

If Whitehead is right, then even a full-color personal memory ought to be translatable into his philosophical language. Olson deliberately chooses to present a genuine memory in a manner that will resist easy translation into the philosophical geometry of a "complex of occasions." The opening line, "I come back to the geography of it," could be the start of a case study using an informant, in this case the author himself. The line appears to take us into the middle of some longer discourse, as if geography had earlier been discussed, had then been dropped, and is now being readdressed. We can easily slide past the familiar idiomatic phrase "I come back to," which like many familiar idioms is semantically and cultural dense with tacit meanings that we rarely articulate to ourselves. Here its significance depends on an interplay of literal and rhetorical meanings: that he regularly revisits the city of his youth; that his thoughts revert to memories of summer play on the parklands of Gloucester; that he is recapitulating an argument about geography he has made before, although exactly what this argument might be is left unstated; and finally that this poem is part of some repetitive process, a coming back to the starting point of self. All these processes, whether activities, thoughts, reasoning, or aesthetics, will have to be accounted for in the theory of occasions. When speakers use this phrase "I come back to," they are often indicating a certain impatience with either their interlocutors or their own powers of expression, as if needing to remind both listeners and themselves that there is always more to be said and that making ideas explicit is at best a faltering process. Is Whitehead in need of such a reminder, and how will his philosophy accommodate the illegibilities of personal expression? What Olson comes back to is "geography," a word with affinities to "geometry" but also resistant to it, since geography is that density of topography and human habitation that can be represented geometrically only by a process of abstraction—that is, by a process of leaving out the roughness of terrain.

The opening memory also resists easy assimilation into the abstraction of process philosophy in its knowing wink at psychoanalysis. Gender identities are strongly marked. Olson and the other boys played "until no flies / could be seen," just heard, and then went home to "where the women / buzzed," making a noise just like those irritating flies. He also fondly remembers his dad as a latter-day Odysseus acting out a castration threat, having seen his father emerge from a tent with a knife between his teeth ready "to take care of" an alleged suitor for Olson's mother. This threat of battle anticipates what follows in the poem, an abrupt switch away from personal expression to a metapoetic register:

This, is no bare incoming
of novel abstract form, this

is no welter or the forms
of those events, this,

> Greeks, is the stopping
> of the battle

When the poet tells us he is stopping the battle, he is usually assumed by readers to be alluding to the halt of an allegorical conflict between Platonists and materialists, just as ancient bards, according to Robert Graves, would use their poetic skills to halt a battle and impose peace.[59] But within the unfolding narrative logic of the poem, the poet also implies that he is stopping the battle in which his father fights, in other words, stepping outside the oedipal framework. If we push the interpretation further still, we could treat this idea of stopping the battle as the moment when the poet discards psychoanalytic temporality in favor of the scientific space-time-line of Whitehead's relativistic metaphysics. Although such an interpretation has some traction, I think the best way to treat the effect of memories on the poem's reasoning with Whitehead is to notice how their complex particularity resists being completely subsumed into Whiteheadian abstractions, leaving a deliberate question mark over the viability of the persuasive geometric concept of a "complex of occasions."

This particularity inheres even in the terminology that Olson has taken from Whitehead's *Adventures of Ideas*. By the time that Whitehead says, "It was the defect of the Greek analysis of generation that it conceived of it in terms of the bare incoming of novel abstract form," he has completely tamed his vocabulary so that it will perform the tricks he wants from it. The attentive reader is supposed to know exactly what these words—"Greek," "novel," "bare," and "welter"—are intended to mean. Relocated into a poem in the Maximus sequence, however, these words start to run wild again. What would Olson's readers have thought was intended by the reference to "Greeks" or the "battle"? Plato and Aristotle were still widely studied in school. Olson could be referring to their intellectual advocates (one of the most prominent literary critical groupings, the circle around R. S. Crane at Chicago, thought of themselves as Aristotelians), but the allusion carries other background resonances, to old-fashioned intellectuals, to the traditions represented by fraternities, and even perhaps to "Greek" love. If the latter seems unlikely, consider what meaning a reader unaware of the Whitehead relation might give to the boast "No Greek will be able / to discriminate my body." Given the enormous charge that the word "discrimination" has in American history, the poem seems to hint at fears around sexuality and identity and perhaps to imply that a healthy manliness is needed. Maybe this is why the poem thinks of the struggle against Greeks as having the ferocity of battle.

The identity of the Greeks becomes even more unstable as the poem proceeds. At the end of the poem it appears to be the poet who is engaged in battle with Gloucester itself:

that forever the geography
which leans in
on me I compel
backwards I compel Gloucester
to yield, to
change
          Polis

is this[60]

Now Olson and Gloucester resemble two Greek men wrestling in the gymnasium. What allegorical battle is this? Is Olson the poet trying on behalf of Whitehead to compel Gloucester "to yield" to the schema of the process philosophy? Is the leaning in of the geography a kind of resistance to the attempt to construct a science of history? "Polis" is an obviously Greek word and therefore might well be the sort of "Greek" discrimination that the poet resists. We can read the final two lines of the poem—"Polis / is this"—not as a proposition to the effect that "this is what Polis is" but as far more open-endedly interrogative, saying something like "Polis—is this really what it means/could this work/is this sufficient? and so forth.

So far I have concentrated on the poem's test of Whitehead's philosophy by identifying various tensions between the smooth abstraction of a theory of history viewed as a complex of occasions and the irreducible rough irregularity, the contingency, of history when viewed from the perspective of its tumbled-together constituents, the endless myriad of personal experiences, actions, and geographies that heap up into history. But the poem itself is not always quite in control of its test bed. The switch to a metapoetic register creates problems that the poem does not resolve. Such an abrupt shift is exactly what T. S. Eliot practices in *Four Quartets* where he repeats a rhetorical device signaling disenchantment with the great tradition of poetry. Again and again he offers a compact, heavily troped lyric followed by a disavowal: "That was a way of putting it—not very satisfactory: / A periphrastic study in a worn-out poetical fashion."[61] Olson's rhetorical move is so similar to that in Eliot's by-then-ubiquitous poem as to risk seeming worn out. Is Olson then offering the memories of his mum and dad as examples of a worn-out style of poetic memoir? This may be slightly the wrong question, not least given that familial confessionalism in poetry becomes a highly visible force only with the poems of Ginsberg and Lowell later in the decade. More plausible is that Olson is unsure whether such personal reminiscence will wear well. Even if Olson's poetic maneuver does not appear to ape Eliot too openly, it does create a problem that bedevils the logic of reflexive statements. Do they also justifiably refer to themselves? Eliot was careful to say "that was a study" because he wanted to avoid reflexively referring to *this* very statement (i.e., "that was a study") itself. When Olson says, "this . . . is the stopping of the battle," he could be referring to the previous section of the poem, or he could be referring to the entire poem, including the statement "this . . . is the stopping of the battle," or he could even be referring to the whole Maximus project. He appears not to want exemptions, but should he? Is a novelistic personal reminiscence capable of bearing the weight of this reflexive inclusivity? The poet may, as he says, be "one / with my skin," but is this convincing enough to make all those other forces yield?

These instabilities in Olson's poem may account, along with Olson's relatively rare opening up of himself emotionally to the reader in the first part of the poem, for the tendency of readers to treat the whole poem as an endorsement of Whitehead. This reception suggests that Olson was right to include the poem in the second volume of *Maximus* once he did indeed endorse Whitehead. When he wrote the poem, he was, however, attempting something much more tentative, to see how far he could trust Whitehead. We can see just how much the poem puts Whitehead to the test by examining Olson's insistence that Whitehead's specialist terminology, with all its scientific overtones, remain valid when used in everyday contexts.

In essence, Olson treats Whitehead's theory as if it were based on analogy. Olson remolds Whitehead's metaphysics into a schema comprising the topological fields of forces that the human scientists modeled for understanding human interactions. We can see this testing in many ways: in the juxtaposition of personal memory and Whitehead's schema, in the deliberate filling out of Whitehead's abstract terminology with echoes of the significance of these words in modern American society, and in the way Olson's poetic usages of Whitehead's discourse incorporate gender into their frame of reference. One of the most memorable ways the poem tests out the viability of phrases and sentences from *Adventures of Ideas* is its handling of the word "occasion." A strong semantic pressure test is applied to Whitehead's bold claim that "the human body is indisputably a complex of occasions which are part of spatial nature." Hoeynck reveals that the early version Olson sent in a letter to Duncan had altered Whitehead's phrase substantially: "the seriality of occasions / the slow westward motion of etc."[62] Hoeynck suggests that Olson reverted to the phrase "a complex of occasions" because Duncan had pointed out that in both poem and existence, "time is synchronistic."[63] Olson does, however, make other alterations to Whitehead's phrasings. He drops the intensifier "indisputably" and makes two other small alterations to Whitehead's formulation, substituting "an American" for the "human body" and interpreting "spatial nature" as implying that these occasions can be represented mathematically via geometry and, by extrapolation, geography: "An American / is a complex of occasions, / themselves a geometry / of spatial nature." Instead of a claim about being, Olson offers a claim about the historical and geographical constitution of American identity—it is not all of humanity that are complexes of occasions, as Whitehead (the Englishman) says, but the historically contingent, typical American who has emerged from the long process of nation building. Olson has taken a long step away from Whitehead. Perhaps we should hear tacit interrogatives in these lines: "is an American a complex of occasions?" or "how well does this conceptual framework help the poet investigator understand the workings of American history?"

Olson probably took the idea of narrowing down Whitehead's idea that geometry is fundamental to the new all-inclusive cosmology (all-inclusive because atoms, stars, minds, and emotions are all interconnected) to the idea that measurement of the kind undertaken by the social sciences is fundamental to human life from the same chapter of *Adventures of Ideas* as he took the concept of occasions. Whitehead offers a definition of "geometry" related to, and as idiosyncratic as, his definition of "occasion": "Geometry is the doctrine of loci of intermediaries imposing perspective in the process of inheritance."[64] Roughly paraphrased (but what would an exact paraphrase of this almost-hermetic language look like?), Whitehead is saying that geometry is a mathematical tool that offers a spatial and therefore structural visualization of causal relationships across some temporal frame. Phase space comes to mind. In this definition of geometry the crucial distinction between Whitehead's use of science and that of the social scientists is particularly evident. Whitehead believes that the world *is* geometric, whereas the social scientists think that when they apply a schema such as topology or phase space to a selected part of the social or psychological world, they are building a useful dollhouse in which they can simulate human lives. And the geometry is in the model—*not* in the people and their history. Whitehead believes the physical sciences are uncovering the workings of the cosmos; their findings warrant the construction of a persuasive metaphysics that includes all aspects of the human.

As Olson struggled in the early 1950s to make provisional conceptual schemes live up to the promise with which earlier social researchers had invested them and came to know Whitehead's philosophy more thoroughly after careful reading of *Process and Reality*, it gradually came to seem the answer to his needs. With a ready-made conceptual scheme of such power, he could stop looking for analogies in physics that when combined with carefully chosen particulars (memories and local histories) could be extrapolated into typologies and abstractions capable of organization into a conceptual scheme that would make possible a study of history with the rigor, and joy, of science. Now he could concentrate on the case study, on the research into his own and Gloucester's history. This came at a cost. Steven Shaviro points out that one consequence of Whitehead's reorientation of philosophy is that epistemology turns into ontology. By turning away from Kantian questions about what is knowledge of the cosmos and how we can form any knowledge of it to questions about how the endless eventfulness of the cosmos brings everything, including human beings, into relation, questions of knowledge fade out. Whitehead, says Shaviro, "thus short-circuits the entire process of epistemological reflection . . . and epistemology collapses back into ontology."⁶⁵ In doing so, Whitehead effectively renders redundant the epistemological questions with which Olson previously engaged. Of course, Whitehead's ontology brought certain benefits. Its compelling account of the entirety of the cosmos, and human relations with it, from the ground up gave Olson confirmation of the centrality of embodiment or, as Joshua Hoeynck expresses it, the "withness of the body" and the importance of "the inheritance of the world as a complex of feeling."⁶⁶ Miriam Nichols persuasively argues that Whitehead helped Olson articulate a more ecological recognition that "the human universe is not a universe centered on the human; rather, it is a human–non-human continuum perceived (inevitably) from the human point of view."⁶⁷ But giving over so much epistemological authority to Whitehead brought losses too.

Epistemology had been a practical mode of inquiry for Olson. Rendering it a minor consideration may have fitted his new circumstances, but it meant that finding poetic modes of inquiry would be harder. Much of what is best in Olson's poetry derives from its enactment of the process of inquiry, starting as it so often does from a moment of disorientation and proceeding to the slow emergence of coherent understanding, which may well be only temporary. I think we should listen to Olson's personal investment when he says in the relatively early essay "Human Universe" that "the trouble has been, that a man stays so astonished he can triumph over his own incoherence, he settles for that."[68] In his most powerful poems Olson does not settle for any easy coherence but lets the dispersing forces at work do their damnedest to pull apart the attempted schemas and conclusions.

Whitehead let Olson off the hook. Whitehead's metaphysical system enabled Olson to stop restlessly trying to develop conceptual schemes based on physics and mathematics, schemes that increasingly looked unsustainable.[69] He could treat events from different periods of history as manifestations of eternal objects, he could assume that his own authorship would be understood in Whiteheadian terms, and above all he could take the system for granted and literally pace out— as he does in several poems—the archeology of America. The result was mixed, however. Inquiry and conceptual analysis split apart. His later poems tend to be either demonstrations of Whitehead's cosmology or records of inquiry without conceptual analysis.

Good as it is, the later, much-admired poem "MAXIMUS FROM DOG-TOWN—I," which heralds his later style, lacks the urgency of active inquiry that shaped "The Kingfishers" and earlier Maximus poems. Here the poet is glimpsed for a moment as a father out walking with his son when they notice a black duck on a flooded road, but otherwise, the poet is present only as the impersonal, confident agent of the narrative. No uncertainty intrudes. The poem indulges in a folksy version of Whitehead's cosmos: "strong like a puddle's ice / the bios / of nature in this / park of eternal / events is a sidewalk / to slide on, this / terminal moraine."[70] Perhaps for a moment we hear the poet treating Whitehead's idea that forms are eternal events with playfulness (something to slide about on), but this does not go anywhere in the poem. Instead, the poem offers us a mythic dimension to a local legend and reminds us that even in quite ordinary, if locally newsworthy, events we can find more general patterns, but it doesn't show us how we might test out our hypotheses about individuation or the possibility that a particular event is also an eternal event. Contrasting with that poem is the fieldwork poem "Letter, May 2, 1959," in which he surveys traces of Gloucester's origins by pacing out distances himself, using his own body as a measure. He generates information whose significance is never clear, because like so much else in this poem, the big Whiteheadian scheme has no grip on this local case study. The diagrams, figures, the dense fabric of historical allusions (which require almost twenty pages of Butterick's guide), and the venting of scholarly frustration may be "interfused / with the rubbish // of creation," but they are not fused into a broader research synthesis. The result is a jumble of signs of inquiry without the substance. The poem is most persuasively read not as a record of triumphant historical analysis but as a confession of the failure of analysis, represented by the ending in which records of the threat of Indian attack are followed by the silence of numbers indicative of local sea depths. Here are the limits of inquiry when severed from locally responsive conceptual frameworks.

Olson's direction of travel away from local conceptual schemes was shared with others working in the human sciences. Over the next decade most thinkers with any investment in this research strategy found themselves concentrating more and more on the social and psychological mechanisms that made conceptual schemes possible. Joel Isaac suggests that when Kuhn wrote *The Structure of Scientific Revolutions*, he thought of paradigms very much in terms of the Harvard interstitial human sciences, as indissoluble amalgams of teaching and practice, of research training and case studies. In arguing this, however, I think Isaac underplays the degree to which Kuhn also worked with a second, more dominant notion by the time he published his history of scientific change. The typical paradigm is a conceptual scheme on steroids. It is true that Kuhn sometimes talks in *Structure* of paradigms as local regimes. Quantum mechanics, for instance, "is a paradigm for many scientific groups, [but] it is not the same paradigm for them all."[71] But his main story encourages readers to identify paradigms with scientific worldviews or "competing paradigms" by concentrating at length in later sections of *Structure* on how the "redefinition" of a science renders it "incommensurable" with what went before. His examples are generally epochal ones such as Newtonian mechanics, Lavoisier's chemistry, or Einsteinian relativity, not balkanized areas of quantum mechanics.

This extension of the conceptual scheme into something much larger, which eventually attracted Donald Davidson's opprobrium, can be traced in the work of other thinkers around 1960. About the same time that Kuhn was introducing the idea of the scientific paradigm, Wilfrid Sellars, who studied at Harvard in 1938–39 for a never-to-be-finished PhD just when Olson was deciding to abandon his studies, presented his own theory of how philosophy and science were interrelated. In "Philosophy and the Scientific Image of Man" (1962) he argues that we live in a world where coexist two seemingly incompatible images, the manifest and the scientific, of the same landscape, "each of which purports to be a complete picture of man-in-the-world."[72] If this sounds suspiciously like an inflated version of the conceptual scheme, it is because Sellars himself thinks of these images as each being a "conceptual framework"; he explains: "the conceptual framework which I am calling the manifest image is, in an appropriate sense, itself a scientific image" except that it does not base itself on "imperceptible entities" such as atoms and genes.[73] The everyday, or manifest, image of the human is in effect a conceptual scheme made up of the researches of the human sciences and the creations of the arts, a scheme controlling consciousness that is so pervasive and invisible as to seem like a sophisticated version of common sense.

More recent philosophers of science have been critical of the grandiosity that overtook conceptual schemes. When Bas van Fraassen took up Sellars's ideas again twenty years later in *The Scientific Image*, he did so in order to downsize Sellars's conceptual framework of science and replace it with a myriad of provisional ones. Van Fraassen argues that what matters in a scientific theory is that it "saves the phenomena"; it is as accurate in its modeling of the phenomena as possible. The principle is pragmatic: "to accept one theory rather than another involves also a commitment to a research program, to continuing the dialogue with nature in the framework of one conceptual scheme rather than another."[74] Sellars and Kuhn both believed that the conceptual scheme was still viable as a means of comprehending the theoretical work of the sciences, if only it could be treated as an underlying principle of valid inquiry. With less self-consciousness about the process, this is what Olson also did in setting his later poetry almost always in Whitehead's universe. And somewhat like the recent philosophers, post-Olsonian poets have tended to resist such expansion of conceptual schemes, instead assembling their own more provisional models, often using spare parts taken from theories of language, to "save the phenomena."

DUNCAN'S AMBIVALENCE TOWARD SCIENCE

Robert Duncan always admired Olson's handling of science, whether he was investigating its place in Western philosophy, in non-Western systems of thought, or in American historiography. Duncan believed that Olson convincingly showed his postwar readers that now "metaphysics only proves true as it is imagined as physics," perhaps thinking of the "projective" conceptual scheme derived from the physics of energy fields that Olson called "composition by field."[75] Olson also realized that physics is only a part of the history of science. He disinterred from oblivion "all the sciences of Man outside of the post-Christian rationalizing science of 'EUROPE' [that] have been declared to be out-of-bounds."[76] And

Duncan attributed to Olson a historical perspective on science similar to that which animated Rukeyser's biography of Gibbs, a perspective that treats the sciences as part of a much wider American cultural history by analyzing "the Westward movement of scouting, exploration, traffic with the natives, first settlements, raids and massacres, exploitations, scientific observations as a great psychic happening, a drive into the mythopoeic."[77] Duncan's image of Olson as a great poetic philosopher and historian of science owed a lot to the admiration for Olson's advocacy of Whitehead, which reached its height in 1956 when Olson gave a series of lectures on the philosopher to an audience of poets in San Francisco. One fruit of this period of Duncan's life was his major poem "Apprehensions," arguably one of his most Olsonian poems, about which he wrote several letters to his friend as he was composing it. In one notable passage Duncan explains excitedly that he is "in the midst of a poem again, projecting a construct of five sections that allows me to come alive," and that he finds in what he calls Whitehead's "scheme" a sense of a possible "new universe . . . an unknowable unknown."[78] Duncan accepts the connections between projective verse, constructs, schemes, and Whitehead.

In this final part of my discussion of the legacies of midcentury attempts to co-opt physics for poetics, I want to contrast Duncan's interests in science with Olson's in order to highlight the consequences of the choices Olson made as he gradually relinquished provisional epistemologies and his joy of firsthand science waned. I shall discuss the logic and the passions at work in Duncan's discursive responses to public knowledge of the sciences and then look in detail at two poems, "Apprehensions" and "The Fire: Passages 13." To begin with, I want to make brief observations about two points where these poet friends diverge in their understanding of the significance of physics and the natural sciences for poetry. The first is evident in Duncan's letter to Olson.

When Duncan casually refers to Whitehead's ontology as a "scheme," he is echoing Olson's understanding of Whitehead, yet the consonance is deceptive. Instead of committing himself as Olson did wholly to this one scheme, Duncan treats the idea that Whitehead's cosmology, or "new universe," is a scheme as a license to think of it as another possible object of poetic interest alongside other mythopoeic universes. It is important to remember that Whitehead definitely did not think that he was unveiling some hidden reality or unknowable unknown; he was offering a truer account of the actual world by integrating subjectivity into an account of material reality and presenting the resulting plenum, not as a set of objects moving through space and time, but as a network of events that coalesce into the solid realities in which we live. The whole point of Whitehead's philosophy is that this is not a new universe but a more accurate picture of our familiar one. Duncan treats Whitehead's philosophy as one more conceptual scheme alongside many others from theological, mystical, and poetic traditions that he invites his readers to consider as subjects for aesthetic admiration of their wonders and mysteries. If conceptual schemes are thought of metaphorically as tools for studying the universe, Duncan wants us to study the tools, not the supposed facts they might enable us to grasp.

One of his main poetic strategies for immersing readers in the aesthetics of the scheme is to create poetic zones where two or more epistemologically dissonant conceptual schemes or mythologies appear together with the result that their epistemic force is neutralized, somewhat in the manner that citation of a proposition neutralizes its truth claims. "Apprehensions" repeatedly performs this poetic cancellation of the epistemic force of such templates for belief. In the following lines, angels and the biological products of evolution hover over the same citational space:

> Angels of light! raptures of early morning!
> your figures gather what they look like
> out of what cells once knew of dawn
> first stages of love that in the water thrived[79]

Spirits, genes, physics, and process philosophy are all modes of constructing new universes, all rhetorical cosmologies of great power, to be celebrated and feared but not to be treated as fixed natural facts, theological realities, or metaphysical truths. Duncan is a pluralist to the very tips of his fingers.

Although Olson and Duncan shared many intellectual heroes, on one scientist they differed strongly. As I said in chapter 1 (see note 14), Olson admired J. Robert Oppenheimer so much that he even incorporated Oppenheimer's own words, as reported by Lincoln Barnett in a 1949 *Life* magazine profile, into his own affirmation of belief in the value of research in poetry: "I am, then, concerned as any scientist is, with penetrating the unknown." Duncan saw the unknown differently. He was critical of Olson's belief that "everything can be known" about the universe.[80] And he was excoriating about Oppenheimer's use of knowledge of the unknown for mass murder.

In "The Fire: Passages 13" Satan looks out of the eyes of all the world leaders of the world war in the 1940s, Roosevelt as well as Hitler, and works his evil particularly deviously in the visions of the nuclear physicists:

His face multiplies from the time of Roosevelt, Stalin,
Churchill, Hitler, Mussolini;   from the dream
of Oppenheimer, Fermi, Teller, Vannevar Bush,

brooding the nightmare formulae — to win the war! the

inevitable • at Los Alamos

plotting the holocaust of Hiroshima •

Teller openly for the Anti-Christ[81]

In the lines from "Apprehensions" we watched angels and cells momentarily coexist in the poem, despite their ontological opposition in everyday life, where the theory of evolution and angelology are mutually exclusive realities. Here we have a much darker coexistence of nuclear physics and Satan, but once again it is only in the poem that these two different forms of understanding can so readily cohabit as to enable the devil to stare out at us through the eyes of a physicist.

Before I continue I want to note just how shocking these lines appear to be, since they seem to equate the work of political saviors like Roosevelt and Churchill with the terrible murderousness of Nazi and Fascist leaders. Even the physicists are treated as morally equivalent to the rulers of the concentration camps because of their contribution to the atomic bomb's slaughter of vast numbers of civilians in Japan. Whatever the rationale for war, these lines insist that ordering mass slaughter is always a moral atrocity. Oppenheimer's Da Vincian range of talents and his resolute leadership of physics research into the unknown all count for nothing when weighed against his alleged Satanism. No physics, however remarkable, could be enough to exonerate Oppenheimer in Duncan's eyes. But I deliberately say that these lines *appear* to be shocking. I shall argue in a moment that just as the point of poetically depicting angels hovering over the genetic heritage of the cell is to invite reflection on the epistemological and affective investments we make in both forms of understanding, so too these lines of denunciation are not spoken by a poet on a soapbox ranting hoarsely at the injustice of the world but are framed by the poem to elicit reflection on the implications and costs of seeing the world as an inferno. Indeed, it would be a mistake to think of Duncan as unable to recognize the force of the kind of argument that we might attribute to Olson's apparent indifference to Oppenheimer's role in building the bomb, the argument that we should not conflate the marvels produced by physics with the moral weakness of the man. There can be a devil's physics just as, according to Duncan, Jakob Boehme perceives a "Devil's chemistry" at work in worldly evil.

Before looking in more detail at this poem, however, we also need to look more widely at Duncan's own interest in science both to better understand the role of science in his poetry and to be able to compare this with Olson's interest in transposing scientific methods of inquiry into poetic projects. To begin with, it might seem as if Duncan was even more enthusiastic about science than Olson. Lisa Jarnot starts her biography of Duncan by saying that he cuts a "curious figure" in modern American poetry because "he possessed a voracious appetite for the ideas of modern science and psychology, yet also gleefully claimed kinship with those who 'practiced the seasons' and adhered to a 'practice of the gods.'"[82] A "voracious appetite" for new scientific ideas sounds even more intense than Olson's joy.

This joyfully voracious appetite is on full show in a fascinating, long interview that Duncan recorded in April 1969 with two Canadian poets, George Bowering and Robert Hogg. Among the typically wide range of topics that Duncan weaves together are impromptu reflections on the significance of science for poets. His questioners begin by asking him about his friendship with Olson. Duncan muses about the usefulness of conceiving poetic form as a "field of time"—a fascinating merger of Whitehead and Olson—in the landscape of eternity, then makes an abrupt turn to physics: "And now of course I think we have got a different picture with particle physics, in which happenings are really that form, I think expanded beyond the field I was picturing in which there could even be a path; because they don't even see those pathways there and they think of it as events."[83] Duncan is excited by the possibility of yet another new universe, "a different picture." Olson was as capable of such excitement at encountering new knowledge as Duncan, but he would never have become so momentarily rapt by the possibilities of particle physics. Whatever Olson might say about his joy in science, it was new knowledge from historians and archeologists that lit him up.

The language of physics has flooded Duncan's discourse because he has just heard on the radio "some guy who got the Nobel Prize this year in whatever, physics," explaining "that particles are individuals; this means that you don't have 32 particles." Duncan instantly wants to construct analogies between physics and poetry, particles and language: "Every event you see in the particle level is an individual event, that just blasts wide open the picture even of the field I guess, so what do we do then in language, and why does language change, why does our concept of a form of a poem change?" Later, reading back over the transcript of the interview, Duncan adds a note to say that "this scrap, pickt up from a transient news report and not to date (July 17, 1969) researcht further, shows at least how immediately news in science is heard of by this poet as belonging to the lore of poetry" [Duncan's spelling].[84] There may be a mistranscription of "lore" for "law," or a deliberate pun, because for Duncan laws were always lore, always norms, models, schemes, or myths. Olson's interest in the early 1950s in the possibility that poetry might be involved in the discovery of laws of the human universe was quite alien to Duncan. He wanted to discover new universes, not tie down the dominant vision of reality with the chains of a set of laws.

From the information that he gives it is fairly certain that Duncan has just heard on the radio the 1968 Nobel Prize winner Luis W. Alvarez. The year 1968 was a remarkable one for Nobel Prizes—the award for biology went to Marshall Nirenberg, whose research, noted the *Scientific American*, "led to the decipherment of the genetic code."[85] Alvarez was the discoverer of many fundamental particles using his special instrument, the bubble chamber, a successor to the cloud chamber, for visualizing and analyzing particle tracks generated by the new accelerators.[86] He was an altogether-unusual physicist, both what he half-jokingly calls a "plumber," someone who actually builds equipment for nuclear research, and a researcher who then uses the equipment to make discoveries.[87] His team at Berkeley made a remarkable number of major discoveries as they tracked the paths of possible new fundamental particles, or "resonances" as they were also known because they manifested first as energy peaks on a graph. Alvarez was close enough to the messy idiosyncrasies of the actual phenomenology of energy measurements to wonder whether, as Duncan reports him saying on the radio, each event was a distinct particle, perhaps by analogy with the endless variety of snow particles. Alvarez was also in the news in 1969 for his use of cosmic ray detectors to map the interior structure of the giant pyramid at Giza looking for the mummified remains of the pharaoh Khefren.[88] He was just the sort of scientist who would appeal to Duncan. The *New York Times* said of Alvarez when his Nobel Prize was announced, "it is probably impossible to build a bubble chamber that could track all the ideas that spin through the fertile, far-reaching and spark-producing mind of Dr. Luis Alvarez."[89]

Ideas also spun through Duncan's mind after hearing Alvarez. Duncan tells his interviewers that "science is always advancing new pictures of what the universe is," and the more far-reaching of these, such as the Copernican revolution, not only alter our sense of what it is to be human, and what a planet might be, but the resultant new feelings about "whereness" are influential "enough to convert literary form."[90] The history of science for Duncan is a history of conceptual schemes or pictures of what the universe is like. Duncan is not fazed by the complexity of the science; even partial comprehension is valuable. In *The H.D. Book*, after explaining the influence of Stravinsky's musical structures on his composition of "The Venice Poem," he adds that through this process he "derived certainties of my own aesthetic, and then of a poetic, of a theory of forms and of the nature of making itself, as I have derived understandings from sciences I do not 'really understand.'"[91] Duncan is willing to forgo conceptual rigor in order to align the cosmological visualization possible in poetic form with the cosmological visions of the new physics. In one of the final interviews he gave he is still keen to do this, telling David Melnick that "science and poetry are identical as pictures of what is happening."[92] How can science and poetry offer identical pictures? This is of course not quite what Duncan is saying. He is saying that the aim of both science and poetry is to offer pictures of what is happening, not necessarily the same one; indeed, that would be very unlikely.

His motive for insisting that science and poetry both produce conceptual schemes is that he is concerned that the sciences too often attempt to monopolize the "real." As Graça Capinha says, Duncan consistently challenges the "subaltern" position of the humanities in relation to "the dominant scientific/rationalistic paradigm."[93] Duncan repeatedly insists that the best poets "work toward the Truth of things" because they are attempting to achieve "the most *real* form in language" that they can.[94] The difficulty for twentieth-century poets is the fierce competition from the sciences for exclusive rights over the real. Discussing Yeats in *The H.D. Book*, Duncan writes: "In an age when what we commonly call Science, the evocation of the use of the world, the presumption of mechanical imaginations in place of all other imaginations, defined its own realm as the sole Real and all other worlds as unreal, there were men in the arts too who attempted to define realistic claims, working purely in terms of semantic or cultural values, at war with unrealistic or animistic feelings of language."[95] Duncan's history is fuzzy here, for it is the relatively recent ascendancy of logical positivism that Duncan is remembering, the logical-positivist commitment to a blunt naturalism, glossed by a survey of the philosophy of science as "the view that all phenomena are subject to natural laws, and/or that the methods of natural science are applicable in every area of inquiry."[96] Many scientists, Oppenheimer included, did not believe this. Nevertheless, his underlying point—that any attempt to claim one realism as the only reality is mistaken—is not affected by this characterization of science. Although, in Kevin Johnston's words, Duncan's "writing challenges the scientific or rationalist categories of subjectivity derived from the Cartesian heritage," he has plenty of room for newer scientific categories as long as they are not exclusive of the reality of other forms of subjectivity.[97]

Where the logical positivists were extreme naturalists who believed that the universe was knowable as a simple unity, Duncan is what Hilary Putnam would call a conceptual pluralist. Just because Duncan's essays and poems appear to celebrate hermetic correspondences antithetical to science therefore does not mean that his poetics is irreconcilable with modern science. In Putnam's words, Duncan is a thinker who believes that "the various sorts of statements that are regarded as less than fully rational discourse, as somehow of merely 'heuristic' significance, by one or another of the 'naturalists' . . . are bona fide statements" answerable to norms of truth and validity.[98] This conceptual pluralism not only shapes his poems but also marks a divergence from Olson. This conceptual pluralism is especially evident in "Apprehensions," one of the four poems that Peter O'Leary claims provide the pillars of Duncan's poetic oeuvre, poems that share a "religious . . . personal-mystical-narcissistic modality"—poems that, I would add, also insist on the aesthetics and ethics of all perspectival schemas for viewing the world.[99]

To return now to our close reading of "Apprehensions," we can begin by observing that the divergence of its pluralism from Olson's poetics is made all the more striking by the strongly Olsonian features of this poem. O'Leary's excellent close reading of "Apprehensions" highlights the enactment of very personal research, which is "not merely a record of occult dread or an evocation of the knowledge of that secret dread" but also "the experience of grasping after information purposefully kept in the dark, so as to be dug and rooted for."[100] Duncan reaches back to childhood to transform "knowledge from an infantile dream" in the light of adult wisdom. The poem trades in knowledge, but O'Leary's terminology insists that this is not simply a report on preexisting knowledge and actions completed elsewhere. In making the poem itself the site of inquiry, a "grasping after information" that is hidden either because it has been kept secret by human action or because it is one of nature's secrets of the kind that scientists try to uncover, Duncan is working closely with Olson's poetics. Although he thought that there might be unknowable unknowns, Duncan shared Olson's interest in discovery. He wrote in a notebook at the time he was planning the poem: "Reality, then, I mean, is something more than what we know. I write a poem as part of the process of realizing what I experience."[101] This realization of the experience of coming into understanding can require a search into the unknown. The poem is the locus of this research.

"Apprehensions" is structurally similar to the poetic template that Olson first used successfully in "The Kingfishers" and frequently redeployed. Both poems announce themselves with an opening aphorism that establishes a *doxa* against which the poem will push hard. Olson begins with the closed-in magisterial pronouncement "What does not change / is the will to change," while Duncan begins with what sounds like the title of a set of instructions for a magic spell: "To open Night's eye that sleeps in what we know by Day."[102] His opening effectively mingles an aphorism—"Night's eye sleeps in what we know by Day"—with an invitation to the reader-adept to awaken to knowledge of the night. This opening injunction uses very similar imagery to Olson's exhortation to his readers to wake up to the new political conditions signified by Mao's new China in part 2 of his poem:

> The light is in the east. Yes. And we must rise, act. Yet
> in the west, despite the apparent darkness (the whiteness
> which covers all), if you look, if you can bear, if you can, long enough

Compare the two injunctions to the reader: in Olson's poem whiteness covers all despite the apparent darkness; in Duncan's, Day covers all, lulling Night's eye or self to sleep. Duncan's poem abstracts from the political message the metaphorical scheme of night/day, sleep/waking, I/eye.

In Olson's poem, the next thing to happen is that a protagonist awakes from a dream in which beautiful birds escape from their cages. Duncan's poem similarly recounts a hypnagogic moment of confusion instigated by reading an essay on "Renaissance Cosmologies" and Marsilio Ficino, which triggers a daydream. But where Olson's protagonist, in the hope of gaining a better knowledge of the violence of American history, begins to search for knowledge of why the kingfisher's feathers ceased to be treasure, Duncan's protagonist stays with the metaphorics of hidden treasure as he visits a haunting excavation. This eerie image of a mysterious underground cavity prompts wide-ranging speculations: about mythic images of Earth, the sources of language's communicative power, the sources of human violence, and various theosophical, historical, scientific, and metaphysical representations of origins. Here too there are parallels with "The Kingfishers," which also ponders communication, in this case cybernetic information theory, and such instances of human violence as cannibalism and war. At the same time the differences begin to mount up. Olson's protagonist concludes as an archeologist hunting among stone ruins in Central America, looking for insights that will help others to rise and act. Duncan's protagonist would never be satisfied by an archeological dig into the caves of his vision. Instead, he searches for and finds correspondences.

The title "Apprehensions" not only signals fears for the future but also alludes to Whitehead's jargon of "prehensions," as if the poem were a demonstration of these mergings of objective and subjective. "Apprehensions" concludes with "Second Poem," entering its final phase with the confident assertion "Wherever we watch, concordances appear."[103] This is deliberately ambiguous; it could be either a Symbolist credo or the working hypothesis of the natural scientist looking for underlying connections between different phenomena. Duncan amplifies his point in the next two lines: "From the living apprehension, the given and giving *melos*, / melodies thereof — in what scale? / Referring to these" ("these" refers to the various orders that are then enumerated). Apprehension, or imaginatively engaged perceptions of the universe in all its heterogeneity, will provide poetic and melodic inspiration. "Apprehensions" continues with a series of anaphoric lines beginning with "the orders of," culminating in another intimated bond between science and poetry: "the orders of stars and of words." In the final lines of "Apprehensions" Duncan creates his signature ambiguity of semantic and syntactical form to hint at the epistemological and emotional confusions underlying this appeal to "orders," which are both the nomological structures that scientists hope to make intelligible and metaphysical commands from unseen powers in the theosophical universe. Having finished his list of the orders, Duncan adds a final three lines that resolve the poem unexpectedly:

There is no life that does not rise
melodic from scales of the marvelous.

To which our grief refers.

The final line is deliberately cryptic. An earlier line describes the excavation as "grievous" (often used in the phrase "grievous bodily harm"), and throughout the poem there are stories and images that could be sources of grief. But there is also a more theosophical point being made. Grief is as inevitable as joy, and knowledge of the elements is never affect free.

Life as music arising from a sense of wonder—Olson could never have written these lines. It is not just the echo of a credo that Olson would have thought otiose, Walter Pater's "all art continually aspires to the condition of music," it is the emphasis on the new, the marvelous, the new universe revealed by a radio interview with a physicist or by the pages written by a Renaissance visionary. Such differences are explored in the poem that recoils from Oppenheimer, "The Fire: Passages 13." "The Fire" follows a very different template from "The Kingfishers" model of an inquiry that begins in confusion, gradually assembles provisional conceptual schemes, and ends with the hope of further enlightenment. The departure from Olson's poetic mode is signaled from the outset by the bookending grids of the same words (in reverse order in the final one) largely associated with the water and greenery around a harbor. Though I am tempted to call each grid a 6×6 *matrix*, thinking of the central role that matrices played in the physics of the period, I must resist, not least because mathematical matrices are always accompanied by a set of rules for calculating the relations between the elements.[104] Here there is none. The opacity of these grids is impressive given the familiarity and evocativeness of words like "shadow," "plash," "downstream," and "purl." We have no way of identifying any intent for, or any statement inferable from, the words we are given. It is therefore also plausible to see them as walls around the poem built from ruined fragments of the figures of a legible language rendered illegible by their jumbling in this new construction. When the poem asks, "Do you know the old language?," a question quickly glossed as "Do you know the language of the old belief?," we have already had the experience of not knowing a language, though whether it is a language of belief we cannot tell. These grids resist any performative or speech-act-based poetics.

Then the poem offers four words—"blood disk / horizon flame"—that herald the theme of the middle of the poem, the destruction of the pagan world of belief by our modernizing Western culture, a theme represented in Piero di Cosimo's painting *The Forest Fire* (ca. 1505), which was based on Lucretius's account of the origin of the world. First, the poem is careful to establish, as Duncan almost always does, a location for the authorial subjectivity in time and space: here a domestic location—"The day at the window." Then the poem ruminates on di

Cosimo's images of "our animal spirits" fleeing the fire and, in doing so, living in momentary enforced harmony "as if in Eden," before presenting the "opposing music" of Hieronymus Bosch's depictions of hell. As the poem ponders Bosch, the speaker himself is gradually caught up in what the poem has previously called "chords and melodies of the spell that binds / the many in conflict in contrasts of one mind." Interpreting Bosch's "faces of the deluded" subjected to "voluptuous torment" in contemporary terms leads to the frightening vision of Satan staring out of the faces of Hitler, Churchill, and Oppenheimer, with "glints of the evil that one sees in the power of this world." As if for reassurance that these glints are not distortions of vision, the poem cites Whitman's satirical denunciation of American politics after the Civil War and then plunges back into the hell of contemporary America, saying, "My name is Legion and in every nation I multiply." It is important that we are at least momentarily uncertain who is speaking here, the poet or Satan or, troublingly, Satan looking out of the poet's eyes.

Duncan's poem takes the risk of allowing the speaker to be so caught up in the horror at the destruction of wilderness and the violence of war that he becomes possessed by the very forces that he opposes. In other words, the denunciation of Oppenheimer is made from the perspective of hell, not the perspective of di Cosimo. Now there is no Christ, no Gadarene swine, to draw out the devils who possess politicians, physicists, and even possibly the poet. The final grid of words begins "cool green waver," as if trying to lower the temperature of the prophetic rage with a "plash" of extinguishing water over the fires of destruction.

Olson treats scientific knowledge as an inherent good; Duncan treats scientific knowledge as being as fallible as any human creation. After reading Duncan it can seem a limitation of Olson's poetry that he almost never questions the trustworthiness of his own poetic voice. It is almost always his own, and more than this, the idiosyncrasies of his voice, its breath patterns so scrupulously indicated by spacing and layout, are guarantees that it is Olson who is speaking. When "Letter 27" says "I am one with my body" or cites unattributed phrases from Whitehead as if they were Olson's own thoughts, we are intended to think of such utterances as carrying their author's warrant. In Duncan's poem, even his own voice can be co-opted, because it is not only pictures of the universe but sounds of the universe, what the eye sees and the mouth speaks, that are shaped by the specific picture of the universe that one momentarily inhabits, whether scientific or mystical. His vision of science and poetry has affinities with Rukeyser, whose later interests in poetry and science could be said to lie somewhere between Olson and Duncan. When poems about the mysterious transformations of nuclear matter and energy appear in the midst of the dialogue in *One Life* that represents the Senate hearings on the Atomic Energy Commission, they are a reminder of the emotional and spiritual depths of the often very prosaic political arguments, of the stories that accompany the atoms. She is always attentive to what one of these poetic irruptions calls not energy but the "work": "The work in the loss of mass. / The work in the lifetimes of the fixed stars. / The work in ideas of unstability."[105] Work, like action, is a concept that presupposes meaning. Yet this awareness of the ethics of science is nothing like as cutting or as relativist as Duncan's. She goes on trusting the methods of science. When she writes her great poem "The Speed of Darkness" and cuts back on the rhetoric that so often swamps her writing with almost as much terse affirmation as can be found in George Oppen's poetry, she still finds it useful to allude to Einstein when she describes the silence out of which she speaks: "But this same silence is become speech / With the speed of darkness." The erotic body that this poem celebrates as a fundamental good is complementary to the scientific body.

PART III
SCIENTIFIC AMERICANS

✳ 6 ✳

SCIENTIFIC
AMERICAN POETRY

Rae Armantrout, Jackson Mac Low,
and Robert Duncan

SCIENCE FOR DEMOCRACY

Halfway through his poem "Osiris and Set" Robert Duncan refers his readers to
a specific issue of the *Scientific American* in the knowledge that they are likely to
have ready access to the magazine, even to have read it. By 1960 the new series of
this mass-market magazine had become a taken-for-granted authoritative source
of knowledge about developments in the sciences as these were seen by scientists
themselves. Behind Duncan's citation is the assumption that reading the *Scientific
American* is as important as reading the newspaper if you want to be informed
about the world. Books offer a summary of research that is over and done with.
A magazine can offer an authentic glimpse of unfolding scientific research more
nearly as it happens. To give his Black Mountain students an example of good scientific writing, Charles Olson reads from a *Scientific American* article on the kid-
ney.[1] In *Gunslinger*, Edward Dorn's character Dr. Flamboyant reads the *Scientific
American*. Jackson Mac Low applies his acrostic proceduralism to a whole series
of articles from the *Scientific American* in 1960. Nearly twenty years later Rae
Armantrout takes phrases from it for poems in *The Invention of Hunger*, as she
mentions in her retrospective essay "The Lyric." Other undocumented allusions
and borrowings by other poets are likely too. In the next chapter I argue that the
magazine created a context for poems by Gary Snyder, George Oppen, and Amiri
Baraka, although with the partial exception of Oppen, it is not possible to trace
direct references. Most poets did not emulate Duncan.

In the early and mid-1950s, the question of where citizens could find reliable information on new scientific developments if they lacked the expertise to distinguish between competing media reports about the sciences was a pressing one. Fred Decker, a leading meteorologist, set out the problem facing the general reader in an article in *Science* complaining that false scientific claims were widely employed by companies to promote their products and services. "Science," he writes, "is represented to society not only by qualified impartial scientists guided by objective logic but also by specialists and former scientists whose scientific conscience may have been more or less eroded by commercial interests, by specialists not broadly enough trained to speak with adequate perspective, by promoters operating on the fringe of science with little or no concern for the long-range growth of science, and by outright charlatans totally unqualified in science but yet accepted by many laymen as 'scientists.' Journalistic practices tend to aid the spectacular claims more than the cautious, qualified reports." How, for instance, can a layperson evaluate a company's claim to be able to seed clouds for rain? Scientists have a duty to remind the public that "in their conclusions there is always a margin of uncertainty." Decker believes that it is essential to provide "authentic, complete, prompt, and understandable reports of scientific developments" to the citizenry because "accurate information [is] needed for progress."[2]

Others were worried that deficits in the public understanding of science threatened not just economic progress but democracy itself. Alan Waterman's report on his first ten years as director of the National Science Foundation insists that "in order to participate fully in the democratic process through intelligent voting, citizens must have at least a general knowledge and understanding of the nature of science and its implications for the national defense and welfare."[3] A year earlier, Eisenhower's Science Advisory Committee report, "Education for the Age of Science," made a similar plea: "A democratic citizenry today must understand science in order to have a wide and intelligent democratic participation in many national decisions."[4] But where was such citizen knowledge of science to come from? Writing in 1955 Gerard Piel, the publisher of the *Scientific American*, expressed concern that humanities intellectuals knew far too little about science:

> Science still occupies the House of Magic in which it was exhibited at the New York World's Fair in 1939. In the popular view science is first of all our most securely established body of knowledge. It is a rich mine of hard facts that have a tougher consistency and more utility than revelation. How these facts originated and got put together, nobody inquires. . . .
>
> I believe you will agree that this is a fair rendering of the image of sci-

ence as it is held in the public eye and mirrored in our press. . . . It contributes to the antirational, antiscientific mood presently in our culture. It promotes the almost complete estrangement of arts and letters from the sciences, which explains why our humanists largely miss the insights science now offers into so many of their habitual concerns.[5]

He must have been pleased when Glenn Seaborg, the Nobel Prize–winning physicist who helped discover many of the transuranic elements, gave a talk a dozen years later affirming that Piel's own magazine had gone a long way toward fulfilling this need for better public understanding of science: "Occupying the middle ground is the unique *Scientific American*, serving highly motivated, intelligent laymen as well as scientists, engineers and other technical people."[6]

The *Scientific American* was published by Scientific American Inc., a company founded in 1948 by Donald H. Miller and two former science editors at *Life* magazine, Gerard Piel and Dennis Flanagan. Flanagan was responsible for the policy of publishing articles like the profile of Oppenheimer that made such an impression on Olson. Scientific American Inc. had purchased the title to a magazine dating back to 1845 with the aim of creating a wholly new format that would "meet the interest, in the first place, that almost all scientists have in fields outside their own," fields where they too were in effect laymen.[7] Between 1948 and 1986, when the magazine was sold to a conglomerate and became the lightweight popular magazine it is today, the *Scientific American* worked hard to accurately convey the live quality of science-in-the-making that books cannot easily convey. Between the late 1950s and the late 1980s, many poets joined those "highly motivated, intelligent laymen" and turned to the *Scientific American* as an authoritative guide to what was happening in a wide range of sciences, whether as a one-stop provision of scientific knowledge or a point of departure for reading further into a topic via more specialist books and articles. Readers could feel confident not only in the reliability of the science but also that the magazine's articles on a topic would provide sufficient common ground for informed debate with others. This sense that everyone read the magazine was crucial. If you read an article in the *Scientific American* on astrophysics, sociobiology, or the city, you could be sure that other poets and intellectuals were likely to be ready to engage in discussion of the issues it raised.

Although the magazine would go on to have an enormous circulation far beyond the scientific professions, when it started it was written by scientists and aimed at both active scientists and the layperson. As Bruce Lewenstein observes, however, the magazine's definition of "layman" was narrow and was accompanied by a relatively narrow conception of the magazine's mission: to report on new developments in the sciences.[8] A prospectus for the new magazine was quite open about its role as a cheerleader for the sciences: "We certainly have a point of view. It is that we are for science. With the men of science, we agree that human want is technologically obsolete."[9] This confidence in technological solutions to social problems would trouble poets. George Oppen challenges them in his long poem "Of Being Numerous." Lewenstein argues that the *Scientific American* interpreted its role in bettering the public understanding of science as entirely a matter of "appreciation" rather than critique. Somewhat harshly he comments that "the magazine was, in essence, a monument to the vision of science as saviour of the world."[10] The magazine certainly had little to say about many pressing issues arising from the increasing dominance of the sciences in American culture, issues such as the changing epistemologies of scientific method, relations between the sciences and other forms of intellectual inquiry, and the ethics of science.

Authorship was a more complicated matter than would appear at a glance. Although an article might have Kinsey Anderson's or Edward O. Wilson's name as author, in-house editors rewrote as much as necessary in order to ensure intelligibility and wide appeal, as well as conformity with the magazine's sense of its own mission. Piel treats this process as if it were quite transparent: "We enlisted the scientists who were authors of work that caught our interest to write the articles about that work," and "in their concern to reach the wider audience interested in science, they accepted our invitation and then our editing."[11] In practice, the magazine subtly concealed its editorial and rhetorical management by presenting the main articles as the sole work of the scientist author, who is therefore placed as the originating authority behind what is asserted. Editorial control resulted in a quiet rhetorical underpinning that ensured that the editors' point of view was consistently maintained, and crucially distanced the articles from the turbulent communication protocols of informed exchanges in the specialist journals. This is the sort of smoothed out communication from the laboratory that Oppenheimer questioned in his American Association for the Advancement of Science speech discussed in chapter 2.

In the remainder of this chapter and in the next I shall offer close readings of several very different poetic interactions with the *Scientific American*. In this chapter I shall concentrate on documented allusions and recyclings of ideas and language from identifiable articles in the magazine. In the next I shall focus on cases where the magazine's influence on the intellectual climate of the time resonates with poets' concerns about nuclear-energy policy, practical sociology of urbanism, and "race" science. I shall conclude the present section with a brief discussion of the problems of tracing the impact of scientific journalism on specific poems. Rae Armantrout's poem "Natural History" does not give any internal sign that it alludes to articles in *Scientific American*, and so raises questions about how knowledge of a scientific source alters a poem, and whether there might be other unidentified poems, by other poets, derived from scientific texts.

Armantrout is now well known for her interest in science. In the collective autobiography of Bay Area Language Writing, *Grand Piano*, she mentions the origins of this early lyric poem, published in her second book of poems, *The Invention of Hunger* (1979), in the course of discussing the short lyric: "it emphasizes the boundary, the mysterious, arbitrary division between outside and in (or now and later). This idea has always fascinated me. It was the focus of my poem 'Natural History': 'Discomfort marks the boundary. / One early symptom was the boundary.' In that poem I drew on an article about termite mounds in *Scientific American*."[12] Close scrutiny of the poem and the magazine suggests that she also quoted from other *Scientific American* articles. Unlike Duncan's "Osiris and Set," or Jackson Mac Low's acrostics, nothing in the poem itself, written more than thirty years before the memoir was published, alerts the reader that she might be citing the *Scientific American*.

While taking part in a roundtable discussion about poetry and science for the online magazine *Jacket*, Armantrout reflects on why, throughout her career, she has brought scientific discourse into her poems: "We all get our view of the world and the universe partly from what we're told about science. I tend to bring in various things that contribute to the way we construe reality: creation myths, political 'news,' and scientific information. . . . I think one can question the metaphors used by scientists without necessarily doubting that they are describing something real."[13] To question poetically the metaphors is not necessarily to doubt the validity of the science. Armantrout respects scientific epistemology, because she sees it, as Lydia Davis argues, as needing to be made a part of one's lived experience: "She reads philosophy, she reads psychology, she reads science, and what she observes is the everyday, the ordinary, 'our' behavior: she tries to make sense of it all, or tries to verify that there is no sense in it; she seeks answers, or at least more questions."[14] Davis could be describing a scientist. "Natural History" brings in material for questioning, material indicated sometimes but not always by quotation marks, from the *Scientific American*. But without the later note in *Grand Piano*, readers would probably not have identified this source. Her poem is an early example of the questioning of the scientific metaphors she discusses, though it already demonstrates the skill at negotiating epistemological hierarchies that so distinguishes recent collections such as *Versed* (2009). It is a poem in five numbered sections, structured as a series of one- or two-line statements and questions in a dialectical mode not unlike that of George Oppen's work. The poem constellates ideas about the human body as a gendered *system*, inviting skepticism toward the tempting analogies or conceptual schemes that encourage the use of such an all-purpose concept. Although the construction of conceptual schemes based on analogies with physics was becoming increasingly unwelcome in sociology, biologists studying social systems in animals were still drawn to them. Armantrout's poem questions the likening of humans to biological systems or animal species thought of as systems, because such analogies can support conservative ideologies such as the belief that there is a natural human tendency to polarize gender identity into warrior masculinity and predatory femininity.

The poem therefore invites readers to wonder if we can really treat human beings as "elaborate systems in the service of / far-fetched demands." Or do such interpretations depend on suspect analogies, such as that between a human system and a termite colony? In the first three sections of Armantrout's poem, the idea of "system" is specifically linked to the idiom "scheme of things," which is one vernacular root of the epistemic tool of a conceptual scheme:

I

Discomfort marks the boundary.

One early symptom was the boundary.

The invention of hunger.
"I could *use* energy."

To serve.

Elaborate systems in the service of
far-fetched demands.

The great termite mounds serve
as air-conditioners.

Temperature within must never vary
more than 2 degrees.

2

Which came first
the need or the system?

Systematic.
System player.
Scheme of Things.

The body considered as a functional unit.
"My system craves calcium."

An organized set of doctrines.

A network formed for the purpose of . . .

"All I want is you."

3

was narrowing their options to one,
the next development.

Soldiers have elongate heads and massive mandibles.
Squirtgun heads are found among fiercer species.
Since soldiers cannot feed themselves, each requires
a troupe of attendants.[15]

The article on termite mounds that Armantrout mentions was actually an article on biomechanics by Steven Vogel, who helped create this field. His article depends heavily on the semantic continuity between different usages of the word "system" to create unexamined homologies between otherwise-heterogeneous structures or networks that include a "plumbing system," a "wind-driven system," a "biological system," and an "ecological system."[16]

Armantrout picks away at the linguistic sleight of hand used by this would-be systematic science. Lines six and seven above nominalize an expanded notion of system to the point where it threatens to collapse: "Elaborate systems in the service of / far-fetched demands." She devotes the second section of the poem to asking more about the conceptualization of organisms and their technologies and cultures as systems. Is the scientist a "System player," someone who, in the everyday idiom, plays the system? Vogel applies engineering principles to various habitats constructed by sponges, limpets, birds, prairie dogs, termites, and other animals to show how they take advantage of the physics of air and water flow to manage their environments. Termite mounds even cleverly create ventilation systems that simulate air-conditioning.[17] Vogel's conclusion that "careful and imaginative consideration of the physical world is imperative when investigating the adaptive strategies of living organisms" is aimed at fellow scientists who too readily seek biochemical explanations when simpler mechanical ones will do, but its tacit invitation to the poet is obvious.[18] Love and sexuality can, for instance, be considered adaptive strategies used by persons. Can imaginative consideration of the physical systems where they occur really deliver new insight?

Armantrout's poem also pushes hard at the metaphorical implications in recent articles in the *Scientific American* by Edward O. Wilson: one strikingly titled "Slavery in Ants" and another, coauthored with Berthold Hölldobler, titled "Weaver Ants."[19] Her grotesquely overdeveloped warrior ants in section 3 are derived from Wilson's martial portraits. In borrowing his language and images she alters his description so that her soldier ants have "a troupe of attendants" rather than the slaves from whom, in Wilson's account, they beg for their food, and her soldiers have "massive mandibles" rather than the "saber-shaped mandibles" of Wilson's vividly martial ants.

Sociobiology evidently caught Armantrout's attention, especially in moments when Wilson's caution about comparing insects and humans is momentarily set aside and rhetorical affinities between different spheres of the "natural" are opened up by the poetic interactions of words ranging from "system" and "energy" to "soldiers" and "queen." Arguments developed in Wilson's textbook *Sociobiology* (1975) inform his *Scientific American* articles. In the first of these, Wilson slips in a considerable number of sociological comparisons between ants and humans, saying at the very start of the essay, "The institution of slavery is not unique to human societies." Ants capture ants from other colonies and, according to Wilson, "enslave" them. His rhetoric's covert moralizing also hints at a discourse of degeneracy when it suggests that ant species may decline as well as advance: "The evolution of social parasitism works like a ratchet, allowing a species to slip further down in parasitic dependence but not back up toward its original free-living existence."[20] In an America where civil rights were still a hot issue, this language of slavery, freedom, and social parasitism was loaded. Wilson is definitely aware of what he is doing, because he concludes his final paragraph with a series of tactical disavowals of the evolutionary progressivism that dominated biology throughout much of the twentieth century: "The slave-making ants offer a clear and interesting case of behavioral evolution, but the analogies with human behavior are much too remote to allow us to find in them any moral or political lessons."[21] The science of ant societies tells us just how important American ideals of freedom and self-reliance remain. Armantrout was alert to the leakage of ideology into scientific writing. In section 5 of the poem she puts part of a sentence in quotation marks—"As soon as it became important / that free energy could be channelled"—derived from a *Scientific American* article entitled "Chemical Evolution and the Origin of Life" ("As soon as it became important for a limited supply of free energy to be channeled into one or a few of the many possible reactions, directed catalysis by enzymes would have become essential").[22]

In the remainder of this chapter and the next I shall offer close readings of several very different poetic interactions with the *Scientific American*. In this chapter I concentrate on two cases of documented allusions and recyclings of ideas and language from identifiable articles in the magazine. In the next I concentrate on cases where the magazine's influence on the intellectual climate of the time resonates with the concerns of poets about nuclear-energy policy, practical sociology of urbanism, and "race" science.

ALEATORY INQUIRY IN JACKSON MAC LOW'S *STANZAS FOR IRIS LEZAK*

On April 10, 1960, the poet Jackson Mac Low moved in with his lover, the artist Iris Lezak, and began writing her a long series of poems. These are love poems but not as we know them. They are rigidly procedural poems, mostly constructed around computational algorithms making acrostic use of the titles of a wide range of sometimes popular and sometimes more technical or specialist source texts. Each line of a poem is normally generated by selecting words whose first letters will spell out in turn the respective word in the title. For example, Mac Low uses the title of Kinsey Anderson's article "Solar Particles and Cosmic Rays" from the

June 1960 issue of *Scientific American* as an acrostic template for a poem of the same title. Anderson compares the workings of the new high-energy linear accelerators for studying nuclear particles to those that already exist in nature—"galactic accelerators" such as supernovas and the vast dust clouds and magnetic fields within the Milky Way—and that create the very high energy particles known as cosmic rays.[23] Mac Low's poem "Solar Particles and Cosmic Rays" uses only language from this article to compose the twenty-one stanzas, each of which has five lines, one for each word of the title, so that the first letters of the words in each line spell out the respective title word. The words in each line of the poem are normally found by searching for the next word in the article with the necessary starting letter.

At first glance, the whole volume of *Stanzas for Iris Lezak* can look like the work of a Flarfist who has traveled back in time a few decades. The poetic language appears to lack any integration, whether semantic, syntactical, logical, or narrative, because of the violently disruptive effect of the generative procedures. "Solar Particles and Cosmic Rays" was one of many poems written over a few months in 1960, then published more than a decade later in the volume *Stanzas for Iris Lezak* (*SFIL*, 1972), a large book running to over four hundred pages. The collection contains about a hundred and seventy poems constructed entirely from words and phrases in the books and articles circulating in 1960, words chosen according to a range of programmatic selection procedures. In "An Afterword on the Methods Used in Composing & Performing *Stanzas for Iris Lezak*," Mac Low explains that his acrostic proceduralism drew upon "practically everything I happened to be reading from May thru October of 1960." He then lists the books and journals at some length, so that we know that in addition to an eclectic mix of books on "Zen & Tibetan Buddhism, politics, poetry, & botany," he was also reading newspapers and magazines, including the *New York Times*, the *Catholic Worker*, and bulletins for the War Resister's League, as well as the *Scientific American*.[24] The eclectic character of this list conveys very effectively how well the *Scientific American* fitted into intellectual culture. During this period Mac Low wrote at least eleven poems drawing on eleven or more articles from four issues of the magazine, and probably used the magazine for parts of a number of others.[25]

In the extensive information about the composition of these poems that Mac Low provides, we learn that the starting point for composition was, not poetic theory nor a fascination with chance or procedural methods, but passion: "we were deeply in love, & our sexual life was very happy."[26] He recalls starting on the series of poems by using works by Rabindranath Tagore as source texts for an acrostic of "My girl's the greatest fuck in town. I love to fuck my girl." So far this doesn't sound very different from the motives and practice of any besotted young lover fascinated by writing verse, a state of mind hinted at by a few reassuringly familiar indications of the genre of the love poem, scattered among the greater number of peculiar titles such as "Insect Assassins" or "Drive on Malaria Covers 92 Lands." But as soon as we look at the content of these poems with conventional titles, the strangeness returns. What does the content of the poem "Solar Particles and Cosmic Rays" have to do with Iris and his love for her?

We need to ask more questions about this drive to transform the printed matter that flowed through his life into love poems. Did he think that what his love for Iris demanded was bulk poetry, as many love poems as possible, and that these formal procedures were the best way to supply such quantity? Surely this is too reductive an explanation. Did he want to show to himself and her that erotic passion could be found everywhere, that love was making the intellectual world go round? This explanation is slightly more compelling, though it ascribes to the poet a delirious reductionism devoted to revealing a pan-erotic fundamentalism that is itself reductive. Even an intoxicated lover sooner or later realizes that the search for universal confirmation of one's own state of mind is ultimately self-defeating. But we don't have to think of the project in quite such narrow terms. When Mac Low decided to mine these texts for the writing of what he calls "purely acrostic-stanzaic chance" poems, he was assaying or mining the sort of texts from contemporary culture that he and other intellectuals were most familiar with. He may have been driven by a desire to find the place for his love in this vast heterogeneous textual culture; he was also trying to work out how to communicate this love to his reader, Iris Lezak. These poems are composed in the first place for a single known reader, the author's lover, and therefore, these poems, and what they say, are gifts for her. It is as if instead of reading out something interesting from a magazine to her, he would present her with a procedural extraction from it.

In writing with such a strong sense of personal relation to a specific reader, Mac Low is part of a long poetic tradition, and also close to another, more local one. Other New York poets of the time were exploring the possibilities of writing not for a generalized audience but for the actual group of readers with and for whom they composed, read, and discussed poetry amid all their other entanglements, amicable and passionate.[27] These poets developed new textual strategies for binding together mutual recognitions with affective and symbolic forms that could be actively challenged and reworked *within* their poems. By composing poems around the actual passions and tensions within their immediate readership, these poets also make possible more generalizing observations about poetry as a form of social interaction. Although Mac Low's poems look and sound very different from those of the New York School, he too is very interested in poetic investigations of what Lytle Shaw calls the "social rhetoric" by which it is possible "to reimagine the social logics that allow group formations in the first place."[28] These poets write poems as part of the continual enactment of their coterie, whereas here Mac Low is writing for one person, but he too is endlessly interested in what it means to be a person in the midst of these many discourses that tell him and Lezak what it is to be human. Andrew Epstein says of the New York poets that when they "write about friendship, community, and selfhood, one can sense that they have internalized and are wrestling with larger debates about the nature of individualism and the scourge of conformity."[29] By writing procedural poems as intimate communications to Iris Lezak, Mac Low is locating his discourses of politics, religion, and science both within larger debates about the nature of individualism and by what knowledge it can be understood and within the intimate converse of a passionate relationship. Today, in an important sense, *Stanzas for Iris Lezak* is a book of poems that now has neither writer nor reader, only a meta-author and meta-reader, yet this has made the poems more interesting, because through this unpromisingly hermetic metapoetics, Mac Low's poems draw out otherwise-obscured features of the scientific discourse of his time. Under the procedural microscope, those stained, selected words reveal latent patterns of embedded social relations.

Commentators have had much to say about Mac Low's principles for using randomness. Brian McHale takes Mac Low largely at face value and argues that his writing is an example of "prosthetic poetry," that is, poetry that shares the poet's "authority with a machine," where the machine in this case is Mac Low's selection process: "When the poet submits to dictation from the Muse, whether supernatural, internalized, or the 'virtual muse' of the machine, we understand this act as a figure for submission to the genius of language itself—or to the language-machine."[30] Because the poems are instances of a mechanical procedure, it is to the circuit diagram of the procedure that we should look for meaning. The poems themselves are instances of diminished agency. Such readings of conceptual art are current in the visual-art world, where the formula behind a facture can be the main interest of the work. The claim that Mac Low's poetry is a repeated enactment of the machine-like character of writing and language troubles McHale, however, and near the end of his essay he backs away a little, introducing Roy Harris's conviction that language is utterly nonmechanical because it is an irreducibly social mode of action. Maybe Mac Low's poetic language is not the machine it seems.

I think that we should be wary of attributing too much automated chance to the poems. In the lines—"Sun of long a rays. / Particles. A rays. They in cosmic long sun"—Mac Low appears to have altered the rules and chosen to select the indefinite pronoun "a" even though in earlier stanzas he avoided such minimal words. This change of the rules indicates that the composition is not quite as machine-like or computationally pure as we thought. Some readers have therefore concluded that the Mac Low machine is more of a chess-playing Turk than a poetic version of Big Blue. Ellen Zweig rescues the Mac Low authorship by detecting a playfulness that no machine intelligence could manifest: "Mac Low tends to use both strict chance methods and looser flexible rules subject to his whim."[31] To invoke "whim" is to hint that his poetry is suffused with a chancy nonepistemic intentionality, both at the level of the generative structure and at the level of individual words and their placement. The poem may not offer knowledge the way its essay counterpart does, but it does provide textual pleasure.

As he proceeds with his poetic experimentation, Mac Low feels his way into an original mode of inquiry by which he exposes often quite-hidden strata, or compositional vectors, in the text fragments whose verbal material he transposes into poems. The poem "Solar Particles and Cosmic Rays" engages with a particularly powerful epistemic institution, and yet if we assume the poem is aleatory, it appears to have no outside, and we can look only to the poem for interesting abstract sonic and visual effects coupled with curious semantic accidents. These effects are by no means trivial; the poem offers some compellingly original sound patterns and visually intriguing punctuation. "Sunspot on late appearing regions" plays a neat game with its fricatives, dentals, and the deft internal rhymes that also hint at a submerged reflexive punning self-congratulation, "spot on." From the fifth stanza onward most lines have one or more long dashes to indicate where an acrostic word could not be selected, and these too can create a kind of concrete poetry in which extended dashes mimic particle tracks like those being recorded in bubble chamber photographs.

"Riometers. As ——————— something." (*SFIL*, 123)

But is this all the meaning we can look for? I want to suggest that the acrostic computational engines do not preclude the intentionality of a significant critique of the science in the article. To begin with, "Solar Particles and Cosmic Rays" well represents the intransigence, or what Craig Dworkin has called the *illegibility*, of modernist poetry.[32] Dworkin argues that the resistance to interpretation of some modernist poetry is itself part of the poem's meaning and not simply a barrier to be erased as far as possible in order to reveal meaning. In Mac Low's poem, repetition of technical nouns such as riometers and accelerators has the effect of making their lumpy resistance to assimilation stand out.[33] But the poetry does considerably more than this; it manages to assay both the epistemological and the ontological authority of the scientific language.

Marjorie Perloff offers a helpful way into thinking about the significance of this capacity to test a scientific discourse through sometimes-violent disruption of its initial conditions. She argues that Mac Low is one of a number of experimental poets whose work would be "unimaginable" without the prior experiments of Marcel Duchamp, which explore, in Duchamp's own words, "a *reality which would be possible by slightly distending* the laws of physics and chemistry" (Perloff's italics).[34] Although Perloff does not make the following claim herself, we could describe Mac Low's poetic strategy as the distension of a scientific discourse in order to see what realities emerge, what new phenomena this new schema might project. To see how this distension works, we can compare the first two stanzas of "Solar Particles and Cosmic Rays," with passages from the article that gives the poem its title, lexicon, and acrostic architecture:

Step origin least assured rays
Past assured rays Throughout investigators concerned least enormously step
Assured not detect
Concerned origin step more investigators concerned
Rays assured years step

Sun, of learned accelerators recent
Particles accelerators recent the invest cosmic-ray learned energies sun,
Accelerators notably design
Cosmic-ray of sun, might invest cosmic-ray
Recent accelerators Year, sun[35]

The source texts for these stanzas are the first few sentences from the opening paragraph and the second half of the next paragraph of Anderson's article:

(1) Throughout the past 35 years investigators concerned with the origin of the enormously energetic particles called cosmic rays have been assured of progress along at least one line. Each time they have contrived

experiments to <u>detect</u> more energetic cosmic rays they have found them. With each upward <u>step</u> in energy, however, it has become <u>more</u> difficult to imagine where cosmic rays come from or to conceive of a process by which they acquire such enormous energies. It is true that the sun showers the earth <u>not</u> only with electromagnetic radiation but also with particles; during periods of high sun-spot activity gusts of energetic solar particles frequently rush into the earth's atmosphere along the lines of force of the earth's magnetic field.

(2) Thus one <u>of the</u> most hopeful theories about the genesis of cosmic rays bears similarities to purely terrestrial experience with the <u>design</u> of the great <u>accelerators</u> by which men have <u>learned</u> to <u>invest</u> atomic <u>particles</u> with <u>energies</u> in the <u>cosmic-ray</u> range. Thanks to the <u>recent</u> advances in man's knowledge <u>of</u> the <u>sun,</u> <u>notably</u> during the International Geophysical <u>Year,</u> that theory can now be subjected to further observational test. It was the late Enrico Fermi who suggested that a "galactic accelerator" <u>might</u> push the cosmic rays up to their immense energies.[36]

Comparing poem and article shows that in the opening stanzas there has been considerable manipulation, whimsical and otherwise. Mac Low darts back and forth in his source to pick choice words that will make striking arrangements, avoiding dull prepositions in favor of more interesting words with what the "Afterword" calls "lexical weight." He sometimes goes back to words he has already used if they create promising effects (e.g., he uses "concerned" instead of "conceived"). His filtration method also has odd consequences: although the opening paragraph of the article is dense with the buzz-word of nuclear physics, "particles," this word does not appear in the first stanza at all, trumped instead by the use of another word beginning with "p," "past."

When these particles do start appearing, so too do long dashes and considerable emphasis on epistemological decision-making by the scientists. Here is stanza 7:

Suggested of large aloft ————
Particles aloft ———————— the instruments carried large ————
   suggested
Aloft 1958, direct
Carried of suggested ———— instruments carried
———————— Aloft ———————— suggested (*SFIL,* 123)

Whether or not the dashes simulate particle tracks, they do visually mark epistemological uncertainty, inviting the reader to ask how what is missing can be known. We read these word-length dashes as invitations either to substitute our own insertions or to guess at what word has been elided, and this reworking pulls attention away from the empirical material to the processes of evaluating and interpreting it. The rays are "assured," the accelerators are "learned," something has been "suggested," and so forth. In stanza 15, one of the few later stanzas without dashes, the word "could" appears three times:

> Studies output later and rays
> Protons and rays These is could later established studies
> And no destined
> Could output studies much is could
> Rays and Y. Studies (*SFIL*, 125)

The fourth line cleverly (but can we attribute an intentional value to it?) conflates the statement that it might be possible to produce ("output") research with an epigram—"much is could"—which aptly represents the "margin of uncertainty" that Decker thought the public should understand about science. The result is an elliptical statement that is also circular: the could studies the could, or to paraphrase this distended conditional further, probability researches how much is probability. If we hear in the next line the synonym of the letter "Y" as "why" so that the line reads—rays, why?—then this caps the accumulating disjoint meanings. The answer to the question "why?" is more research, another circular answer, responding to the why with another why.

"Solar Particles and Cosmic Rays" also makes evident a feature of the standard specialist scientific paper that is often tucked out of sight. A typical specialist article has to convince its expert readers that the results can be trusted, that the data and methods are reliable in the face of many uncertainties and unknowns. Joseph Harmon and Alan Gross call the rhetorical devices (those "coulds" and "suggesteds") that experts use to manage trust "hedges." A hedge is a phrase such as "we wish to suggest the possibility that . . ." which helps readers "gauge the author's epistemic commitment" and can also "communicate doubt within the boundaries of a prose where the absence of the personal conveys the impression of authority and neutrality."[37] One of the most significant consequences for poetry of the spread of such scientific writing across the culture has been the prominence and status of this rhetorical management of propositional claims.

Strategies such as the disjunctive speech act, the paratactic incompletion of sentences, and the insulation of phrases from the completion of the subject-verb-predicate circuit can be thought of as counterparts of the explicit hedges used in the science articles. Some of this rhetorical maneuvering beneath the surface of the article is picked up by Mac Low's writing through of the text, which elicits the subjectivities embedded in such repeated words as "concerned," "showed," "could," and "coincidence."

Mac Low's acrostic rewritings of *Scientific American* articles are hard to love. They look unpromising at first sight, lacking in the cognitive and affective possibilities of a guiding intentionality, lacking in poetic elegance or precision, and composed as part of a hermetic love pact. As I have shown, they deserve attention because their interpretive violence exposes a set of interlocking norms of poetic communication around readership, discourses, probability, and knowledge that remain of vital interest. Their intelligibility is opaque, and their unintelligibility transparent. His poems do surprisingly carry out valuable inquiries into the uneasy certainties of science and the easy uncertainties of desire.

"OSIRIS AND SET": ROBERT DUNCAN CITES THE *SCIENTIFIC AMERICAN*

One of the most striking poetic references to the *Scientific American* occurs in Robert Duncan's poem "Osiris and Set." In the middle of this poem from *Roots and Branches* (1964) Duncan abruptly switches from a theosophical account of Set, Osiris, and Horus, avatars of a psychospiritual conflict of the faculties, to refer to "a graph in *Scientific American*, September 1960," that "shows the design of sensory and motor intelligences."[38] From Egyptian gods to physical anthropology: this is a big leap in time and knowledge. Poets rarely give sources in their poems, and when they do, they even more rarely make explicit reference to scientific writings. Duncan's reasons for this citation are complex, not least because he tacitly alludes not only to the whole article in which the illustration appears but also to the entire issue, including the extraordinary amount of weaponry displayed in its advertising. He can confidently cite the magazine because by 1960 the new series of the *Scientific American* had become the journal that intellectuals read for the latest reliable and intelligible information about the sciences. His readers were likely to have seen this or similar issues of the magazine.

Until it explicitly mentions the *Scientific American*, "Osiris and Set" is one of his less memorable poems, placed at the end of a series of poems that reflect on the "mandrill," or primate, origins of humanity alongside memories of the loss of his birth mother, speculations about the puzzling persistence of mass human violence, and archeological ideas of how cultural change in ancient times "brought the wilderness into the heart of Egypt."[39] Osiris and Set are characters in Egyptian narratives of a long familial conflict between the gods Set and Horus that started with Set's murder of Osiris, husband to Isis. These narratives are generally assumed by archeologists to be mythological versions of geopolitical strife between the north and the south of the country. Duncan also treats Set as an ancient name for the most basic motor instincts of the human animal, its "primitive terror" and "first knowing," and similarly treats Horus as an image of the raw senses.[40] This psychological-neurological interpretation of Set and Horus spurred Duncan to refer to the striking illustration in an article by Sherwood Washburn, professor of physical anthropology at the University of California, entitled "Tools and Human Evolution" in the September 1960 issue of *Scientific American*. Washburn reprints two of the neurosurgeon Wilder Penfield's heuristic images of the human nervous system as homunculi overlaid on a sketch of the cerebral cortex. The relative size of the different body parts of the homunculi is indicative of the relative intensity of the nervous signals to the brain from that region. These compellingly weird pictures depict two grotesque male figures with enormously enlarged mouths and hands relative to their bodies, tensely gripping the outer arc of a curved, cross-sectional diagram of the human brain. One has closed lips; the other appears to be screaming. It would be easy to interpret these figures as in some way demonic. When Duncan imagines Set "striving there, at the edge," he is recalling how the homunculus appears to be straining to hold on to the edge of the diagram.

Although Duncan directly alludes only to the illustrations of homunculi, his poem takes its bearings from the entire magazine, its images, articles, and even its advertisements. The hefty 280-page annual special issue that year was devoted to "presenting a series of articles on the human species, with special reference to its origins," and overall amounted to a particularly telling example of the immense authority of the natural sciences. Use of the indefinite pronoun "its" rather than "our" or "human" underlined the objectivity of the scientific approach to its object, humanity. The range of topics—evolution, society, language, agriculture, cities, and the history of science itself—demonstrated the power of scientific method to produce knowledge about every area of human existence, including many areas of experience that might have seemed to a poet to be the prerogative of artists and writers. Indeed, sometimes the magazine appears to go out of its way to provoke such readers. In the article Duncan refers to, a study of the vital part played by the use of tools in human evolution, Washburn blames the inadequacies of language for his difficulties in explaining how the mutual reinforcement of tool use and growing bipedalism created a feedback that helped accelerate the evolution of early hominids toward our modern human form: "English lacks any neat expression for this sort of situation."[41] When Duncan decided to make an explicit reference to the article, perhaps he was also tacitly responding to this invitation to provide a "neat expression."

More certain is that he was responding to the powerful manner in which this entire issue of the magazine projected a compelling, and sometimes-disturbing, image of science. Contributors are explicit that science can be both a force for good and a potentially destabilizing one. The leading historian Herbert Butterfield's article, "The Scientific Revolution," sketches the enormous improvements in the quality of life brought about by the "scientific-industrial revolution" and even hints that writers should thank scientists that they have so many readers: "great literature is perhaps more widely appreciated at the present day than ever in previous history."[42] On the other hand, a brief report in the magazine from the Committee on Science in the Promotion of Human Welfare, a subcommittee of the American Association for the Advancement of Science, warns that "disparity between scientific progress and the resolution of the social issues which it has evoked . . . threatens to . . . disrupt the history of man" (ellipses in original).[43]

The disparity between scientific progress and social issues is assumed to be in favor of science, but an attentive reader would notice disturbing signs that the scientists themselves could be blind to racial issues and what we now call biopolitics. In Marshall Sahlins's essay "The Origin of Society," the anthropologist sometimes known as a stone-age economist, argues that "the decisive battle between early culture and human nature must have been waged on the field of primate sexuality." The "overthrow" of primate sexuality led to a form of human society that "substituted kinship and co-operation for conflict, placed solidarity over sex, morality over might."[44] Although Sahlins emphasizes the disjunction between primates and what he calls the "primitive life" of humans, the illustrations tell a more overtly racist story. The series of pictures starts with black-and-white photographs of baboons and then switches to similarly monochrome images of !Kung hunter-gatherers, as if primates and Bushmen were visually equivalent.[45] A reader like Duncan did not have to look far for tangible examples of the disruption feared by the American Association for the Advancement of Science and could find it in the heart of what was supposedly the most advanced objective research.

An essay on human evolution, for instance, speaks of the need for genetic selection of human traits to avoid degeneracy. A perceptive reader would have noticed that although its famous author Theodosius Dobzhansky, a founder of the modern evolutionary synthesis, starts off by disowning eugenic principles, saying that "a society that refused, on eugenic grounds, to cure children of retinoblastoma would, in our eyes, lose more by moral degradation than it gained genetically," he does so only to reintroduce them under new guises. These are the discredited eugenic motifs that helped provide the ideology for Nazi treatment of Jews, homosexuals, Romany, and the disabled.[46] Dobzhansky worries that medical treatment may have so diminished the effect of previously fatal genetic weaknesses that these genes will be passed on in increasing numbers to future generations. And he fears, like so many supposedly well-meaning eugenicists before him, that if people of low intelligence continue to have more children than high achievers, these inferior people will genetically entrench a dumbing down of humanity. Eugenics of a kind may be needed after all: "It may well be, however, that the social cost of maintaining some genetic variants will be so great that artificial selection against them is ethically, as well as economically, the most acceptable and wisest solution." Man, he writes, "should be able to replace the blind force of natural selection by conscious direction, based on his knowledge of nature and on his values."[47] Duncan's poem ponders the wisdom of trying to "replace the blind force of natural selection by conscious direction," which the poem pejoratively calls "subduing the forces of nature."

I said this was a hefty issue of the magazine. Much of this heft consists of striking advertisements that hint that decisive battles between culture and human nature may not yet be over. At a distance of more than fifty years, the advertisements can seem shockingly frank in their celebration of power and warfare, a potent reminder of one of the main uses of physics at the time. The back-cover advertisement for General Dynamics is titled "Dynamic America," and its account of submarine history beginning with the recent submarine circumnavigation of the planet and the launching from a submarine of a Polaris missile capable of carrying a nuclear warhead is accompanied by color drawings of early attempts to build submersibles. Images of submarines and missiles recur so often that the magazine might be a catalog for the defense industry. The Hughes Aircraft Company claim to be "putting Polaris on a precision path"; "Minuteman, a three-stage solid pro-pellant intercontinental missile, is moving through its early development" at Space Technology Laboratories Inc.; "Kollmorgen Starfinder Polaris missile submarines use the Kollmorgen-designed and built Type II periscope"; "The airborne data processing system for the EAGLE missile concept uses advanced digital data han-dling complemented by the intelligence of the human operator"; Avco is "arming America's Pentomic Army";[48] Laboratory for Electronics Inc. "is experienced . . . in Electronic Warfare"; the Martin Company presents color photographs of "Five major U.S. missiles developed and built by Martin"; and in an approximation of a centerfold, a color spread for RCA shows a "Strange 'fish' under the polar ice," a new-generation submarine using a magnetic videotape recorder that cleverly fits into a torpedo housing.[49] For a poet there must have been a special pathos in the Autonetics pitch that replaces T. S. Eliot and Scheherazade (what a wonderful conjunction) with a missile: "Somewhere in a wasteland, the Air Force Minute-man will keep its lonely vigil all through a thousand nights."[50]

Here made visible to anyone picking up the magazine were evident the "grave implications" of the "permanent armaments industry of vast proportions" with its "task forces of scientists in laboratories and testing fields" that President Eisenhower would call in his farewell address a few months later the "military-industrial complex."[51] The defense of science as the science of defense was an effective weapon around 1950 for persuading policy makers to fund scientific research and became an ideology that would shape how the arts and humanities would respond to the sciences, both positively and negatively, for the following decades. Was the scientific imaginary now given over to destruction? For Duncan this became a very real fear, that "chemists we have met at cocktail parties, passt daily and with a happy 'Good Day' on the way to classes or work, have workt to make war too terrible for men to wage."[52]

The advertisements tell another story too. The corporations and their scien-tists lay claim to creativity, life, and mystery. Lockheed is "minding the future" perhaps, like the Martin Company, believing, "The creative capacity of man is equal to the challenge of space."[53] The Electron Tube Division of RCA bragged that their tubes were "measuring the growth of the living cell" and "shedding light on the mysteries of life."[54] The Rand Corporation, a think tank devoted to Cold War military strategy, took a full-page ad in a position opposite the contents page, thus making its message pretty well unavoidable to any reader of the maga-zine, and filled it with a portrait of the biologist Ernst Haeckel accompanied by a quotation from his book *Riddle of the Universe* (1900).[55] How might poets of the time have responded to the following claim in Haeckel's words that is implicitly endorsed by the Rand Corporation and to some degree by the journal itself?

> An entirely new character has been given to the whole of our modern civilization, not only by our astounding theoretical progress in sound knowledge of Nature, but also by the remarkably fertile practical application of that knowledge in technical science, industry, commerce, and so forth. On the other hand, however, we have made little or no progress in moral and social life, in comparison with earlier centuries: at times there has been serious reaction. And from this obvious conflict there have arisen, not only an uneasy sense of dismemberment and falseness, but even the danger of grave catastrophes in the political and social world. It is, then, not merely the right, but the sacred duty, of every honorable and humanitarian thinker to devote himself conscientiously to the settlement of that conflict, and to warding off the dangers that it brings in its train.[56]

Duncan possibly recalls this passage in his poem: "Osiris-Kadmon [was] into many men shatterd [*sic*], / torn by passion." Rand chose the quotation presumably because of its Cassandra-like prescience about the Holocaust and the world wars of the twentieth century—and yet here too is an irony. Were readers expected to recognize that Haeckel's ideas, however unintentionally, like those of the eugenicists, colluded with the very catastrophe he predicted, because the Nazis regularly employed his statements, such as the claim that "politics is applied biology," as part of their ideology?[57] Duncan was always insistent that there can be no innocent position.

Duncan's poem brings several conceptual schemes into tension with one another: Egyptian mythology, scientific neurology, social anthropology, a nationalistic vision of Cold War campaigning. These systems of knowledge compete fiercely for epistemic authority. Duncan's ethical poetics invites the reader to recognize other possible relationships between knowledges than outright conflict for epistemic supremacy. Set's cock, the phallic missiles of American defense, the genitals on the graph of neurological intensity, the lordly exercise of masculine power—Duncan's poem does not choose between these modes of imagining the ethical dilemmas of his time. English may lack a "neat expression" as the *Scientific American* writer has it. Poetry in Duncan's hands is all about the tensions between neat and disorderly expression.

This poem enacts the strangeness of the incommensurability between different knowledges and epistemologies. In the three lines "the great boat of the gods / penetrates the thick meat, / sending quick nerves out that are tongues of light," three forms of contemporary knowledge—theosophy and the sciences of biology and psychology—poetically transform from one to another in an ontologically dizzying sequence. The first and most fantastic transformation begins with the metaphorical excess of "boat" (a reference to Ra's solar boat), which as a physical object could in principle "penetrate" the "meat," a vivid word for the medical, and consumable, material body. As we read, however, we are conscious of the impossibility of scale. A model boat with a sharp prow might enter flesh, but not one "great" enough to accommodate gods. The literal impossibility of this physical entry makes the reader recoil. Metaphorical interpretations are also elusive, leaving us with a sense of some miraculous, poetic transubstantiation. Whatever it is that happens to enable the great boat to penetrate flesh is beyond adequate, intelligible representation. This entry of the god's boat stimulates the nerves and they produce signs for consciousness, crossing the body/mind barrier, which still resists explanation by scientists and philosophers. The poem's continuous syntax in these three lines equates the act of the gods entering the body with the ordinary act by which the flesh of the nervous system somehow produces consciousness— here "tongues of light" rather than the Bible's Pentecostal "tongues of fire." These three lines insist on a poetic commensurability between knowledges that remains only a poetic possibility. And it is this hope that is invoked again near the end of the poem when the sistrum sounds through "us" and at least momentarily displaces the suspect neurological naturalization of violence.

This first part of the poem is an attempt (as so often in Duncan, his statements are provisional) to talk about Sahlins's primate brain in terms of a struggle between faculties represented by a narrative of contending mythic gods, a story not dissimilar to Plato's parable of the horses, or as Peter O'Leary suggests, a "Freudian primal scene."[58] How then might this violence be redeemed? The poem introduces the shadowy feminine figure of Isis, described punningly as "She-That-Is, / our Mother" who "puts it all together." After her restorative ministrations, Horus is identified with the homunculus, while Set is associated with the "forces of nature." As he does in "Apprehensions," Duncan creates a poetic space in which two competing models of human psychology, the sensorimotor homunculus and Egyptian gods, confront one another. In the resulting rapture the neurological system no longer tries to suppress what the gods represent, and a redemptive transformation supposedly occurs. Instead of the screams of battle, "the sistrum /

sounds through us." The lyric poetic spirit inherent in the Egyptian myth redeems division and conflict. At such a moment we transcend our ordinary limitations embodied by the nervous system, and allegorically poetry transcends the limited epistemological imaginary of contemporary science represented by the *Scientific American.*

In the 1950s, Cold War sociology strove for value-free scientific objectivity. What values it did admit to were nationalistic and purely epistemological. Imagination was replaced by instincts. Any residual ethics was embodied in utilitarian technocratic planning. The articles in the September 1960 issue of *Scientific American* are social-scientific studies of human evolution and raise questions about whose discourse about the human has epistemic and ethical credibility. Were the many pictures and cheerful endorsements of the power of American weapons signs of the achievement of science or signs of catastrophic disruption to come? Had primate nature been overcome, or should it, and what was going on in that vast brain? What kinds of intervention in human development could or should be made? Duncan's poem "Osiris and Set," one of a series of poems that formed part of a lifelong correspondence with the poet H.D., used the speculative revivification of mythological thought to ask difficult questions about the scientific America revealed in this and other issues of *Scientific American*. In his quite-different procedural poetics, Jackson Mac Low also asked how the sciences were now implicated even in the most intimate relationships. Both poets could assume that the magazine was a part of the intellectual diet of the time. As late as the end of the 1970s, Rae Armantrout could assume the magazine's ubiquity to the extent that she felt no need to identify a representative source for the current discourses about gender and system from which her poem tweezers into place egregious phrases for its linguistic microscope.

7

DEFYING
SOCIAL SCIENCE

George Oppen and Amiri Baraka

SPECIAL ISSUES

George Oppen was as curious as Mac Low, Duncan, and Armantrout about what the scientists were publishing about their work, and as critical of the ethics and politics behind the often-shadowy negotiations between scientific knowledge and other knowledges. In the notebook entries published as "Daybook III," he expresses concern that high school graduates too often emerge "with a concept of themselves here, and knowledge as something out there from which they may [be] handed facts as if over a counter." This epistemological naivety can be remedied only if "they somehow come to get a glimpse of the production floor, of the workshop, where everything is a sort of mess, and everything might easily have been made quite differently." As an example he gives the concept of "nature," saying that it is freely used to mean "physical existence, physis, things," as well as knowledge of them through "natural science—physics—the study of things," by people who do not recognize that the etymologies of the words "nature" and "physis" are so different as to be almost incompatible.[1] He himself was scrupulously careful to talk of the abiotic world as "mineral" rather than as nature. Knowledge mattered to Oppen. When Paul Vangelisti sent him a poetic rite for a community united in celebration of its confidence in the "burning sword" of Jesus, Oppen wrote supportively: "I tend to respond to the Jordan's candid tongue [referring to a phrase in Vangelisti's poem], to the whole Western and Judaic philosophy of Being, as the true sword of knowledge—Of course, like you, recognizing this as a response, a poetics, not a political program."[2]

But where did Oppen think those students could go to see what is going on in the workshops of knowledge? His argument implies that a good dictionary would help and so would an epistemologically sophisticated poetry. To make use of such assistance the students would still need to be able to witness the production of knowledge itself, yet public presentation of science in the mass media tends to edit out the mess. Even popular-science writing at its best, as in the typical article in the *Scientific American* in the period when Oppen was writing, would present the tidied and polished product of research, not the messy experimental scene that precedes the final graphs and firmed-up theories. Nor would the typical popular-science article normally reveal dissent about the meaning of the "mess," dissent that can result in sometimes-sharp conflicts between different specialists in a scientific field. As it happens, the *Scientific American* did in fact have a partial answer to this dilemma.

Starting in 1950, the *Scientific American* devoted an issue each year, usually in September, to a single specific area of science—planet Earth, scientific imagination, radiation, the cell, mathematics—with the side effect of making visible to the nonspecialist how differences of method, conflicts, and uncertainties arise from the complexity of actual scientific practice down on Oppen's production floor. Disagreements between the authorities contributing to each symposium sometimes reveal that scientific knowledge is not just monolithic fact. The magazine was aware of this possibility from the start. In J. Robert Oppenheimer's introduction to the very first special issue, "The Age of Science: 1900–1950," the article in which he makes the sideswipe at poets that I mentioned in the introduction, Oppenheimer tacitly acknowledges the importance of letting the public see how knowledge is made. He refers to the "errors and byways" along the route to scientific knowledge and uses a metaphor similar to Oppen's to explain the dual character of science. On the one hand, science provides a "serene and active workshop of the human spirit," and on the other, science goes out into the world to try to achieve "a vast extension of human resources, of man's power to control and alter the environment in which he lives, works, suffers and perishes." Oppenheimer hopes that the ten articles in the special issue will give the layman to whom they are addressed a better sense of both the "unity of science" in the past half century and the disunities marked by those research cul-de-sacs and the "extraordinary diversification and specialization of the several sciences." Even social science is given a nod. In his conclusion he observes that "in these days of growing crisis, men have talked with earnest desperation of the application of scientific method to new areas, to problems of man's behavior and to human society."[3]

These special issues repeatedly caught the interest of poets, especially when the science encroached on human affairs. Engagement with a special issue of the *Scientific American* enabled the sort of triangulation of readership, poetry, and science that Gary Snyder attempted in *Turtle Island* (1974). "Let no one be ignorant of the facts of biology and related disciplines," he wrote in this often-didactic, Pulitzer Prize–winning collection of poems and essays. Otherwise, the "great achievements of science" will be wasted on "software and swill."[4] As he was writing these poems and texts, two *Scientific American* special issues, on ecology and then on energy, took readers behind the scenes of the attendant epistemological and policy debates.

Meteorologist Burt Bolin, who would later become the first chairman of the Intergovernmental Panel on Climate Change, warns: "Tampering with the biological and geochemical balances may ultimately prove injurious—even fatal," and therefore, humanity should attempt to "leave them close to the natural state that existed until the beginning of the Industrial Revolution."[5] Snyder echoes such arguments in *Turtle Island*: "Man's careless use of 'resources' and his total dependence on certain substances such as fossil fuels (which are being exhausted, slowly but certainly) are having harmful effects on all the other members of the life-network." As a result, "pollution is directly harming life on the planet." Those engaged in a "revolution in consciousness" need "to take over 'science and technology' and release its possibilities and powers in the service of this planet."[6] Just a year later, the September 1971 special issue of *Scientific American* on energy revealed that far from everyone in the science world was convinced of such threats. The special issue begins with an introductory essay by Chauncey Starr, dean of engineering at UCLA, which opens with the arresting prediction that "between now and 2001, just 30 years away, the U.S. will consume more energy than it has in its entire history," and so it is not surprising given this starting point that he concludes his article by saying that the country must concentrate on developing breeder reactor technology, "which should make it possible for nuclear fission to supply the world's energy needs for the next millennium."[7] But a small number of articles oppose such arguments. Notably, Roy Rappaport summarizes his famous comparative energy audit of subsistence farming in New Guinea and the extravagant fossil fuel agriculture of the West. His conclusion that the resulting "ultimate ecological problem" could be "the most difficult to solve, since the solution cannot easily be reconciled with the values, goals, interests and political and economic institutions prevailing in industrialized and industrializing nations," is similar to Snyder's concerns in *Turtle Island*.[8] By exposing the production floor of scientific knowledge to view, the special issues of *Scientific American* allowed writers to position their interventions more readily alongside current epistemological and political divisions in the sciences.

SOCIAL SCIENCE AND THE CITY: GEORGE OPPEN'S "OF BEING NUMEROUS"

The 1965 special issue on urban planning would have appeared to fulfill Oppenheimer's prophecy particularly successfully: it contained a dozen articles by social scientists "on cities: how they arose, how they are evolving in various circumstances, how they shape themselves."[9] Articles on the history of the metropolis and on several specific cities, including Calcutta and New York, are followed by discussions of how a scientific approach can successfully address problems of land use, transportation, services, renewal, and design that arise in modern cities. On display was the confidence of social scientists that the problems of urban humanity could be addressed scientifically. They would have been aware that social science was not prominent in the topics usually selected for the general issues of *Scientific American*. The publication of the cities issue was a chance for sociology to make known to the general public its hard-won epistemic authority that positioned it alongside the well-established natural sciences. It has often been said that the ultimate proof of scientific knowledge is that it works. Airplanes stay up, lights stay on. Now social science could demonstrate that its science also worked; it was solving urgent problems in the new urban metropolises around the world, including New York.

Talking to L. S. Dembo, George Oppen said that "'Of Being Numerous' asks the question whether or not we can deal with humanity as something which actually does exist."[10] Readers of Oppen have tended to think of this as a purely philosophical preoccupation bound up with his study of the works of philosophers such as Maritain, Hegel, and Heidegger, who were so important for his later poetry. There is, as far as I am aware, little direct reference to sociology in his published and unpublished notebooks and letters, which has led commentators away from considering whether he was at all interested in it. I think he was. We should not treat Oppen's persistent reading and rereading of the philosophers as a sign of a lack of concern with sociological issues. Reading philosophy was in part his means of divesting himself of those aspects of Marxist sociology that he now found repugnant, in order to retain what was still viable of the social and political idealism that had led him to Marxism in the first place. A return to Hegel, the philosopher who most influenced Marx and who continued to influence modern political theory of the state, was an obvious rational move, even though in the 1950s logical positivists reviled Hegel as a totalitarian thinker. At the risk of exaggerating Oppen's philosophical aptitude, I would suggest that his engagement with Hegel can be understood as a strategy similar to that undertaken by the British philosopher Gillian Rose. The wittily blunt title, *Hegel contra Sociology*, of her brilliant critique of the sociology associated with Talcott Parsons sums up her belief that the "search for a general logic for the exact or sociological sciences" has been a mistake.[11] Only by returning to Hegel and his thinking about the speculative concept might it be possible to move beyond what she felt had become a sterile opposition between hermeneutic and scientific sociology. Oppen also wanted to use Hegel to counter a positivist sociology as well as a rigidly Marxist one. Philosophy was a form of social theory for him.

In practice, most commentators recognize Oppen's investment in sociological themes even if they avoid such discourse. One example will have to suffice. It occurs in Charles Altieri's ambitious attempt in *The Particulars of Rapture* to show why we should redirect critical attention to the affective forces of sympathy, identification, and "intensities." This turn to affect is not a turn to passivity, sentimentality, or indulgence in highly spiced emotion. This is a muscular, investigative affect. Sections 32 and 33 from "Of Being Numerous," for example, show how "an aesthetic perspective invites us to *ask* what states, roles, identifications, and social bonds become possible by virtue of our efforts to dwell fully within these dispositions of energies and the modes of self-reflection they sustain" (my italics). Altieri gives as an example section 33 of George Oppen's "Of Being Numerous," in which one "can *test* the possibility that the shift in section 33 to simple syntax and direct affirmation can provide a distinctive sense of peace" (my italics).[12] Altieri is pointing to the apparently straightforward statement in section 33 that follows a complex meditation on beauty in section 32, which ends with a portrait from the man's point of view of the passionate desire that binds a couple for whom there is "Not truth but each other // The bright, bright skin, her hands wavering / In her incredible need." Section 33 reflects on the implications of this twining of beauty, passion, love, and need:

Which is ours, which is ourselves,
This is our jubilation
Exalted and old as that truthfulness
Which illumines speech.[13]

Altieri argues that the unexpected emphasis on need draws attention to "what people might hold in common," an interpersonal perspective that in section 33 becomes "the sense of sharing that the poem asserts."[14] This jubilation is a "social emotion" made possible by recognition of mutual need, which then enables us to understand our mutuality as the virtue of truthfulness. Though this typically incisive interpretation of the effect of framing a single extended statement as an independent quatrain perhaps underreports the masculinity inherent in making the woman an emblem of mutual passion, it offers a persuasive understanding of how affect can be a mode of inquiry. Altieri's interpretation also demonstrates that even when Oppen is eliciting strong affects in order to test possibilities, he sees them as social issues.

The question that so exercised Oppen, of how to think about the relation of the individual to society, was also very live in the sociology of the time. Kingsley Davis, whose article "The Urbanization of the Human Population" introduces the cities issue, was already well known as the author, with Wilbert Moore, of a functionalist theory of social inequality proposed in the 1940s. In their landmark paper Davis and Moore are at pains to insist that theirs is a theory of groups and structures, not actual individuals: "the discussion relates to the system of positions, not to the individuals occupying those positions."[15] Like most of their contemporaries, these social scientists had "chosen the meaning / Of being numerous" (*NCP*, 166). Hamilton Cravens characterizes the dominant trend in Cold War sociology as based on such a theoretical choice of numerousness over singularity: "There was, in short, no such thing as an individual apart from a group in social and behavioral science thinking." He puts it even more forcibly: "the 'I' and the 'We' were tightly fused together."[16] The meaning of this fusion varied. Davis and Moore led the way toward what became the dominant theory in sociology in the 1950s, structural functionalism. Their theory presupposed that all features of a stable society must have "social functions" in maintaining the social order. When they examined economic and social inequality, they concluded that it was justified because "every society, no matter how simple or complex, must differentiate persons in terms of both prestige and esteem, and must therefore possess a certain amount of institutionalized inequality."[17] Unfairness works. Modern societies need to use "unequal economic returns" to regulate their functioning hierarchies.[18] This type of rationalization of the status quo could be described in Oppen's terms as trying to "define / Man beyond rescue / of the impoverished" (*NCP*, 185).

By the end of the 1950s Davis had had second thoughts about the viability of associating functionalism with a defense of inequality, and in 1959 he published what purported to be a critique of the "myth" of structural functionalism. This was not quite a recantation, however. He was not espousing some alternative form of sociology; he was cleverly defending functionalism by arguing that the concept of functionalism as defined by its various proponents is so baggy that in effect all scientific sociology is rightly functionalist. To ground his arguments he has to go back to basics and justify the scientific credentials of sociology, and as he does so, it soon becomes clear that he takes for granted that a social theory will be based on a conceptual scheme. "Every science describes and explains phenomena from the standpoint of a *system* of reasoning which presumably bears a relation to a corresponding *system* in nature." Later, in a footnote that reminds us of the enduring influence of the notion of the conceptual scheme, he clarifies the thinking behind the cautionary adverb: "A body of theory includes a conceptual framework which, not being in the form of evidential propositions, is not subject to verification." Davis has a specific worry that functional analysis has not broken the hold of the semantic chain of historical usage. He charges his functionalist peers with relying too heavily on "the purposive and moralistic reasoning of ordinary discourse" when a scientific sociology ought to be "doing the opposite of moralistic reasoning." "Such words as 'function,' 'dysfunction,' 'latent,' 'needs' are treacherous" because they have their origins in "ideas much older than sociology or anthropology, they are susceptible of easy expansion by knitting together ready-made intuitions, connotations, and ambiguities."[19] His argument amounts to the injunctions: forget the roots of words, concentrate on the present, and suspend the ethical.[20] That these are all preoccupations of Oppen's poem is a reminder of just how attuned Oppen was to the intellectual debates of his time.

After 1945, two departments of sociology dominated American sociology: Harvard under Talcott Parsons and Columbia under Hank Lazarsfeld and Robert Merton. Columbia's structural functionalism provided the most influential template for sociology by configuring "methodological positivism," "the validation of the scientific method," and "value neutrality (while at the same time engaging in applied sociology, which was problem solving)."[21] The cities issue of *Scientific American* aimed to demonstrate the successes and potentials of such problem-solving applied sociology. But the tide was turning. Craig Calhoun and Jonathan VanAntwerpen show how Davis's theory of the "functional value of inequality became a shibboleth for insurgents in the 1960s, who sought to show how functionalism legitimated the existing order and neglected the role of power."[22] These insurgent sociologists were being radicalized by the political campaigns of the

new social movements of black activists, feminists, and the New Left. In a tren-
chant brief sketch of the transformation of sociology in the late 1960s and 1970s
from a would-be scientific discipline to cultural study, Immanuel Wallenstein fills
out the picture. These new political ideas displaced earlier beliefs that cultural
formations could always be traced back to economic determinants with a new
emphasis on the centrality of identity, on how different social groupings confer
identity on the singular individual. Value neutrality also came under fire, which
led to new arguments against a rigid demarcation between science and the hermeneutic inquiries of the humanities. Sociology could no longer offer itself as a
"superdomain" acting as mediator between the two.[23] Wallenstein's analysis enables us to see that Oppen's poem "Of Being Numerous" can be read as a timely
intervention in a culture-wide shift of intellectual norms.

Oppen was particularly interested in the whole issue of what he called "the
problem of the concept of humanity." Peter Nicholls shows how Oppen specu-
lated in his notebooks on how best to think about this "problem," writing for instance that "Of Being Numerous" "does not mean to solve such a problem, but
to permit the problem to remain a problem, while giving MEANING, if I can, to
its terms."[24] Such thinking appears to inform the elusive reasoning of section 35:

> . . . or define
> Man beyond rescue
> of the impoverished, solve
> whole cities
>
> before we can face
> again
> forests and prairies (*NCP*, 185, his ellipses)

Solving problems was what the urban technocrats highly valued, as the cities
issue repeatedly makes plain. Abel Wolman, an emeritus professor of sanitary
engineering at Johns Hopkins, treats the city as a superorganism with "metabolic
problems that have become more acute as cities have grown larger and whose
solution rests almost entirely in the hands of the local administrator." This planner "must ultimately provide solutions fashioned to the unique needs of his own
community."[25] Part of the value of the cities issue was its exposure of dissension
around such claims. Another contributor to the issue, Nathan Glazer, a professor of sociology who assisted with the research and writing of David Riesman's

The Lonely Crowd (1950), is sympathetic to critics of urban renewal who argue that "it has done little for the poor" and points out that although most planners still support urban renewal, "they are now in large part the professionals trained to fill needs created by the urban renewal program itself." He concludes that the biggest weakness of urban renewal is that it focuses on the city as a whole or totality and loses sight of local diversity: "All the criticisms of urban renewal point to the fact that, whereas the program speaks of the whole city and all the ways in which it must be improved, provisions are made to influence only one aspect of the city—the physical nature of a given locale." Nevertheless, Glazer concludes that urban renewal is not bust, and "under the pressure of a number of gifted critics, urban renewal has become an instrument that any city can use to develop policies well suited to its needs."[26] Solving the problems of whole cities remains possible.

The cities issue also foregrounded two themes of great interest to Oppen: the general assumption that New York was what Benjamin Chinitz calls "the archetypal metropolitan region" and the pervasive technocratic assumption that the problems of the modern city can be solved by planning supported by careful calculation employing scientific methods.[27] The contributors are clearly fascinated by New York. Davis points out that Greater New York is "the largest city in the world and nearly twice as large as New York City proper."[28] Chinitz, a professor of economics at Pittsburgh who specialized in urban and regional economics, describes New York breathlessly as the city of cities: "The clichés are true: the height and breadth of the city are overwhelming; the nervous energy of its people, the variety of their origins and attitudes and goals are stimulating; its output of ideas, books, articles, paintings, plays and music, of fads and styles, of protest and restless dissatisfaction influences the entire country and the world." A hint of self-congratulation on behalf of his profession enters the next sentence: "For years the city has pioneered in social advances and the use of government to humanize an industrial society, and in spite of profound difficulties its government is remarkable for the range of its services and the expertise of its personnel."[29] For a poet who called this city home and was one of its creators of new ideas and books, a man with a history of political protest, these clichés about "humanizing" industrial society could appear patronizing and diminishing, as if the political and creative activity of the individual living in the archetypal city was inevitably dissolved in the multitude. No doubt Oppen was also aware of the legends that had grown up around the city's foremost planner, Robert Moses, much in the news in 1964 as his power ebbed away—a towering figure whom his biographer describes when he was an idealistic young man as looking out at New York and "dreaming dreams of public works on a scale that would dwarf any yet built in the cities of America."[30]

The other theme foregrounded everywhere in the cities issue of *Scientific American* is the value of planning as a mode of calculating utility and managing the urban economy. In a discussion of how to make "a frontal attack on the *problem*" of equitable distribution of tax revenues between New York districts, Chinitz asserts that an answer "must lie in improved techniques for *calculating* the costs and benefits to individual communities that arise from their accommodation to any new order" (my italics). On the same page he talks expansively of "*planning* as a vital influence in shaping the patterns of land use" (my italics).[31] Other contributors to the cities issue are more cautious, conscious that they cannot any longer take for granted public confidence in planning. Hans Blumenfeld's essay, "The Modern Metropolis," treats the notion of planning as badly in need of decontamination because although the idea of urban "planning" has a good ancestry in Patrick Geddes, "the Scottish biologist who was a pioneer in city planning," in recent times Soviet "centralized planning and ownership" has made the concept appear suspect to many. Blumenfeld's redemptive strategy is to dress up the concept of planning with positive rhetorical associations, to equate it with "*rational* distribution" and to talk of a proposal to improve transport connections as an "*ideal* of planning" (my italics). His concluding remarks try to resolve the conundrums of planning by making explicit the tensions between market control and public management: "Any plan that seeks to control the growth of the metropolis rather than leaving it to the play of market forces will require the setting up of new forms of control."[32] Such a plan would be "radical." A poet aware of the etymology of the word in the Latin *radix*, or "root," a word that might well characterize his own political history, might be justified in believing that "The Roots of words / Dim in the subways" (*NCP*, 172).

This type of social-scientific reasoning was one target of an essay by Heidegger, "Identity and Difference," that meant so much to Oppen a sentence from it appeared to him in a dream and was then incorporated in an altered form in the poem "Route," which was published alongside "Of Being Numerous" in the 1968 New Directions edition *Of Being Numerous*: "'Substance itself which is the subject of all our planning' / And by this we are carried into the incalculable" (*NCP*, 201). In a note to himself written in June 1966, published in his *Selected Letters*, Oppen gives a detailed circumstantial account of how he came to write these lines. They were among several sentences from a dream he had the night after reading Heidegger's *Essays in Metaphysics: Identity and Difference* (1960) "without being able to understand it clearly."[33] Oppen then reread the essay in search of the day residue around which the dream version had presumably co-hered, only to find he was unable to locate the instigating sentence. Yet as Peter Nicholls points out in his thorough account of this incident, any reader going to the Kurt Liedecker translation of "Identity and Difference" can readily identify the sentence that inspired the lines Oppen used in the poem: "To the extent that Being is challenged, Man is likewise challenged, that is to say, Man is 'framed' so he will safeguard the Existence which concerns him as the very substance of his planning and calculating, and thus pursue this task into the immeasurable."[34] The search for the instigating words was not wasted however. Oppen records that although he felt great resistance to Heidegger's idealism, the essay "was of such importance as to alter the subjective conditions of my life, the conditions of my thinking, from that point in time."[35] Nicholls provides a compelling psycho-logical reading of this dream rigmarole as a process by which Oppen was able to borrow Heidegger's authority for what was really his own idea that the modern preoccupation with the calculative rationality of planning could be halted by the absolute impenetrability of matter. I would add that Oppen's confusion may have been the result of his tendency to conflate sources and quotations, in this instance Heidegger and the sort of pervasive social-scientific discourses of his time that so often exemplified Heidegger's complaint that "our whole existence is challenged everywhere . . . to plan and calculate everything."[36]

IS THE CITY A WORK OF ART?
OPPEN'S CRITIQUE OF KEVIN LYNCH

Everyone who writes about Oppen sooner or later admits to bafflement about a particular phrase, possible allusion, or poetic structure, admissions that con-tribute to a widespread belief that his poetry, especially "Of Being Numerous," is often elusive or downright puzzling. This is surprising given the relative sim-plicity of his vocabulary, the absence of learned allusions demanding elucidation that make the *Cantos* or *The Maximus Poems* obviously challenging, the famil-iarity of his subject matter, and, above all, the recognizable modernist shape of the poetic form itself. The poetic form, which at first glance looks to be a variant on the modernist poem as written by Williams, Pound, Olson, Duncan, Levertov, Creeley, and many others, is deceptive. When we look closer, we find that Oppen does several things differently. He does so because as a poet he experiences a crisis surrounding poetry's capacity to speak about the contemporary world: the wars, conformism, consumerism, the seeming ineffectuality of the individual, and the difficulty of distinguishing what is real, true, or accurate knowledge. The pressure from social science is symptomatic of this crisis.

In his insistently dialogical poem "Of Being Numerous," which, in Michael Davidson's apt words, is "created out of conversations with others" (NCP, 280), Oppen, like Pound, Olson, and Duncan, quotes others. But unlike those other poets, Oppen usually anonymizes his sources, often hiding them so well that critics cannot decide if he is citing a specific passage, and he habitually alters even those quoted passages that he expects at least some readers to recognize. In a poetic climate that fiercely defended the originality of a specific poet's own style against copying or even critical imitation, he repeatedly comes close to doing this.[37] Oppen violates another familiar norm, the bracketing of direct affirmative utterances so that they are not read as the poet's settled position or statement of fact. The New American poets treat propositions in their poetry with reflexive caution. Creeley is a master at this, and almost all the prominent New American poets distinguish their work by similar strategies.[38] Oppen permits himself direct, nonreflexive affirmations of his own commitments, ranging from expressions of personal sentiment to the wisdom or "philosophical statements" that Henry Weinfield believes lift the poetry from being minor and eccentric to the status of major work.[39] These direct statements unsettle the norms of the modernist poetry of the time. Yet these statements are not to be taken at face value either, since almost invariably they are countered by other statements in the poem or subjected to a corrosive skepticism. To understand how Oppen's handling of direct statement works in the poem, we have to look at the various forms his expression, allusion, critique, quotation, and verbal echo take because it is in this wide configuration that their significance can be found. One place to start is with his interest in knowledge.

Oppen's commentators tend to agree that his preoccupations with epistemology and knowledge are central to "Of Being Numerous," though they approach this interest from different angles. Weinfield begins with the proposition that "the form of the poem, the fact that it is a sequence poem, emanates from the poetic/epistemological problem that generates its central theme and that is contained in its title: the problem of finding unity in multiplicity."[40] Oren Izenberg makes the large claim that Oppen is "reconceptualizing what it means 'to know' in a way that cuts it loose from our objects of sense, and that demands a reconsideration of what forms of attention toward others might be adequate."[41] Nicholls begins by citing Oppen's retrospective judgment that in the poem he was exploring "the condition of 'What it is rather than That it is'" and then goes on to specify the types of inquiry the investigation of the "What" entailed: "to test what 'existential certainties' might exist" at a time of social crisis, "to understand the generational conflict," and, above all, to ask the "question whether or not we can deal with humanity as something which actually does exist."[42] Susan Thackrey argues that Oppen passionately believed that the poem should not "participate in any sort of pre-determined meaning," and this led him to a rigorous investigation of perception understood in the broadest terms as including any kind of thinking and feeling.[43] She cites in support of this line of reasoning a well-known passage from a letter Oppen wrote in 1966, the year he finished "Of Being Numerous": "I think that poetry which is of any value is always revelatory. Not that it reveals or could reveal Everything, but it must reveal something (I would like to say 'Something') and for the first time. The confusion of 'must not mean but be' comes from this: it is a knowledge which is hard to hold, it is held in the poem, a meaning grasped again on re-reading—One can seldom describe the meaning— but sometimes one has stumbled on the statement made in another way."[44] All of the commentators in their different ways agree that the sort of knowledge that Oppen values is "hard to hold."

"Of Being Numerous" itself has a considerable amount to say about what knowledge means for the poem, for its author and readers, and for the society in which it is written. The first section begins with what at first appears to be an epistemological foundation for the sense of self: "There are things / We live among 'and to see them / Is to know ourselves'" (NCP, 163). Section 6 warns that the kind of knowledge about our collective condition that is offered by the media is suspect just because of its affective theatricality: "the discovery of fact bursts / In a paroxysm of emotion / Now as always" (NCP, 166). In section 12 the poem cites, with a slight alteration, a sentence from Whitehead that the philosopher intends as a starting point for his process philosophy: "'In these explanations it is presumed that an experiencing subject is one occasion of a sensitive reaction to an actual world'" (NCP, 169).[45] Epistemology is often assumed to begin with such supposedly incontrovertible observations, but the poem puts all the concepts in this sentence under the pressure of its contextual irony. Can there be any such explanations; is genuine experience even possible in this metropolitan world; what occasions might there be other than a singular sentience reacting to stimuli; just how sensitive are these subjects; and just what is actual about the actual world?

These internal questions about the scope of any inquiry into the conditions of present-day knowledge find their most intense expression in section 31, which concentrates on the acts of knowing that knowledge requires:

Because the known and the unknown
Touch,

One witnesses—.
It is ennobling
If one thinks so.

If to know is noble

It is ennobling. (*NCP*, 182–83)

Both the poetic sound and the poetic line sharply interrupt what without this interference would be a fairly smug affirmation of epistemological confidence. The line "If to know is noble" easily sounds to an American ear to be "If to know is 'no bull,'" or, more explicitly, not bullshit. By leaving this half-uttered demotic expletive latent in the sonic pattern, the poem compels a reader who notices this to hear a curious kind of class-conscious tension about knowledge, whose value may be that it is either upper class and noble or lower class and no bull. The syntax is so torqued by the lineation that we are left with circularities (it is ennobling if one thinks it is ennobling) and logical inferences without predicates or premises (we don't know what one witnesses as a result of being on the borders of knowledge, nor whether the act of knowing is noble). These twists create a crisis of meaning centered on complacencies about the value and methods of the many sorts of knowing that the whole poem discerns in contemporary life.

These challenges to the reader to consider what it means to know what is actually there in the city and the world are part of a wider pattern of activity in "Of Being Numerous" whereby, in the words of the art critic Thomas Crow, who is speaking in the context of visual art, "the violent acts of displacement and substitution entailed in making any object intelligible are already on display in the art."[46] Oppen's poem is constantly reflecting on how its own art should be interpreted. The violence it does to everything from simple affirmations to philosophical statements by authoritative thinkers entails an unusual degree of attention to how interpretation comes about. The poem not only requires self-consciousness about epistemological issues but also asks questions about art and aesthetics, about literary and philosophical tradition, and about the limits of poetry in its time, to name just some aspects of its testing of the boundaries. It is in the midst of this constant unsettling of the norms and expectations of modernist poetry that the poem engages with the rising ambitions of social science in its time. To see further how this questioning of the possibilities of poetic meaning works to unsettle existing assumptions about the relations between poetry and science, it will help to examine in more detail how Oppen uses the words of others.

The anonymous statements in quotation marks—we are not likely to find a specific source for the commonplace scientific phrase "A state of matter" (*NCP*, 172) in section 17, used several times in the *Scientific American* in the early 1960s—and the unattributed quotations such as "There is nobody here but us chickens" (*NCP*, 172), which is also in section 17, and the phrase "sad marvels" (*NCP*, 163, 187), which may or may not be a quotation, contribute to the collective effect of representing not the authorship of singular individuals but, as it were, utterances caught rising momentarily above the hubbub of contemporary culture. At worst these utterances may seem to be "a ferocious mumbling" (*NCP*, 173), but Oppen's silencing of attribution is not usually intended as disrespect for the author's contribution. His point is that "A state of matter" and the tidied-up line from Louis Jordan's 1946 song "Ain't Nobody Here but Us Chickens" are part of the common culture, half or quarter remembered, and now uttered and affirmed by many.[47] The song itself demonstrated this, as Oppen probably knew, drawing on an already-long history of comedy around the joke about the thief pretending to be a chicken despite his human voice. Frequently, we cannot quite tell if a phrase is a quotation. Oppen could well have encountered the phrase "sad marvels" in John Cuddon's 1960 account of Istanbul, where it occurs at the close of a bravura passage about the ravages to which cicadas are prone from a vast range of inventively tormenting predators. Cuddon is in an Istanbul cemetery whose melancholy atmosphere triggers this tacit allegory of the many forms of death, or sad marvels, that humans have encountered in the violence of history. He ends with a gesture toward the actual insect: "We may be indifferent to these sad marvels, but it has had its admirers."[48] If Oppen did borrow the phrase, it would be not because he was plagiarizing it or wanted to show it off but precisely because it had been used by someone or some many people before. He wanted his poem to read as if written by that numerousness.

At times the poem measures itself against the poetics that were associated with several immediate contemporaries who were felt at the time to be defining the possibilities of innovative poetics. In these cases too Oppen emphasizes the nonsingularity of the words, thoughts, and feelings in the poems to which he alludes and usually accompanies this emphasis with ruminations on the poet's epistemological responsibility to "what there is." Allen Ginsberg begins *Howl* with fierce certainty: "I saw the best minds of my generation destroyed by madness, starving hysterical naked, / dragging themselves through the negro streets at dawn looking for an angry fix."[49] Oppen introduces epistemological doubt to this type of claim: "How shall one know a generation, a new generation? / Not by the dew on them!" (*NCP*, 178). One will not, in other words, know it by any early-morning dew collected by those who were staggering through the dawn. But Oppen doesn't refute Ginsberg's poetic claim to recognize and represent his generation. Instead, he introduces epistemic criteria for knowing the new generation that include "torn" earth, "wounds untended," and "voices confused." In asking for such criteria, however, Oppen is opening up a distance between speaker and proposition that Ginsberg tried to close.

Oppen also alludes to another well-known poetic opening gambit. Robert Duncan announces *The Opening of the Field* with a poem that celebrates the power of creative fiat of the poetic imagination in a phantasmagoric display of different kinds of "field," including a Neoplatonic "eternal pasture folded in all thought," an underworld that issues forth as a "disturbance of words within words," and a children's grassy playground:

> It is only a dream of the grass blowing
> east against the source of the sun
> in an hour before the sun's going down
>
> whose secret we see in a children's game
> of ring a round of roses told.[50]

Section 34 of "Of Being Numerous" toys with similar motifs while quietly resisting any such poetic idealism. These are people and grass he can actually see: "Children and the grass / In the wind and the voices of men and women // To be carried about the sun forever" (*NCP*, 184). Oppen is explicit that instead of the idealism of eternities and dreams, the place for poetry to begin is the "beautiful particulars," beautiful even though the scene includes papers blowing on sidewalks, the sort of particulars missing from Duncan's idealized world. For Duncan, the auratic scenes are an "everlasting omen of what is," while for Oppen, the children, the grass, the voices, even the sun, are not primarily omens or symbols of the imaginative potency evident in conceptual schemes but simply what is there and what poetry should attend to. In these and other allusions to contemporary poetry, he is taking readers through reflections on what it means to realize, "It is now difficult to speak of poetry—" (*NCP*, 180), and more generally, he is in each case suggesting that poetry faces an epistemological crisis and somehow needs to wrest back some epistemic authority from the encroaching sciences and mass media.

One of the most striking of these encounters with contemporary poets occurs in a passage that challenges an attractive idea that was becoming popular at the time: that the city could be a work of art. In section 11 Oppen appears to oppose any aestheticization of the city. He concludes a series of images of the city (strangely intercut with what most commentators believe is an allusion to *Hamlet*) with the portrait of a friend, Phyllis, whom he recalls speaking about the happiness she felt on her first day at work at simply being in the city. He introduces her with a glancing reference to the Phyllises of pastoral, of whom the most visible recent example appeared in Williams's *Paterson* in book 4, where Phyllis is a young woman courted by the married poet Paterson, against backgrounds of "tall buildings," cheap hotels, and the East River.[51] Oppen's Phyllis is not just a poet's fantasy; she exists off the page. He remonstrates with an imagined unsympathetic reader who is repelled by the apparent sentimentality of the picture he sketches of Phyllis about to exit the bus—"So small a picture . . . it cannot demean us" (*NCP*, 169). And then, as if freed by this gesture, he bursts into a far stronger affirmation of his own similar affection for the city:

I too am in love down there with the streets
And the square slabs of pavement—

To talk of the house and the neighborhood and the docks

And it is not "art" (*NCP*, 169)

He could mean that such talk of love for the city is not artful, but the final line's use of scare quotes gives priority to the firm negative: streets, pavement, house, neighborhood, docks—in other words, the city—are not art. But who would be calling it art anyway?

Kevin Lynch did just this in his article in the cities issue: with the right planning the city could be "a work of art."[52] Back in 1960 this urban planner, a student of Frank Lloyd Wright, had published a groundbreaking examination of urban planning, *The Image of the City*, still today required reading on most degree courses in planning or architecture, in which he announces that urban planning is a "temporal art."[53] At times Lynch talks as if the city itself is art, itself a poem: "we need an environment which is not simply well organized, but poetic and symbolic as well. It should speak of the individuals and their complex society, of their aspirations and historical tradition, of the natural setting, and of the complicated functions and movements of the city world."[54] *The Image of the City* argues that the modern city offers a sense of the infinite, that "at every instant, there is more than the eye can see, more than the ear can hear," in experiences that form "sequences of events" reaching back deep into memory.[55] Oppen uses similar language to talk about the city and the world as being "Occurrence, a part / of an infinite series," though he would appear to be unimpressed by any claim that the city is special in its capacity to be endless, since "Surely infiniteness is the most evident thing in the world" (*NCP*, 184). Lynch suggests a theatrical metaphor for the variety of the city: "We are not simply observers of this spectacle, but are ourselves a part of it, on stage with the other participants."[56] The city makes possible what Oppen calls "audience-as-artists" or performers, in a reference that Nicholls identifies as being aimed at the Living Theatre.[57] Lynch is especially interested in the "clarity or legibility" of the city that is held by individuals as a "generalized mental picture" of the urban environment.[58] One of his key conceptual innovations is to argue that we should not think of these mental images as singular but as "public images," pictures of the city shared by "large numbers of a city's inhabitants."[59] This concept underwrites his proposals to improve the legibility of the city through enhancing the potential for visualizing its manifold complexity.

The public image of the city is constituted by five elements, "paths, edges, districts, nodes, and landmarks." All of these are shaped by visual perception, so that landmarks such as high buildings, domes, or towers might effectively be "'bottomless'" because they have a "peculiar floating quality."[60] What matters is what Walt Whitman says in section 40 of Oppen's poem about the statue atop the new capitol building in Washington, DC: "you can see it very well" (*NCP*, 188). Lynch's central argument is therefore that "a distinctive and legible environment not only offers security but also heightens the potential depth and intensity of human experience."[61] The moral plea implicit in this and other similar statements is a sign of a conceptual overstretch that has often caught the attention of later commentators. Throughout Lynch's writings there is a sleight of hand operating around the question of what it means for the city to be "imageable," legible, or perceptually cooperative. He may say in *The Image of the City* that he is simply concerned with "the need for identity and structure in our perceptual world" of the city, but as this formulation's use of the charged terms "identity" and "structure" indicates, he cannot keep himself from suggesting that more is at stake than just visual harmony and readable landscapes, that, as he expresses it in the *Scientific American*, the legible city is "the physical basis of an open society."[62] Lynch wants us to assume that when we read the city, we are not just being entertained but reading the whole history signified by this urban environment, learning the city's potential from its signs of humanity's achievements.

This latent assumption in *The Image of the City* that to read the city is to read its achievement becomes more visible in the article Lynch wrote five years later for the cities issue of *Scientific American*. Here Lynch is polemical. The city can be "a work of art, fitted to human purpose."[63] His essay veers between this utopian vision and a dystopian one of a future "single world city" where there is no escape from the "pressure of other people," as he sees the ugliness of this looming future fate and asks a question that Oppen might have asked: "What could we do to make it a more human place?" His answers are limited to correcting environmental concerns, the "perceptual stress," the "lack of visible identity," the illegibility, and the lack of flexibility. At present, "the language of the cityscape is as baffling as a news release."[64] He recognizes the need to conserve heritage, perhaps including such sights as the year-marked stone in the pylon of Brooklyn Bridge that Oppen sees "Frozen in the moonlight" (*NCP*, 165), but Lynch does not want the planners to "freeze the patterns of use." "Freezing the city" would be a mistake.[65] As the resonances between these quotations and Oppen's poem suggest, this is the one article in the cities issue that we might reasonably speculate that Oppen read, given the many echoes of its discourse in his poem.

The *Scientific American* article exposes to a skeptical reader the limitations of Lynch's sociological approach to urban surfaces. In *The Image of the City* he argues that "the very naming and distinguishing of the environment vivifies it, and thereby adds to the depth and poetry of human experience."[66] As city dwellers we "must be able to read the environment as a system of signs" and so be able "to relate one part to another and to ourselves, to locate these parts in time and space, and to understand their function, the activities they contain and the social position of their users."[67] To see accurately the things we live among in the city is to know who we and others are. We have to remind ourselves that this is a very literal seeing, however, as is evident in Lynch's proposal in the same article for a system of paths that would make the city legible:

This is their observation platform for seeing the city, their principal means of comprehending it. It is from the path network that the city dweller sees the relations among the city's parts, recognizes its organization, becomes familiar with its landmarks and develops a sense of being at home instead of lost in the city's immensity. . . . The views from the system would expose the city's major physical parts, its dominant functions and its principal social areas.[68]

The logic of this and other similar passages results in paradox because Lynch is led astray by the misleading analogy between, on the one hand, the interaction of the individual mental image of the city and the public image (constructed from overlapping individual images) and, on the other hand, the interaction of singular individual (the city dweller) and the community (the people). The viewer on the observation platform is really an abstract figure of the collective who is viewing the same collective living in the city, but by talking in the singular about the "city dweller" seeing the relations among the city's parts, Lynch invites the absurd image of a viewer watching herself or himself. If Lynch's logic of urban legibility is correct, in Oppen's words,

> You could look from any window
> One might wave to himself
> From the top of the Empire State Building— (NCP, 168)

Near the end of the article Lynch claims that the sense of "being at home" in the city depends on "an active relation between men and their landscape, a pervasive meaningfulness in what they see."[69] We might wonder how being able to see the city via the path network and other planned structures will ensure that what is perceived is also conceived to be meaningful. Is the meaningfulness of a cityscape in which we can identify specific buildings and routes among them really a precondition for, let alone similar to, the larger sense of meaningfulness in a life lived in the city? Can we look at the things of the city and know ourselves?

We do not know whether Oppen read the cities issue of *Scientific American*, whether he heard about its contents from friends or family, or whether he was simply unaware of it. Although its appearance in September 1965 and its pervasive use of terms that occur in his poem—other examples include "populace," "flow," "locality," and the repeated use of "numerous"—make it possible that he encountered it while he was writing the poem, the verbal resonances that I have indicated, especially the strong ones with the work of Kevin Lynch, may nevertheless have been picked up by Oppen from conversations of those around him or from other reading. The debate about cities and the role of sociological experts that was presented in the cities issue reflected widespread intellectual concern with such matters, concern that would have been inescapable for a poet living in New York. Reading "Of Being Numerous" in conjunction with these sociological essays enables us to understand the stakes in his reasoning about singularity, about what it means to "see" the city, and about how we might begin to imagine ways of ameliorating its problems without using calculative rationality. His philosophically rich language in the poem is charged with the density of a myriad allusions to contemporary debates about the city and its history.

AMIRI BARAKA AND THE
PERSISTENCE OF RACE SCIENCE

You are an American poet recently self-identified as a *black* poet, living on the East Coast in that year of international cultural revolution, 1968. Your attention is drawn to a study in the August issue of *Scientific American* of not only your ethnic group but your own city of Newark. A few years earlier you wrote ironically, "There are facts, and who was it said, that this is a scientific century."[70] Here is more uncomfortable evidence of that assertion, for alongside facts about lasers and leukemia, respiration and infrared light, archeology and the optics of the eye is a lead article offering "a study of ghetto rioters," with the unintended but nevertheless obvious to you effect of treating you and your fellow black Americans as objects of study equivalent to machines, bodily organs, old bones, and cancer. As if that were not enough, a provocative subtitle makes clear in racial language just who these rioters really are: "Why do negroes riot? An analysis of surveys made after the major riots of 1967 in Detroit and Newark indicates that some of the most familiar hypotheses are incorrect."[71] In your eyes, the *Scientific American* has a charge sheet already in such matters, since back when you first came to New York in 1954 it published an article by a professor of zoology entitled "The Biology of the Negro." Then here in this 1968 *Scientific American* article, despite the magazine's liberal support for civil rights, the perspective emphasizes that this is an article written by white scientists for whom African Americans are other ("they"): "Negroes are still excluded from economic opportunity and occupational advancement, but they no longer have the psychological defenses or social supports that once encouraged passive adaptation to this situation."[72]

That was a thought experiment. We are unlikely ever to know if Amiri Baraka did actually sit down to read this specific *Scientific American* article on "race," but we can reasonably consider this article and others like it as potential contributions, not to a liberal awakening to the extent of racist oppression, but to what seemed to many African Americans an already extremely hostile intellectual climate in which science could sometimes be on the wrong side. By considering just how hostile science could appear, we gain a better understanding both of Baraka's poetry and of the ideological risks of the epistemological competition for scientific status that led sociological research or social science to be presented alongside the natural sciences. Framing the investigation into the attitudes of rioters toward different social values as a science equivalent to physics or medicine could call the objectivity of them all into question.

Baraka himself had firsthand experience of these riots; he had been arrested, and he had recorded his impressions of the conflict in a number of poems. He wrote angrily and sarcastically about the "black crime hearts" of those who were caught up in these events, as well as of the threats they experienced:

> In the black crime hearts. Where they beat firecracker blue sidewindow
> nights the ladies wait for you in bars, and people laugh, the people
> try to sleep, or stay up all night locked up in something. These nights
> each night, another face, or stray bullet . . .[73]

In this poem "Newark Later" he captures the shared anxiety about the fate of friends—"where are you?? call us?? You alright??" His theme is the "not yet complete redisintegration," a typically incisive multiple pun on the many sociological terms used to objectify his community—redistribution, desegregation, reintegration, and destruction—a pun that represents the complex driving forces behind the rioting.

Looking back in his autobiography on this period he observes that there was "no science we could relate to."[74] Perhaps the article on ghetto rioters spurred him to say as much in the preface to the collection *Black Magic* (1969), which brings together three books of poetry written between 1961 and 1967, *Sabotage*, *Target Study*, and *Black Art*: "Black Art was a beginning, a rebeginning, a coming in contact with the most beautiful part of myself, with our selves. The whole race connected in its darkness, in its sweetness. We must study each other. And for the aliens we say I aint studying you."[75] Otherwise, these "aliens" study you and call it science. The stereotypical American is a white scientific American who believes that race science is a picture of reality. This is why Ishmael Reed's poem "badman of the guest professor" includes the founding scientist of genetics research, Charles Darwin, among the poets and novelists (T. S. Eliot, William Faulkner,

Ernest Hemingway, and Shakespeare) who are, "you know, whitefolkese / business."[76]

Under these conditions, an African American poet might well feel like defying science altogether, physics as well as genetics and sociology. This is what Baraka attempts in several poems written at this time, notably the well-known poem "Ka 'Ba," whose title refers to what is usually called the Kaaba, an ancient stone temple in the shape of a cube inside the great mosque in Mecca. In this poem, he depicts "black people" cleverly "defying physics in the stream of their will."[77] Probably written before the publication of "A Study of Ghetto Rioters," this poem rejects the very idea that any science, biological or social, could play a valid political role as a science when its treatment of "race" was so corrupted by unexamined racial ideologies. Baraka, I shall argue, can be read as applauding fellow African Americans for defying the epistemic claims of the sciences that appear to support oppressive beliefs about "race." He strikes out at the discourse made current by these mostly well-meaning articles about African Americans because publication in journals such as the *Scientific American*, with its mission to promote scientific knowledge, turned African Americans into the objects of white scrutiny.

The *Scientific American* saw itself in the forefront of the attempt to discredit misuses of genetics and "race" science to justify discrimination against African Americans. Its preoccupations are immediately evident in this list of its main articles on "race" issues between 1950 and 1980: "The Biology of the Negro" (1954); "Attitudes on Desegregation" (1956); "Metropolitan Segregation" (1957); "Attitudes toward Desegregation" (1964); "Residential Segregation" (1965); "The Social Power of the Negro" (1967); "A Study of Ghetto Rioters" (1968); "Intelligence and Race" (1970); "Attitudes toward Racial Integration" (1971); "Attitudes toward Racial Integration" (1978). The *Scientific American* was liberal in its treatment of "race," and most articles emphasized the "powerful forces of segregation" and the pervasive damage caused by whites' racial discrimination against African Americans, whether the issue was intelligence, crime, or culture. When Nobel Prize–winning physicist William Shockley, one of the inventors of the transistor, wrote to the Letters page of the January 1971 issue of *Scientific American* to argue that the authors of "Intelligence and Race" were "irresponsible" because they neglected to investigate whether the possession of Caucasian genes raised the IQ of African Americans, the magazine responded robustly. Shockley's eugenicist thesis that "welfare programs may be encouraging dysgenics—retrogressive evolution through disproportionate reproduction of the genetically disadvantaged"[78]—was challenged first by a spoof letter fulminating against articles included solely for their "emotional 'relevance'" from "Alfred E. Bigot" (alluding to *Mad* magazine's American everyman Alfred E. Neuman) and then from one of the authors of "Intelligence and Race," Walter Bodmer, firmly critiquing Shockley's reasoning on eugenics, exposing his misuse of the statistics of heritability, and calling for action on the economic and political discrimination against African Americans. An even more reform-minded approach is evident in the third of the articles on desegregation published in 1971. Its authors conclude that an increasing tendency for northern white Americans to "say that blacks should not intrude where they are not wanted is a measure of negative response to black militance."[79]

Nevertheless, like those "Caucasian genes," traces of race science lingered on. Racial science is still hanging around in the 1954 article "The Biology of the Negro," which now reads as troublingly racist despite its ostensibly liberal attitudes and the credentials of its author, Curt Stern, a leading geneticist who pioneered a nonracial approach to human genetics and was a signatory to the important UNESCO report *The Race Question*. His *Scientific American* article gives the impression that "race" is a firm, genetic fact: "within the same country live white and Negro races" and so the question is "what is the probable genetic future of the whole population of the U.S." given that "there has been a great deal of miscegenation" and the "African genes" are mingling with the "Caucasian genes"? And for Stern, "race" can be measured by traits such as "dark skin" and "thick lips." Even when the article says that "for the long run, there seem to be no inherent biological weaknesses in the Negro which place him at a disadvantage," the qualifiers ("for the long run," "seem") and the negative construction hint at the opposite.[80] An African American reader would very likely find these figures troublingly apophatic. Baraka, or Everett Le Roi Jones as he was then called, was discharged from the Air Force that year and went to live in New York, where he became a part of the literary scene. Would he have been reassured to be told that there appeared to be no "inherent biological weaknesses" standing in the way of his being a poet?

Baraka knew very well that black Americans engaged in urban protests for historical, political, economic, sociological, and even ethical reasons. Science, however, had what purported to be a scientific answer based on treating black persons as objects of scientific inquiry without any apparent self-consciousness of how this might appear to be a denial of the humanity and subjectivity of black Americans. In a rare example of momentary self-awareness that there might be a problem with white sociology itself, one of these authors, James Comer, hastily transfers the problem onto young, white, female idealists: "The thong sandaled, long-haired white girl doing employment counseling may be friendly and sympathetic to Negroes, but she cannot possibly tell a Negro youngster (indeed she does not know that she should tell him): 'You've got to look better than the white applicant to get the job.'"[81] So it is not the fault of the more serious, presumably male, social workers and their scientific approach that there can be tension between researchers and their subjects. Science has no causal role in discrimination.

The sociologist Richard America identifies a particular type of "racist [who] is also represented as an expert and a scientist, an objective and dispassionate analyst who presents findings and conclusions backed by data and evidence."[82] The problem in the 1960s was that (white) scientists were indeed "studying" the "race" of black Americans, studying their intelligence, the "negro experience" and the "inflammatory" concept of "black power," desegregation, and supposedly scientific questions such as "why do negroes riot?" A poisonous document like the report commissioned by the governor of Alabama, *The Biology of the Race Problem* (1962), was not taken seriously as a scientific publication itself, but its author, Wesley Critz George, who is identified on the cover as a "PhD biologist," could draw extensively on serious scientific publications as evidence for the report's explicitly eugenicist thesis that, in the words that he cites from Judson Herrick's *The Evolution of Human Nature* (1956), "unquestionably, racial and individual differences in mental capacities and attitudes are correlated with corresponding differences in the bodily organization."[83] George therefore concludes that political steps are needed to avoid "deterioration in the quality of our genetic pool."[84] Each time the governor's report cites an American scientist saying something racist, such as "the road to social deterioration runs by way of continued breeding from inferior stock" (a statement attributed to "Professor James C. Needham, biologist of Cornell University"), the entire scientific profession is associated with such a dehumanizing approach.[85] Never mind that this Professor Needham had actually retired in 1935. Indeed, George's book starts with a four-page list of scientists under the heading "Some Authorities Cited" and is careful to list their institutional affiliations. Of course, much of this is sham, and many of these scientists did not in any way intend to support racial theories, but to an African American reader the impression would have been that science did still tacitly support such pernicious ideologies. George could even cite articles from the same issue of *Scientific American* that Duncan utilized in his poem "Osiris and Set" and assume these articles supported his racial theory. According to George, the archeologist Robert Braidwood's article on the agricultural revolution shows that this milestone "involved the white man primarily, although it appears that Mongoloid people of China and America were not far behind in time."[86] George immediately adds, no doubt with the *Scientific American* in front of him, that "in recent centuries, the Scientific Revolution, too, must be credited to the genius of the white man, with some contributions by the Mongoloids."[87] This is a distortion of what the scientists in the *Scientific American* were saying, but as we saw in the case of Dobzhansky, it did not need much effort to twist the original argument in this direction. Braidwood's cautious qualification that his account is based on his well-known work in Iraq, and that there are "vast areas of the world still incompletely explored archeologically," is easily overlooked alongside his extensive discussion of the Fertile Crescent as the origin of the agricultural revolution.[88]

Scientists were, however, becoming increasingly aware of the need to challenge such practices by racial theorists, as was demonstrated by the debate following the appearance in *Science* in 1964 of an article entitled "Racial Differences and the Future." Hiding behind the title's rhetorical mix of pseudoscience and policy was an egregious essay advocating the sterilization of black Americans for eugenic ends. Its author was the endocrinologist Dwight Ingle, a professor of physiology at the University of Chicago, who had been advocating such ideas out of the public eye in the journal he created and edited, *Perspectives on Biology and Medicine*. By publishing these appalling proposals in *Science*, the leading American general-science journal, published by the American Association for the Advancement of Science and read mainly by scientists themselves, Ingle was able to give public voice to a belief that a eugenics program is urgently needed. Just as his title cloaks his intentions, so too does much of his rhetoric. He laments that he can see "no possibility that a comprehensive program to upgrade the genetic and cultural heritage of all the races will be undertaken for several decades."[89] He protests repeatedly that he is not a racist. As a scientist he would "accept the principle of equal legal and moral rights for the individual regardless of race or religion," and as a citizen he would recognize that "many Negroes are good schoolmates, good neighbors, and good citizens."[90]

But it soon becomes clear that however good they may be as citizens, Ingle believes black Americans are "handicapped by a substandard culture" that is genetically determined.[91] He therefore not only opposes desegregation but also opposes what he name-calls in biological language "interbreeding." He also repeats the eugenicist canard that "the histories of the Negro and white races show that the latter have made greater contributions to discovery and social evolution."[92] Then

Ingle attacks just the sort of political and cultural movements that Amiri Baraka was committed to: "the more militant Negro leaders who now dominate the civil rights movement, having been told that there are no genetically determined racial differences in drives and abilities, are demanding equal representation in jobs and in government at all levels of competence." To Ingle the dangers are obvious: a deterioration of standards at all levels of society. To avoid these "biosocial" problems, he argues explicitly for a eugenics program that would entail sterilization, not a prospect to be concerned with since "the procedure for sterilization of either sex is now simple," as well as a "program of conception control."[93] This would have been strong stuff in the decades before the genocide of the Second World War discredited eugenicist ideology; in the 1960s it was inflammatory. Yet what is surely most shocking is that the editors of a major scientific journal would consider it suitable for publication. They even gave Ingle several pages to reply to his critics and allowed him to make more racist statements such as this: "We look in vain for a country which is governed wisely by Negroes."[94]

The article provoked a storm of responses published across two issues of *Science* in December 1964, a backlash that was probably one of the reasons such articles largely disappeared from the serious scientific press around this time. Several correspondents make an ad hominem attack. Jules Rabin says, "I consider Ingle to be manifesting bad culture himself when he launches forth on this awesome set of proposals in the detached spirit of a sanitary engineer." Alan R. Beals, an anthropologist, contributes a savage satirical critique based on the alleged fact that "it is well known that Bigots possess a substandard culture," a problem that requires us to "find a biological definition of race which conforms to the prejudices of the paranoid and the uneducated."[95] Paula Giese, who later became a well-known American Indian activist, identifies the core issue as being that "publication in *Science* confers a cachet of some minimal scientific respectability," and she challenges the editors to take some responsibility for having published such "mischievous suggestion[s]" during an election campaign in which civil rights were a hot issue.[96]

Moral repugnance and the ethics of editorship are not, however, the only major concern in the responses to Ingle's article. Again and again the contributors protest at its speculative, prejudicial, *unscientific* character. They take issue with Ingle's science and its arguments, saying that "intelligence is a socially-acquired ability," not a genetically determined capacity, and that there is no firm evidence that what is ordinarily called intelligence is only found in advanced civilizations, since, for instance, a hunter probably requires a high intelligence to spear an antelope.[97] Morton Fried, an anthropologist at Columbia, makes an argument with special resonance for our reading of the significance of physics in Baraka's poem:

> A line in Wigner's Nobel laureate address states that the "specification of the explainable may have been the greatest discovery of physics so far" (*Science*, 4 Sept., p. 995). If we substitute "science" for "physics" in that statement and apply it to Ingle's "Racial differences and the future" (16 Oct., p. 375), we find that article sadly lacking in basic scientific orientation. An investigator does not have to know everything about the subject of his research before beginning, but it is a minimum essential that he sufficiently delineate the subject of his inquiry so that he knows when he is looking at it and not at something else.[98]

Fried assumes that physics and science can be treated as interchangeable; physics stands for science. He endorses Eugene Wigner's definition of science as the specification of what is explainable. When Baraka uses the term "magic" to describe African American potential, he is also saying that it is not explainable by science, and in doing so, he is not so distant from social scientists such as Fried, since Fried's point is that Ingle fails to be scientific because he wrongly designates a domain of social life as explainable.

The Ingle incident is a reminder that science already had an image problem when it came to black Americans. With this troubling history of faltering attempts at a pro–African American sociology still shadowed by a few racist survivals in the scientific literature in mind, we are in a position to grasp the impact of the 1968 *Scientific American* article on Newark and Detroit. The study was based on research for the National Advisory Commission on Civil Disorders, popularly known as the Kerner Commission. Fieldwork by black researchers using questionnaires led to the conclusion that the rioters were motivated overwhelmingly by a sense of "blocked opportunity" and not, as many media commentators had claimed, because they were criminally minded lowlifes or because they felt economically deprived.[99] "Both the Detroit and Newark surveys indicate that rioters have strong feelings of racial pride and even racial superiority."[100] The article reaches the relatively positive conclusion that African Americans have relinquished "the traditional stereotype that made non-achievement and passive social adaptation seem so natural" and now have "a sense of black consciousness," and that the riots were a protest against "continued exclusion of Negroes from American economic and social life."[101] It contains information that in the context of such exclusion, white America needed to hear, such as evidence that rioters were not outside agitators and the widespread belief among those questioned that African Americans are "'braver' and 'nicer' than whites."[102] The problem is the same with earlier articles: framing this discussion as scientific creates an unintended disenfranchisement. The understandable aspiration of sociologists to be recognized as scientists comes at a price in this area of inquiry, because in Baraka's own words when explaining the background to his essay "Jazz and the White Critic," "the sociological context was race."[103]

From Baraka's perspective the sociological context was race; from the standpoint of many social sciences the key context was science. Just as correspondents objected that Ingle was not scientific, a correspondent to *Scientific American* objected that "A Study of Ghetto Rioters" did "not come up to [the] usual standards of completeness and correctness of scientific method," not because violent political protests are discussed as if they were a form of dysfunction amenable to scientific investigation but because the methodology is not scientific enough.[104] Why? The study needed more thorough statistical evidence than provided by the several accompanying graphs and diagrams, and it should have disallowed even the limited intrusion of political and ethical judgments evident in the authors' conclusion that "none of these factors [poverty, unemployment, lack of education, and deviancy] sets the rioter off from the rest of the community in a way that justifies considering him a personal failure or an irresponsible person."[105] The authors respond to their critic by saying that the only purpose in explicitly denying the role of these factors would be to make "a moral condemnation of riot behavior" and presumably therefore be unscientific. This, of course, is just the problem an African American poet would have with this sort of "scientific" assessment—that moral judgments were being justified as scientific—as we are now in a position to see in a close reading of the poem "Ka 'Ba."

Morton Fried found no difficulty substituting science (race science) for physics in an aphorism about scientific method. This capacity for physics to represent scientific method is the starting point for my discussion of Baraka's striking picture in "Ka 'Ba" of African Americans "defying physics." This frequently reprinted, six-stanza poem celebrates the African roots of black Americans and recalls the Middle Passage and slavery as the continuing existential condition of black Americans.

A closed window looks down
on a dirty courtyard, and black people

call across or scream across or walk across
defying physics in the stream of their will

Our world is full of sound
Our world is more lovely than anyone's
tho we suffer, and kill each other
and sometimes fail to walk the air

We are beautiful people
with african imaginations
full of masks and dances and swelling chants
with african eyes, and noses, and arms,
though we sprawl in grey chains in a place
full of winters, when what we want is sun.

We have been captured,
brothers. And we labor
to make our getaway, into
the ancient image, into a new

correspondence with ourselves
and our black family. We need magic
now we need the spells, to raise up
return, destroy, and create. What will be

the sacred words? (*BM*, 142)

Many African American poets of the time were similarly celebrating black culture in the face of the old slurs that it was "substandard." Nikki Giovanni talks "bout those beautiful beautiful beautiful outasight / black men / with they afros / walking down the street."[106]

At the time, Baraka returned again and again to this theme, as in "The Test":

They drive us
against
the wall white
people

do, against
our natures
free and easy atoms
of peaceful loving
ness. Beautiful things. Our sign permits
of the upward gaze. Toward heaven, or haven, in
the spirit reach of black strength, up soaring,
like Gods we are in hell, fallen, pulling now
against the gravity of the evil one himself.
Black streak from sun power. We are Gods, Gods
flying in black space. (*BM*, 188)

The rhetorical exaggeration that ascends from beauty to spirit and to godlike power is counterpointed by the tacit counternarrative to the scientific one: black Americans embody physics and take energy from the sun.

"Ka 'Ba" finds a compelling, familiar image for these ideas. The title of the poem invites an initially implausible analogy between the apartment block with its courtyard and the holy sanctuary at Mecca, a comparison based at first just on the idea of both spaces being enclosed within a larger building. The perspective from the window mimics that of the white majority, whose minds are closed to otherness and "look down" on the black Americans, who are already engaged in the creation of a new magic in their everyday actions by "defying physics." But what could justify the conferral of such spiritual distinction on ordinary social interactions taking place in this "dirty" urban location where people "call across or scream across or walk across" the spaces that divide them? The poem invites us to assume that these exchanges are constitutive of the culture of black people. So the next line, "defying physics in the stream of their will," sounds at first as if it is a poetically original way of saying that they defy the daily oppression due to economic and political discrimination, as if they are always rowing upstream. But we are left with a lingering residue of meanings. How can physics stand for oppression?

Although we could equate physics with fate since the laws of physics are an inescapable condition of our lives, I have been showing how much evidence there is that this defiance is aimed at the scientific "cachet" given to ideologically driven theories of race, because at this point in Baraka's poetic career the most recognizable image of science is provided by its dominant form, physics. Hence, physics has to be defied. But what meaning will the poem give to this idea, other than a poetic exaggeration of some gesture like a dancer's elegant leap (defying gravity, as it were). The next stanza's final line hints that they are defying gravity like the "beautiful things" in "The Test," saying that despite their loveliness black Americans "suffer, and kill each other / and sometimes fail to walk the air." Using the colloquial image of joy as "walking on air," Baraka subtly intimates that in addition to being open to joy his fellow Americans are able to circumvent scientific laws.

The following stanza subtly builds on the lingering questions raised by the first stanza about the apparent dissonance between the spiritual centrality of the Kaaba and the marginality of the apartment complex and courtyard and about the significance of defying physics. First, this apartment is metonymic of a culture, "our world" that is "full of sound," African American music as well as speech. The poem's simple declaration tacitly redoubles its predicates here: our world is full, our world is sound (musical), and our world is sound (healthy, whole, well-founded). The attribution of these qualities is defended even as they are reasserted; there is loveliness as well as suffering, some of it self-imposed. It is the final line of the second stanza that finally resolves the unfinished thinking of the first, by seemingly admitting a further weakness, "and sometimes fail to walk the air," while actually reinforcing the extraordinary claim that these people can indeed defy physics, at least sometimes, and walk on air. We make sense of this by thinking of it as entirely a metaphor, a joyous sensation of being lighter than air sometimes, a jubilation that arises when black culture most fully engages with itself, when the calls, screams, and interactions are most intense. At the back of this thinking are religious tropes of walking on water and angelic flight.

The poem ends with a question that tacitly admits that so far the poem has only defied physics through its poetic vision. The people in the poem still need some sort of antiphysics, some sort of "magic" if they are to be able "to raise up" and walk the air, and the poem imagines these as "sacred words," presumably of the kind represented by the Kaaba itself:

> We need magic
> now we need the spells, to raise up
> return, destroy, and create. What will be
> the sacred words? (*BM*, 142)

African American sociologists were beginning to develop secular answers to such questions. In 1972 Lerone Bennett wrote, "It is necessary to develop a new frame of reference which transcends the limits of white concepts, . . . We must say to the white world that there are things in the world that are not dreamt of in your history and your sociology and your philosophy."[107] What makes "Ka 'Ba" such a powerful poem is that the earlier subtle raising up of the author's people, so that they appear to be able to defy physics by walking on air, lingers as a demonstration of the power of words to "raise up / return, destroy, and create." For a contemporary black reader the poem is an invitation to rearticulate everyday energies in a new form for which the impossible levitation becomes a sign of hope just because it is scientifically impossible. Physics therefore does represent all that needs to be defied. The question the poem leaves its readers is why "defying physics" and hoping for "magic" should be a powerful move into the future. What will be "the sacred words": poetry, African knowledge, or a new science that at present appears to be magic?

By the end of the 1960s, many black sociologists were turning to the last two options as they challenged the reliance of mainstream sociology on what these insurgent social scientists increasingly called "white experience." Their arguments were presented in a landmark collection of essays, *The Death of White Sociology* (1973), by an impressive array of African American intellectuals. These sociologists were part of that wider movement in sociology that Wallenstein describes as attempting to overthrow the dominant scientific methodology of functional analysis.[108] Robert Staples, for instance, argues unequivocally that "the ideological content of structural-functional theory lends itself easily to legitimation of the prevailing form of racial domination."[109] For these sociologists, black magic would mean appropriating science from its ideological misuses by the white establishment, but not abandoning fundamental scientific methods. In a key article on African Americans and science, the sociologist Joseph W. Scott argues that they need to take over science. Scott, who by then had left not only the University of Notre Dame but America itself for the African University of Ibadan in Nigeria, starts with an endorsement of scientific method and then makes no bones about the fact that "the use of science is mandatory for Black people if they are to get and keep control over the physical and social forces that determine their life-chances." The problem is that the existing conceptual schemes are inadequate for use by African Americans: "Today as Blacks apply the existing theories and models in an attempt to make life more predictable and to get control over the external and internal forces that shape their lives, they are finding that the theories and models which were developed on whites do not apply well to Blacks. Scientific knowledge in the social sciences in this country is mostly a body of facts and generalizations about white experience. . . . Scientific knowledge claims not to be culture-bound or race-bound, but the assertion of universality does not make it empirically so."[110]

Over the next decade Baraka himself began to change his mind about science. During what he later called his "third world Marxist period," he became convinced that political analysis requires scientific rigor. In a 1978 interview in *boundary* 2 with Kimberley Benston, Baraka criticizes his book *Blues People* for its lack of understanding of the class system, saying that he previously "lacked an all-around scientific approach to it."[111] A similar comment in a much later interview suggests that Baraka was referring to the claim by Marxist sociologists that science is an indispensable part of Marxist analysis. He is clearly now at home with scientific metaphors and analogies. He praises Duke Ellington for his "constant experimentation" and continues: "Like Mao said, the three major struggles in life are class struggle, the struggle for production, and the struggle of scientific experiment. That's a struggle, trying to find out. The question of scientific experiment is a question of human development."[112] Artistic experimentation is being likened to scientific experimentation, and a famous African American musician is by implication credited with scientific status. Baraka also cites Jesse Jackson's comment that African Americans should treat politics as a science, and he calls W. E. B. Du Bois's *Black Reconstruction* "a great work of science."[113] Just how far his thinking has changed from the time of "Ka 'Ba" can be seen in the poem "When We'll Worship Jesus," which uses its refrain to contrast the various commitments that African Americans will make instead of the unthinking Christian piety expressed in the worship of Jesus: "we worship our selves / we worship nature / we worship ourselves / we worship the life in us, and science, and knowledge, and transformation / of the visible world / but we aint gonna worship no jesus." The value of science is now perceived as its capacity to help "understand what there is here in the world! to visualize change, and force it."[114]

CODA

///////////////////////

AROUND 1975 a dispute about the biohazards of the new technique of gene splicing spilled out of the laboratory into the media in what has become known as the cloning controversy. Two years later a retrospective *Time* magazine article reported that the mayor of Cambridge, Massachusetts, Alfred Velluci, had used the controversy to harass Harvard University's allegedly insecure research facilities, saying that they should stop risky genetic research because "something could crawl out of the laboratory, such as a Frankenstein."[1] Back in 1972, David Baltimore, who won the 1975 Nobel Prize for medicine, was warning other scientists that the dangers of transposing sections of DNA from one organism to another were "commonly analogous to the atom bomb." These fears became known to the public when a group of leading molecular biologists, including Baltimore, circulated a letter in several science journals calling for a voluntary moratorium on cloning experiments until their risks could be better assessed. A conference was arranged at the Asilomar conference center in Monterey, California, where the scientists hoped that a decision could be reached on self-regulation of this potentially dangerous research. At first the organizing committee wanted to keep the press out, but a journalist friend strongly advised them not to do this: "A secret international meeting of molecular biologists to discuss biohazards? If the press isn't allowed, I'll guarantee you nightmare stories."[2] This was good advice, for, as it was, the newspaper stories would be bad enough. On the day after the conference ended, the *Boston Globe* published an alarmist story: "Scientists to Resume Risky Work on Genes: Danger of Andromeda Strain Posed."[3] The Asilomar conference was big news. Even *Rolling Stone* sent a journalist, Michael Rogers, who wrote a detailed, witty, and informed account that has become the best record of the event.[4]

Rogers's report accurately captures the fears of the time, but it also reveals the power of the metaphors and accompanying rhetoric that had become embedded in molecular biological research into DNA. After James Watson and Francis Crick discovered the triplet structure of DNA, Marshall Nirenberg and others had managed by painstaking laboratory manufacture to demonstrate which amino acids were produced by which specific triplets. Scientists began to claim that they could read what Robert L. Sinsheimer called the "book of life."[5] And now it appeared that the scientists were also rewriting that book:

> A single strand of human DNA, microscopically small, contains at least the information of a library of 1000 volumes.
>
> The chemical keys to that library have been hard to find. To translate one volume, even harder. And to write one's own book—impossible.
>
> Until recently. . . . Recombinant DNA engineering is the reason for Asilomar: the discovery of the first rudiments of grammar for that previously unspeakable genetic tongue.
>
> The ancient Greeks believed in a mythologic being called a Chimera—a female monster composed of two or more animals. Molecular biologists now believe in DNA molecules that they call precisely the same thing. Moreover, they make them themselves. Recombinant DNA engineering uses certain newly discovered enzymes to disassemble the long DNA molecule in so orderly a fashion that the loose bits of genetic coding may then be rejoined, grammatically, into coherent sentences. And such a sentence may well describe—and create—the mutual offspring of two altogether different creatures incapable of mating in nature. . . .
>
> If one knows the grammar, one can begin to make up new sentences.[6]

Not only had scientists apparently discovered that organic life had at its core a language, but they were also becoming *writers* of a new type of text, the genetic materials of living organisms. As the reporter says, "If one knows the grammar, one can begin to make up new sentences."

In *Physics Envy* I have shown how midcentury poets joined researchers from many other disciplines across the natural, social, and human sciences in borrowing from physics research concepts and methods of inquiry for use in domains far removed from nuclear materials. Now in the 1970s the direction had been reversed. Molecular biology was looking to concepts of language and linguistic methodologies that had hitherto belonged to the social and human sciences. The entire balance of epistemological power between the natural sciences and poetry had apparently shifted. This was a heady time for writers. You didn't even need to call yourself a poet; you could call yourself a "language writer" and be understood to be engaged at the frontiers of inquiry. Just up the road from Asilomar a group of avant-garde poets in the Bay Area would begin to develop a poetics based on new sentences.

A whole linguistics of cell architecture had grown up over the past two decades, as John Maynard Smith explains: "The colloquial use of informational terms is all-pervasive in molecular biology. Transcription, translation, code, redundancy, synonymous, messenger, editing, proofreading, library—these are all technical terms in biology."[7] Treating molecular structures as a code was a bold conceptual move with wide cultural implications that have been extensively analyzed by the historian of science Lily E. Kay in *Who Wrote the Book of Life?* (2000). Kay argues that the combination of the cybernetic theory of information and the achievements of cryptanalysis during World War Two gave enormous impetus to the linguistic-code interpretation. The resulting "cultural poetics of science" had far-reaching implications: "though remarkably compelling and productive as analogies, 'information,' 'language,' 'code,' 'message,' and 'text' have been taken as ontologies."[8] Analogies taken as ontologies—this is a story we have already encountered before in Whitehead and in many of the ad hoc conceptual schemes derived from physics. For anyone working creatively with language, whether philosophers, structuralists, literary theorists, or poets, this idea that the cell, the basis of the organism and life itself, was organized like a language was deeply compelling. Many cultural theorists have alluded to this language at the heart of the organism; Derrida mentions it approvingly at the start of *Of Grammatology*.[9]

In 1980, a few years after the Asilomar controversy, Ron Silliman gave a milestone talk on poetics to an audience of Bay Area writers in which he named the prose poetry of disjunct sentences that many of them were writing "the new sentence."[10] This branding proved so successful that the poetry of Silliman himself, Lyn Hejinian, Bob Perelman, and a significant number of the much wider group known as the Language Writers has subsequently been read through the poetics of the new sentence. Silliman and his contemporaries could assume that their poetic researches into language had a stronger epistemological foundation than ideology critique, because the new molecular biology helped give timeliness and authority to structuralism's theory that society is structured like a language.[11]

I don't know whether Silliman or other poets read or heard about the *Rolling Stone* report, but the idea that "new sentences" were the basic blocks of new scientific methods for the study of life was very much in circulation. The Asilomar story and the *Rolling Stone* report point to the direction that interrelations between poetry and science would follow over the next two decades or more. A poet who wanted, in Lyn Hejinian's words, to pursue "a romance with science's rigor" was likely to start not with physics but with molecular biology and go on to look closely at other life sciences, ranging from sociobiology to neuroscience.[12] Conceptual schemes would fade into the historical background largely unnoticed, replaced by the idea that language was the generator of subjectivity, social interaction, belief systems, and ideologies, as well as being a model of how other biological systems operated. Language, information theory, and communication would all become central. Thinkers as diverse as Jürgen Habermas, Jacques Derrida, Gilles Deleuze, and Judith Butler would provide different frameworks of ideas about language that would encroach on poetic territory by appearing to dictate to poets how language, their aesthetic medium, should be conceptualized in highly authoritative terms that often appeared to brook no alternative. Instead of the physicists and social scientists telling poets that the world was made of atomic elements and nothing else, or that social sciences now had full rights of inquiry over their bodies or their cities, now a new challenge faced poets. They had to assimilate the new thinking and at the same time find ways to hold on to their own vision of language. Scientific rigor now looked very different from the physical reductionism displayed in the conceptual schemes improvised from physics. Looking even further into this future, from the 1980s onward, a new generation of historicist scholars would return to the study of science and poetry that had briefly flared around midcentury and go on to produce a rapidly expanding literature. American poets of all persuasions would increasingly write from a knowledge of science and would also be more aware of poets writing in other countries and languages. In a sign of how things have changed, in my own country, two poets with growing reputations in America, J. H. Prynne and Allen Fisher, would not only inform themselves about current scientific developments in the specialist literature rather than the popularizations but engage in sophisticated arguments with these ideas. But that really is another book.

NOTES

INTRODUCTION

1. J. R. Oppenheimer, "The Age of Science: 1900–1950," *Scientific American* 183, no. 3 (September 1950): 20–23, here 20.

2. The staff of *Fortune* magazine, "The Scientists," *American Scientist* 37, no. 1 (January 1949): 107–18, here 109.

3. Daniel J. Kevles, *The Physicists: The History of a Scientific Community in Modern America* (Cambridge, MA: Harvard University Press, 1995), 391. He does not give his source.

4. Muriel Rukeyser, *The Life of Poetry* (1949; Ashfield, MA: Paris Press, 1996), 11.

5. The Week in Review, *New York Times*, December 1, 1946.

6. Rukeyser, *Life of Poetry*, 17.

7. John Ashbery, "The Instruction Manual," in *Some Trees* (1956; New York: Ecco Press, 1978), 14.

8. Rukeyser, *Life of Poetry*, 160.

9. Hans Albrecht Bethe, "Within the Atom," in *The Scientists Speak*, ed. Warren Weaver (New York: Boni and Gaer, 1947), 98.

10. Preface to Weaver, *Scientists Speak*, v. Weaver adds that a quarter of a million requests were made for copies of the talks.

11. T. S. Painter, "The Science of Heredity," in Weaver, *Scientists Speak*, 214.

12. James Franck, "Medical Benefits from Atomic Energy," in Weaver, *Scientists Speak*, 281.

13. J. R. Dunning, "Atomic Energy and Structure," *American Scientist* 37, no. 4 (1949): 505–28, here 507.

14. Hans Reichenbach, *The Rise of Scientific Philosophy* (Berkeley: University of California Press, 1951), 141.

15. C. H. Waddington, *The Scientific Attitude* (Harmondsworth, Middx.: Penguin, 1941), 41. I am not sure how available this book was in America during the war itself. The book was revised and reprinted in 1948 for a Pelican edition that was widely distributed in the United States. A short positive review in *Philosophy of Science* began by saying that "it is not too often that we find a scientist who is willing to bare his soul to public scrutiny" and added with a backhanded compliment that "the correctness or incorrectness of individual arguments is not of great consequences in a work like this." Russell L. Ackoff, review of *The Scientific Attitude*, by C. H. Waddington, *Philosophy of Science* 16, no. 3 (1949): 266. Ackoff was then a professor of philosophy and mathematics at Wayne State University.

16. Waddington, *Scientific Attitude*, 47.

17. Michael Polanyi, "Freedom in Science," *Bulletin of the Atomic Scientists* 6, no. 7 (July 1950): 195–98 and 224, here 197.

18. Mark A. May, "The Moral Code of Scientists," in *The Scientific Spirit and Democratic Faith*, ed. Eduard C. Lindeman et al. (New York: King's Crown Press, 1944), 44.

19. Henry Margenau, "The Democratic Responsibilities of Science," in Lindeman et al., *Scientific Spirit*, 55.

20. David E. Apter, *The Politics of Modernization* (Chicago: University of Chicago Press, 1965), 461. The hostile reviews of his book strongly suggest that on such issues he was by then out of line with his contemporaries.

21. Robert K. Merton, "The Normative Structure of Science" (1942), in *The Sociology of Science: Theoretical and Empirical Investigations*, ed. Norman W. Storer (Chicago: University of Chicago Press, 1973), 275–76. Cited in Steven Shapin, *The Scientific Life: A Moral History of a Late Modern Vocation* (Chicago: University of Chicago Press, 2008), 21.

22. Shapin, *Scientific Life*, 88.

23. Ralph W. Gerard, "The Scope of Science," *Scientific Monthly* 64, no. 6 (1947): 496–512, here, 499, 500.

24. Ibid., 496.

25. Ibid., 497.

26. Harlow Shapley, "Status Quo or Pioneer? The Fate of American Science," *Harper's*, October, 1945, 312–17, here 312.

27. Ibid.

28. James Conant, *On Understanding Science: An Historical Approach* (Oxford: Oxford University Press, 1947), 11.

29. James B. Conant, "The Advancement of Learning in the United States in the Post-war World," *Science*, n.s., 99, no. 2562 (February 4, 1944): 87–94, here 92. Despite the title, the paper was given in November 1943 to a meeting of the American Philosophical Society, which helps account for the emphasis on the significant contribution of many fields toward intellectual progress.

30. Ibid., 94.

31. Ibid., 10, 9.

32. Ibid., 25.

33. David A. Hollinger, *Science, Jews, and Secular Culture: Studies in Mid-Twentieth-Century American Intellectual History* (Princeton, NJ: Princeton University Press, 1996), 162.

34. Lily E. Kay, *Who Wrote the Book of Life? A History of the Genetic Code* (Stanford, CA: Stanford University Press, 2000), 128.

35. Talcott Parsons, *Essays in Sociological Theory Pure and Applied* (Glencoe, IL: Free Press, 1949), 44.

36. Kurt Lewin, *A Dynamic Theory of Personality*, trans. Donald K. Adams and Karl E. Zener (New York: McGraw-Hill, 1935), 99.

37. Philip Whalen, "Statement on Poetics," in *The New American Poetry, 1945–1960*, ed. Donald M. Allen (Berkeley: University of California Press, 1999), 420.

38. Luis Alvarez, a pioneer in the use of these evolving devices, says that the "Wilson expansion chamber had two difficulties that rendered it unsuitable for the job": it was very slow and became slower still if you tried for greater precision by increasing the pressure in the chamber. See Luis W. Alvarez, "Recent Developments in Particle Physics," *Science*, n.s., 165, no. 3898 (1969): 1074.

39. Scientific vocabularies and discourses offer significant opportunities for poetry that a num-

ber of contemporary poets have explored. I have discussed these in several essays. See in particular my exploration of the reader impact of specialist vocabulary in "Strips: Scientific Language in Poetry," *Textual Practice* 23, no. 6 (2009): 947–58. Similar themes are investigated in my earlier essay "Can Poetry Be Scientific?," in *On Literature and Science*, ed. Philip Coleman (Dublin: Four-Courts Press, 2007), 190–210.

40. Shapley, "Status Quo or Pioneer?," 314.

41. Ibid.

42. George A. Lundberg, *Can Science Save Us?* (New York: Longmans, Green, 1947), 89.

43. Douglas Bush, *Science and English Poetry: A Historical Sketch, 1590-1950* (Oxford: Oxford University Press, 1950), 151.

44. Lyn Hejinian, "The Person: Statement," *Mirage: The Women's Issue* 3 (Spring 1989): 24.

45. Merton, "Normative Structure of Science," 268.

46. Hilary Putnam, *Meaning and the Moral Sciences* (London: Routledge and Kegan Paul, 1978), 73.

47. Lisa Jarnot, *Robert Duncan, the Ambassador from Venus: A Biography* (Berkeley: University of California Press, 2012), xxi.

CHAPTER 1

1. William Carlos Williams, *Paterson* (New York: New Directions, 1963), 3.

2. George Oppen, "Of Being Numerous," in *New Collected Poems*, ed. Michael Davidson (New York: New Directions, 2008), 163.

3. Ludwig Wittgenstein, *Tractatus Logico-Philosophicus*, trans. D. F. Pears and B. F. McGuinness (London: Routledge and Kegan Paul, 1961), 73.

4. Ronald Bush, "Science, Epistemology, and Literature in Ezra Pound's Objectivist Poetics (with a Glance at the New Physics, Louis Zukofsky, Aristotle, Neural Network Theory, and Sir Philip Sidney)," *Literary Imagination* 4, no. 2 (2002): 191–210, here 208.

5. Alice Fulton, *Feeling as a Foreign Language: The Good Strangeness of Poetry* (Saint Paul, MN: Graywolf Press, 1999), 178.

6. Alice Fulton, *Powers of Congress* (Boston: David R. Godine, 1990), 1.

7. Ronald Johnson, biographical note, in Edward Lucie Smith, ed., *Holding Your Eight Hands: An Anthology of Science Fiction Verse* (New York: Doubleday, 1969), 116–17.

8. Robert Creeley, "Was That a Real Poem or Did You Just Make It Up Yourself?," in *Collected Essays* (Berkeley: University of California Press, 1989), 572. The quotation comes from William Carlos Williams, *Autobiography* (1951; New York: New Directions, 1967), 390–91.

9. Creeley, "Was That a Real Poem or Did You Just Make It Up Yourself?," 575.

10. George Oppen, "The Mind's Own Place," in *Selected Prose, Daybooks, and Papers*, ed. Stephen Cope (Berkeley: University of California Press, 2007), 30.

11. Bernard Williams, *Truth and Truthfulness: An Essay in Genealogy* (Princeton, NJ: Princeton University Press, 2002), 44. In his book, Williams capitalizes these virtues in order to mark his use of them as "terms of art" and explains that they are complexes of activity involving both skills and moral inclinations. "Sincerity basically involves a certain kind of spontaneity, a disposition to come out with what one believes, which may be encouraged or

discouraged, cultivated or depressed, but is not itself expressed in deliberation and choice. Equally, accuracy does involve the will, in the uncontentious and metaphysically unambitious sense of intention, choice, attempts, and concentration of effort" (45).

12. Robert Duncan, "The Truth and Life of Myth," in *Collected Essays and Other Prose*, ed. James Maynard (Berkeley: University of California Press, 2014), 189.

13. Charles Olson, *Collected Prose*, ed. Donald Allen and Benjamin Friedlander (Berkeley: University of California Press, 1997), 155.

14. Charles Olson, "A Letter to the Faculty of Black Mountain College," *Olson: The Journal of the Charles Olson Archives* 8 (Fall 1977): 31. From a note at the end of the draft it appears to have been written on March 21, 1952, and Mary Harris confirms this. Mary Emma Harris, *The Arts at Black Mountain College* (Cambridge, MA: MIT Press, 1987), 172. The editors of *Olson: The Journal of the Charles Olson Archives* append a note to say that there is no evidence as to whether or not the letter was ever circulated.

15. Denise Levertov, "Some Notes on Organic Form," in *New and Selected Essays* (New York: New Directions, 1992), 67.

16. Lyn Hejinian, *The Language of Inquiry* (Berkeley: University of California Press, 2000), 3.

17. Charles Bernstein, "Artifice of Absorption," in *A Poetics* (Cambridge, MA: Harvard University Press, 1992), 18.

18. Ron Silliman, *In the American Tree* (Orono, ME: National Poetry Foundation, 1986), xix.

19. Bryan Walpert, *Resistance to Science in Contemporary American Poetry* (New York: Routledge, 2011), 147.

20. Ibid., 30. Walpert exempts his fourth poet, Joan Retallack, from these strictures.

21. Ibid., 82.

22. Joan Retallack, *The Poethical Wager* (Berkeley: University of California Press, 2003), 143; Joan Retallack, "AID/I/SAPPEARANCE," in *How to Do Things with Words* (Los Angeles: Sun and Moon, 1998).

23. Alison Hawthorne Deming, *Science and Other Poems* (Baton Rouge: Louisiana State University Press, 1994), 58–59.

24. Readers may recognize the allusion to John McDowell's discussion of the fear that if we jettison the belief in direct perceptual knowledge of the world, or the "myth of the given," we raise "the spectre of a frictionless spinning in the void for this region of thought." He repeats this image throughout his study of the irreducibly conceptual character of experience. John McDowell, *Mind and World* (Cambridge, MA: Harvard University Press, 1996), 18.

25. John Guillory, "The Sokal Affair and the History of Criticism," *Critical Inquiry* 28, no. 2 (Winter 2002): 470–508, here 498.

26. Ibid., 506, 498, 485, 486. In referring to the constructedness of quarks, he is alluding to a leading figure in science studies, Andrew Pickering, and his controversial history of the development of the standard model of subatomic physics in *Constructing Quarks* (Edinburgh: Edinburgh University Press, 1986).

27. Paisley Livingston, *Literary Knowledge: Humanistic Inquiry and the Philosophy of Science* (Ithaca, NY: Cornell University Press, 1988), 119, 132.

28. Ibid., 119.

29. John Dupré, *The Disorder of Things: Metaphysical Foundations of the Disunity of Science* (Cambridge, MA, 1993), 10.

30. Peter Burke, *A Social History of Knowledge: From Gutenberg to Diderot* (Cambridge: Polity Press, 2000). Burke identifies a crisis in what constituted advanced knowledge from around 1900 onward: "The subversion of what had been orthodoxy was clearest in physics. Whether or not it was correctly understood, Einstein's famous General Theory of Relativity (1915) encouraged relativism, while Heisenberg's uncertainty principle, formulated (in 1927) with respect to quantum mechanics, undermined certainty more generally." Peter Burke, *A Social History of Knowledge II: From the "Encyclopédie" to Wikipedia* (Cambridge: Polity Press, 2012), 261.

31. Matthew Arnold, *Essays in Criticism, Second Series* (London: Macmillan, 1896), 1.

32. I. A. Richards, *Science and Poetry* (London: Kegan Paul, Trench, Trubner, 1926), 7.

33. John Crowe Ransom, *The New Criticism* (Norfolk, CT: New Directions, 1941), 3.

34. C. K. Ogden and I. A. Richards, *The Meaning of Meaning: A Study of the Influence of Language upon Thought and of the Science of Symbolism* (London: Kegan Paul, Trench, Trubner, 1923), 326.

35. Ibid., 123.

36. Ibid., 111.

37. I. A. Richards, *Principles of Literary Criticism*, 2nd ed. (London: Kegan Paul, Trench, Trubner, 1926), 124, 118.

38. Ibid., 171–72.

39. Ibid., 138.

40. Ibid., 251.

41. Ibid., 266.

42. Richards, *Science and Poetry*, 54, 47.

43. Ibid., 54.

44. Ibid., 6.

45. Ibid., 32, 60. *Science and Poetry* went through several editions. The 1935 edition reads slightly more cautiously. Now it is "most poetry" that is opposed to science, and Richards uses scare quotes for both uses of "truth." I. A. Richards, *Science and Poetry*, 2nd ed. (London: Kegan Paul, Trench, Trubner, 1935), 29. All further citations from *Science and Poetry* are taken from the first edition.

46. Richards, *Science and Poetry*, 56.

47. Ibid., 51.

48. Moritz Schlick, "Meaning and Verification," cited in Oswald Hanfling, *Logical Positivism* (Oxford: Basil Blackwell, 1981), 20.

49. Richards, *Science and Poetry*, 57.

50. Rudolf Carnap, "The Elimination of Metaphysics through Logical Analysis" ("Überwindung der Metaphysik durch Logische Analyse der Sprache," in *Erkenntnis*, vol. 2 [1932]), reprinted in A. J. Ayer, ed., *Logical Positivism* (New York: Free Press, 1959), 62.

51. Richards, *Science and Poetry*, 25.

52. I. A. Richards, preface to *Poetries and Sciences* (New York: W. W. Norton, 1970), 7.

53. Ibid., 57.

54. Ibid., 56.

55. Wittgenstein, *Tractatus Logico-Philosophicus*, 65; Ogden and Richards, *The Meaning of Meaning*, 89.

56. In one of the most thorough recent discussions of the theoretical issues involved in asking whether poetry offers knowledge, Raymond Geuss offers a sophisticated modern version of Richards's arguments about science and poetry. Geuss concludes that it makes no sense to ask the general question of whether poetry can offer knowledge. He does, however, concede in his concluding remarks that poetry can be "implicated" in existing knowledges, which presumably could include scientific knowledge: "The modern world is deeply immersed in and devoted to the acquisition, testing, transmission and application of bodies of propositional beliefs in a highly self-conscious way, and it would be extremely strange if modern poetry, too, were not implicated in various forms of propositional knowledge. Some poetry may contain straightforward assertions or knowledge claims, but not all poetry does, and even in poems that do contain such claims, the cognitive aspects of the claims may not be the most important thing about them." Raymond Geuss, "Poetry and Knowledge," *Outside Ethics* (Princeton, NJ: Princeton University Press, 2005), 205.

57. Roman Jakobson, "Closing Statements: Linguistics and Poetics," in *Style in Language*, ed. Thomas A. Sebeok (Cambridge, MA: MIT Press, 1960), 350.

58. Tillotama Rajan, *Deconstruction and the Remainders of Phenomenology* (Stanford, CA: Stanford University Press, 2002), 25. She goes on to argue that the rise of structuralism in departments of English was made easy because literary criticism "was anxious for scientific and technobureaucratic legitimation" (25).

59. Stephen Greenblatt, *Shakespearean Negotiations* (Berkeley: University of California Press, 1988), 20.

60. Fredric Jameson, *Postmodernism; or, The Cultural Logic of Late Capitalism* (Durham, NC: Duke University Press, 1990), 196.

61. Stephen Greenblatt, "The Touch of the Real," *Representations* 59 (1997): 14–29, here 14.

62. Peter Galison, "Ten Problems in History and Philosophy of Science," *Isis* 99, no. 1 (March 2008): 111–24, here 123.

63. "Thoughts exist as it were in the dimension of meaning and require a background of available meanings in order to be the thoughts that they are." Charles Taylor, *Philosophical Arguments* (Cambridge, MA: Harvard University Press, 1995), 131.

64. The phrase "poetic epistemologies" provides the title of Megan Simpson's valuable study of "language-oriented women writers who conceive of writing as a process of knowing that constantly interrogates its own methods and processes." Her approach concentrates on the consonance between their poetics and ideas from Continental philosophy, notably deconstruction and phenomenology, that have been central to recent literary theory, especially feminist theory. Simpson shows that these poets have made a substantial contribution to these theoretical debates. Megan Simpson, *Poetic Epistemologies: Gender and Knowing in Women's Language-Oriented Writing* (Albany, NY: State University of New York Press, 2000), 121.

65. Cleanth Brooks and Robert Penn Warren, *Understanding Poetry*, 3rd ed. (New York: Holt, Rinehart and Winston, 1960), xiii. By the time they produced the fourth edition in 1976,

they had apparently lost confidence entirely in this formulation, which is not mentioned. The early reflections on the limits of scientific discourse are repeated, but the general emphasis is much more on pedagogy than before.

66. Ibid., 8.

67. Ibid., 9.

68. John Crowe Ransom, "Wanted: An Ontological Critic," in *Beating the Bushes: Selected Essays, 1941-1970* (New York: Norton, 1972), 3.

69. Stephen Wilson, *Information Arts: Intersections of Art, Science, and Technology* (Cambridge, MA: MIT Press, 2002), 3.

70. Philip Kitcher, "The Naturalists Return," *Philosophical Review* 101, no. 1 (January 1992): 53-114, here 56.

71. Charles Taylor, "Overcoming Epistemology," in *Philosophical Arguments* (Cambridge, MA: Harvard University Press, 1995), 8, 14.

72. Lisa M. Steinman, *Made in America: Science, Technology, and American Modernist Poets* (New Haven, CT: Yale University Press, 1987), 2, 3.

73. Ibid., 74, 166, 134.

74. Larry McCaffery, "Matches in a Dark Space: An Interview with David Antin," in *Some Other Frequency: Interviews with Innovative American Authors* (Philadelphia: University of Pennsylvania Press, 1996), 50.

75. Joel E. Cohen, "Mathematics as Metaphor," *Science*, n.s., 172, no. 3984 (May 17, 1971): 674-75.

76. William L. Laurence, "Structure of Life," *New York Times*, January 14, 1962, sec. 4, 7. Cited in Lily E. Kay, *Who Wrote the Book of Life? A History of the Genetic Code* (Stanford, CA: Stanford University Press, 2000), 275.

77. Robert Rosen, *Dynamical System Theory in Biology*, vol. 1 (New York: Wiley, 1970), 247.

78. Cohen, "Mathematics as Metaphor," 674.

79. Philip Mirowski, *More Heat than Light: Economics as Social Physics, Physics as Nature's Economics* (Cambridge: Cambridge University Press, 1989), 357.

80. Ibid. He cites Norbert Wiener, *God and Golem, Inc.* (Cambridge, MA: MIT Press, 1964), 89.

81. Mirowski, *More Heat than Light*, 400.

82. W. H. Auden, *The Dyer's Hand, and Other Essays* (New York: Random House, 1962), 81.

83. Sianne Ngai, *Ugly Feelings* (Cambridge, MA: Harvard University Press, 2005), 128.

84. Although the phrase originates earlier in the twentieth century, it remained current throughout the postwar period covered by this study. In 1971 President Nixon said that the basic aim of government energy policy should be "the blessings of both a high-energy civilization and a beautiful and healthy environment." Cited in Milton Katz, "Decision-Making in the Production of Power," *Scientific American* 225, no. 3 (September 1971): 191-96, here 191. See Richard Nixon, "Special Message to the Congress on Energy Resources" (June 4, 1971), *Public Papers of the Presidents of the United States: Richard Nixon* (Washington, DC: General Services Administration, 1972), 714, http://www.presidency.ucsb.edu/ws/?pid=3038.

85. Ngai, *Ugly Feelings*, 130.

86. John Rawls, *A Theory of Justice* (Oxford: Oxford University Press, 1973), 530-41.

87. Muriel Rukeyser, *Willard Gibbs* (Garden City, NY: Doubleday, Doran, 1942), 81.

88. Muriel Rukeyser, *The Life of Poetry* (1949; Ashfield, MA: Paris Press, 1996), 163.

89. George Levine, "Why Science Isn't Literature: The Importance of Differences," in *Realism, Ethics and Secularism: Essays on Victorian Literature and Science* (Cambridge: Cambridge University Press, 2008), 181.

90. For a fuller account of the history of interrelations between science and poetry and an extended bibliography of studies of science and poetry, see my entry "Science and Poetry" in Roland Greene et al., eds., *Princeton Encyclopedia of Poetry and Poetics*, 4th ed. (Princeton, NJ: Princeton University Press, 2012), 1264–73.

91. Edward Brunner, *Cold War Poetry* (Urbana: University of Illinois Press, 2001), 213. Gordon's poems can be found in Don Gordon, *Displaced Persons* (Denver: Alan Swallow, 1958).

92. Kurt Brown, *Verse and Universe: Poems about Science and Mathematics* (Minneapolis, MN: Milkweed Editions, 1998), xiii. Other anthologies include Helen Plotz, ed., *Imagination's Other Place: Poems of Science and Mathematics* (New York: Thomas Y. Crowell, 1955); Lucie-Smith, *Holding Your Eight Hands*; John Fairfax, ed., *Frontier of Going: An Anthology of Space Poetry* (London: Panther, 1969); John Heath-Stubbs and Phillips Salman, eds., *Poems of Science* (Harmondsworth, Middx.: Penguin, 1984); Bonnie Bilyeu Gordon, ed., *Songs from Unsung Worlds: Science in Poetry* (Boston: American Association for the Advancement of Science and Birkhäuser, 1985); Lavinia Greenlaw, ed., *Signs and Humours: The Poetry of Medicine* (London: Calouste Gulbenkian Foundation, 2007); Maurice Riordan and Jocelyn Bell Burnell, eds., *Dark Matter: Poems of Space* (London: Calouste Gulbenkian Foundation, 2008). Several of these anthologies also include Anglophone poetry from earlier centuries, and in a few of them there is a notable blurring of the border between science and science fiction.

93. William Bronk, *Life Supports: New and Collected Poems* (San Francisco: North Point Press, 1981), 64.

94. Thomas Nagel, *The View from Nowhere* (Oxford: Oxford University Press, 1986), 61.

95. Archibald MacLeish, *Collected Poems, 1917–1952* (Boston: Houghton Mifflin, 1952), 232. The collection won the 1953 National Book Award and was therefore visible to other poets in the 1950s.

96. Edward Dorn, *Gunslinger 1 & 2* (London: Fulcrum, 1969), 76–77.

97. See the discussions in Brian McHale, *The Obligation toward the Difficult Whole: Postmodernist Long Poems* (Tuscaloosa: University of Alabama Press, 2004); and in Robert Von Hallberg, "This Marvellous Accidentalism," in *Internal Resistances: The Poetry of Edward Dorn*, ed. Donald Wesling (Berkeley: University of California Press, 1985).

98. Christian Bök, *'Pataphysics: The Poetics of an Imaginary Science* (Evanston, IL: Northwestern University Press, 2002), 5.

99. Alfred Jarry, "Elements of 'Pataphysics," *Evergreen Review* 4, no. 13 (May–June 1960): 131–32, here 131.

100. Roger Shattuck, "Superliminal Note," *Evergreen Review* 4, no. 13 (May–June 1960): 24–33, here 30.

101. Louis Zukofsky, "A-12," in *"A"* (Berkeley: University of California Press, 1978), 186. The line has no obvious elucidating connection to its context.

102. Mark Scroggins, *The Poem of a Life: A Biography of Louis Zukofsky* (New York: Shoemaker Hoard, 2007), 45.

103. Mark Scroggins states that Zukofsky found the work drudgery and disliked the "rambling, directionless prose" to the extent that he asked not to have his name given as translator. Ibid., 75–76.

104. Anton Reiser, *Albert Einstein: A Biographical Portrait* (London: Thornton Butterworth, 1931).

105. Ibid., 107–8.

106. Louis Zukofsky to Ezra Pound, June 18, 1930, in *Pound/Zukofsky: Selected Letters of Ezra Pound and Louis Zukofsky*, ed. Barry Ahearn (New York: New Directions, 1987), 35.

107. Reiser, *Albert Einstein*, 117.

108. Ibid., 197.

109. Zukofsky, *"A,"* 23.

110. Reiser, *Albert Einstein*, 18, 37, 79, 127.

111. Louis Zukofsky, "Sincerity and Objectification: With Special Reference to the Work of Charles Reznikoff," *Poetry: A Magazine of Verse* (February 1931): 272–85, here 273–74.

112. Louis Zukofsky, *Complete Short Poetry* (Baltimore, MD: Johns Hopkins University Press, 1991), 82.

113. Daniel Tiffany, *Toy Medium: Materialism and Modern Lyric* (Berkeley: University of California Press, 2000), 268, 273.

114. Wallace Stevens, "A Collect of Philosophy," in *Collected Poetry and Prose* (New York: Library of America, 1997), 861.

115. Ibid.

116. Bas C. van Fraassen, *The Scientific Image* (Oxford: Clarendon Press, 1980), 10.

117. Seldon Rodman ed., *One Hundred Modern Poems* (New York: New American Library, 1949), viii, xxv. Citing Robert Graves, *The White Goddess* (New York: Farrar, Straus and Giroux, 2013), 10.

118. Rodman, *One Hundred Modern Poems*, x. The Rimbaud quotation is a translation of "Les inventions d'inconnu réclament des formes nouvelles." The passage comes from a letter to Paul Demeny, May 15, 1871, in Arthur Rimbaud, *Oeuvres complètes*, ed. Antoine Adam (Paris: Pléiade, 1972), 254. Some critics argue that what Rimbaud meant was closer to the inventions of a shaman or magician than to the discoveries of a scientist. What is at issue is not the discovery of *the* unknown but the invention of *an* unknown. See, e.g., Ross Chambers, "On Inventing Unknownness: The Poetry of Disenchanted Reenchantment (Leopardi, Baudelaire, Rimbaud, Justice)," *French Forum* 33, nos. 1–2 (2008): 15–36. Chambers's argument takes its lead from Michael Taussig's work. This line of reasoning also has resonances with the aims of 'Pataphysics.

119. Steven Meyer, *Irresistible Dictation: Gertrude Stein and the Correlations of Writing and Science* (Stanford, CA: Stanford University Press, 2001), 21.

120. See Meyer's discussion in chapter 1 of his *Irresistible Dictation*, esp. 4–5, 9.

121. Gertrude Stein, "Composition as Explanation," in *Writings and Lectures, 1911-1925*, ed. Patricia Meyerowitz (London: Peter Owen, 1967), 27.

122. Ezra Pound, "The Serious Artist," in *The Literary Essays of Ezra Pound*, ed. T. S. Eliot (London: Faber and Faber, 1954), 42.

123. Ezra Pound, *The ABC of Reading* (London: George Routledge and Sons, 1934), 1, 71.

124. Bush, "Science, Epistemology, and Literature in Ezra Pound's Objectivist Poetics," 195.

CHAPTER 2

1. Alice Fulton, *Feeling as a Foreign Language: The Good Strangeness of Poetry* (St. Paul, MN: Graywolf Press, 1999), 53.

2. David Ignatow, "Poet to Physicist in His Laboratory," in *Poems, 1934-1969* (Amherst, MA: Wesleyan University Press, 1970), 188. Included in John Heath-Stubbs and Phillips Salman, eds., *Poems of Science* (Harmondsworth, Mddx.: Penguin, 1984), 290.

3. G. S. Rousseau, "Literature and Science: The State of the Field," *Isis* 69, no. 4 (December 1978): 583-91, here 587.

4. Louise Kertesz, *The Poetic Vision of Muriel Rukeyser* (Baton Rouge: Louisiana State University Press, 1980), 191-92.

5. Edwin B. Wilson, review of *Willard Gibbs*, by Muriel Rukeyser, *Science*, n.s., 99, no. 2576 (May 12, 1944): 386-89, here 387.

6. Charles A. Kraus, review of *Willard Gibbs*, by Muriel Rukeyser, *Journal of the American Chemical Society* 65, no. 12 (1943): 2475-76, here 2475.

7. Elbert C. Weaver, review of *Willard Gibbs*, by Muriel Rukeyser, *Journal of American Chemical Education* 20, no. 5 (1943): 259-60, here 259.

8. Charles Olson, *Collected Prose*, ed. Donald Allen and Benjamin Friedlander (Berkeley: University of California Press, 1997), 169. The "H-mu" Olson refers to is the physicist Louis de Broglie's wave equation that gives the wavelength of a particle as h/μ, or the ratio of Planck's constant to the product of the particle's mass and velocity (its momentum). Olson's editors identify Merritt as H. Houston Merritt, a professor of neurology, but they say that they have not been able to identify Theodore Vann. "Stockpile Szilard" is Leo Szilard, who first theorized the nuclear chain reaction and is generally credited as the instigator of the Manhattan Project. Looking back on the use of atom bombs to force Japan to surrender, Szilard wrote in his autobiography that "the atomic bombs at our disposal represent only the first step in this direction, and there is almost no limit to the destructive power which will become available in the course of their future development." Spencer R. Weart and Gertrud Weiss Szilard, eds., *Leo Szilard: His Version of the Facts* (Boston: MIT Press, 1978), 211. Cited in Richard Rhodes, *The Making of the Atomic Bomb* (New York: Simon and Schuster, 1986), 749.

9. Michael Davidson, *Guys Like Us: Citing Masculinity in Cold War Poetics* (Chicago: University of Chicago Press, 2004), 33. The citations come from "Projective Verse," in Olson, *Collected Prose*, 240.

10. Lundberg writes: "it will always be the privilege of the poet to smash the world to bits and then rebuild it nearer to the heart's desire." George A. Lundberg, *Can Science Save Us?* (New York: Longmans, Green, 1947), 102.

11. Olson, *The Collected Poems of Charles Olson* (Berkeley: University of California Press, 1987), 157.

12. M. Gell-Mann, "A Schematic Model of Baryons and Mesons," *Physics Letters* 8, no. 3 (February 1, 1964): 214-15, here 215.

13. Ibid., 214.

14. Andrew Pickering, *Constructing Quarks* (Edinburgh: Edinburgh University Press, 1986), 114.

15. Helge Kragh, *Quantum Generations: A History of Physics in the Twentieth Century* (Princeton, N J: Princeton University Press, 1999), 323.

16. Gell-Mann, "Schematic Model of Baryons and Mesons," 215.

17. Editorial heading, *Physics Today* 2, no. 2 (February 1949): 6.

18. Muriel Rukeyser, "Josiah Willard Gibbs," *Physics Today* 2, no. 2 (February 1949): 6-13 and 27, here 6, 7.

19. The staff of *Fortune* magazine, "The Scientists," *American Scientist* 37, no. 1 (January 1949): 107-18, here 111.

20. Leo Spitzer, "Part VIII: Summaries and Conclusions," *Reviews of Modern Physics* 30, no. 3 (July 1958): 1095-108, here 1102.

21. Martha J. Schwartzmann and Martin D. Turner, "Nomenclature, Nomenclature . . . ," in the Phimsy column, *Physics Today* 22, no. 11 (November 1969): 19.

22. Pearl Faulkner Eddy, "Insects in English Poetry," *Scientific Monthly* 33, no. 1 (July 1931): 53-73; and 33, no. 2 (August 1931): 148-63, here 158.

23. Frederick W. Grover, "Poetry and Astronomy," *Scientific Monthly* 44, no. 6 (June 1937): 519-29, here 519, 529.

24. John W. Hill and James E. Payne, "Scientists Can Talk to the Layman," *Science*, n.s., 117, no. 3042 (April 17, 1953): 403-5, here 405.

25. Eric Larrabee, "Science, Poetry, and Politics," *Science*, n.s., 117, no. 3042 (April 17, 1953): 395-99, here 398, 395, 397.

26. June Z. Fullmer, "Contemporary Science and the Poets," *Science*, n.s., 119, no. 3103 (June 18, 1954): 855-59, here 856, 859.

27. John V. Hagopian, Herbert M. Hirsch, Ansel Adams, and J. Z. Fullmer, "Contemporary Science and the Poets Reconsidered," *Science*, n.s., 120, no. 3127 (December 3, 1954): 951-55, here 952.

28. Ibid., 953.

29. Ibid., 952.

30. Ibid., 954.

31. C. N. Hinshelwood, *The Structure of Physical Chemistry* (Oxford: Clarendon Press, 1951), 3; Hagopian et al., "Contemporary Science and the Poets Reconsidered," 955.

32. Richard P. Feynman, *"Surely You're Joking, Mr. Feynman!": Adventures of a Curious Character as Told to Ralph Leighton,* ed. Edward Hutchings (London: Vintage, 1992), 66. Cited in Elizabeth Leane, *Reading Popular Physics: Disciplinary Skirmishes and Textual Strategies* (Aldershot, Hants.: Ashgate, 2007), 413.

33. Oscar Riddle, William Newberry, and Joseph Gallant, "Literature, Science, Manpower Crisis," *Science*, n.s., 125, no. 3259 (June 14, 1957): 1212-14, here 1212.

34. Joseph Gallant, "Literature, Science, and the Manpower Crisis, *Science*, n.s., 125, no. 3252 (April 26, 1957): 787-91, here 788.

35. Ibid.; Helen Plotz, ed., *Imagination's Other Place: Poems of Science and Mathematics* (New York: Thomas Y. Crowell, 1955).

36. Riddle, Newberry, and Gallant, "Literature, Science, Manpower Crisis," 1213.

37. Ann Lauterbach, *The Night Sky: Writings on the Poetics of Experience* (New York: Viking, 2005), 71.

38. Leane, *Reading Popular Physics*, 31.

39. So-called because in the early days of quantum physics almost all the leading nuclear physicists in the world gathered at the fifth Solvay conference in 1927 to discuss the new quantum theory of electrons and photons.

40. Horace Freeland Judson, *The Eighth Day of Creation: Makers of the Revolution in Biology* (London: Jonathan Cape, 1979), 244.

41. Charles Olson, *The Special View of History*, ed. Ann Charters (Berkeley: Oyez, 1970), 42. Ralph Maud suggests that Olson read the Anchor edition of *What Is Life?* published in 1956, though he did not then own a copy.

42. Steven Carter makes the ingenious suggestion that Duncan somehow embeds a poetic simulation of Schrödinger's wave equation in his poetry so that the critic can use it "as a tool for uncoding Duncan's self-referential language." In this and other discussions in his study of science and Black Mountain poets, Carter relies on metaphorical interpretations of quantum mechanics, complementarity, and other concepts, interpretations that derive more from popularizers such as Gary Zukav than the science itself. Steven Carter, *Bearing Across: Studies in Literature and Science* (Lanham, MD: University Press of America, 2002), 27.

43. Robert Duncan, *A Selected Prose*, ed. Robert J. Bertholf (New York: New Directions, 1995), 3.

44. Leah Ceccarelli, *Shaping Science with Rhetoric: The Cases of Dobzhansky, Schrödinger, and Wilson* (Chicago: University of Chicago Press, 2001), 87.

45. John Polkinghorne, *Quantum Theory: A Very Short Introduction* (Oxford: Oxford University Press, 2002), 20.

46. Walter Moore, *Schrödinger: Life and Thought* (Cambridge: Cambridge University Press, 1989), 215.

47. Erwin Schrödinger, *What Is Life? The Physical Aspect of the Living Cell with Mind and Matter and Autobiographical Sketches* (Cambridge: Cambridge University Press, 1992), 3.

48. Ibid., 68.

49. Ibid., 77.

50. This epigraph to chapter 2 comes from Goethe's poem "Vermächtnis" (Legacy): "Being is eternal; for laws there are to conserve the treasures of life on which the Universe draws for beauty." The epigraph to chapter 3, beginning "Und was in schwankender," comes from the words of the Lord in the prologue to *Faust II* (lines 348-49). The epigraph to chapter 4, beginning "Und deines Geistes," comes from the prooemium of "Gott und Welt." They are all encouragements to let the imagination flourish.

51. Schrödinger, *What Is Life?*, 87.

52. Cited in Judson, *Eighth Day of Creation*, 245; Schrödinger, *What Is Life?*, 71.

53. Ceccarelli, *Shaping Science with Rhetoric*, 109.

54. Schrödinger, *What Is Life?*, 79.

55. Duncan, *Selected Prose*, 12.

56. Ibid., 3, 11.

57. Paul Dirac, "The Evolution of the Physicist's Picture of Nature," *Scientific American* 208, no. 5 (May 1963): 45–53, here 47.

58. Duncan, *Selected Prose*, 3, 6.

59. Ibid., 3.

60. Ibid., 97.

61. Donald Allen and Robert Creeley, *The New Writing in the USA* (Harmondsworth, Middx.: Penguin, 1967), 24.

62. Robert Creeley, *The Collected Poems 1945–1975* (Berkeley: University of California Press, 1982), 379.

63. Although Heisenberg published widely, only two of his books were widely read in this period: Werner Heisenberg, *Philosophic Problems of Nuclear Science*, trans. F. C. Hayes (London: Faber and Faber, 1952), published in the same year in the United States by Pantheon; and Werner Heisenberg, *Physics and Philosophy: The Revolution in Modern Science* (New York: Harper, 1958).

64. In a sense this is an unfair comparison, since Schrödinger's book was one of the most influential science books of the twentieth century. Heisenberg did influence some important thinkers. Thomas Kuhn's thinking during the development of his ideas for *The Structure of Scientific Revolutions* was clearly shaped by reading *Philosophic Problems of Nuclear Science*. Kuhn's interest in the character of scientific revolutions, as well as his conviction that normal science comprises "puzzles," has antecedents in Heisenberg's constant comparison between classical, Newtonian, and quantum physics and his argument that each of these phases of science centers on a choice of "questions."

65. It seems likely that Heisenberg's invitation was part of the process by which László Bárdossy's new government of what was then still an independent Hungary attempted to strengthen ties with Nazi Germany. Thomas Powers offers a persuasive sympathetic view of Heisenberg's activities during the war, which became the basis for Michael Frayn's play *Copenhagen*—evidence that Heisenberg continues to fascinate writers. Thomas Powers, *Heisenberg's War: The Secret History of the German Bomb* (London: Jonathan Cape, 1993).

66. Heisenberg, *Philosophic Problems*, 67.

67. Ibid., 67, 93.

68. Ibid., 14.

69. Ibid., 15.

70. Robert Creeley, *Collected Essays* (Berkeley: University of California Press, 1989), 502.

71. Robert Creeley, *Tales Out of School: Selected Interviews* (Ann Arbor: University of Michigan Press, 1993), 137. The interview was presumably conducted in 1977, because it was published in *Boundary 2* in 1978.

72. Joel Isaac, *Working Knowledge: Making the Human Sciences from Parsons to Kuhn* (Cambridge, MA: Harvard University Press, 2012), 209.

73. Clyde Kluckhohn, "The Place of Theory in Anthropological Studies," *Philosophy of Science* 6, no. 3 (July 1939): 328–44, 340.

74. Isaac, *Working Knowledge*, 90.

75. Victor F. Weisskopf, "Problems of Nuclear Structure," *Physics Today* 14, no. 7 (July 1961): 18–27, here 20.

76. Michael J. Moravcsik, "High Energy Physics: An Informal Report of the Rochester Conference," *Physics Today* 12, no. 10 (October 1959): 20–25, here 21.

77. Robert E. Marshak, "The Nuclear Force," *Scientific American* 202 (March 1960): 98–114, here 107.

78. Ibid., 101.

79. I. I. Rabi, "Science and the Satisfaction of Human Aspirations," *Proceedings of the National Academy of Sciences of the United States of America* 50, no. 6 (December 15, 1963): 1218–22, here 1219.

80. Snow is reported to have said in a conversation with his son: "That man [Rabi], sitting on the sofa there, is the man who gave me the idea for the two cultures." John S. Rigden, *Rabi: Scientist and Citizen* (New York: Basic Books, 1987), 258.

81. Cited in ibid., 257.

82. R. R. Wilson, "My Fight against Team Research," *Daedalus* 99, no. 4 (1970): 1076–87, here 1086.

83. Adolph Baker, *Modern Physics and Antiphysics* (Reading, MA: Addison-Wesley, 1970), 72.

84. Abraham Pais and Karl Darrow, respectively, cited in Ray Monk, *Inside the Centre: The Life of J. Robert Oppenheimer* (London: Jonathan Cape, 2012), 503.

85. George B. Kistiakowsky, "Introduction," *Proceedings of the National Academy of Sciences of the United States of America* 50, no. 6 (December 1963): 1193–94, here 1193.

86. J. Robert Oppenheimer, "Communication and Comprehension of Scientific Knowledge," *Proceedings of the National Academy of Sciences of the United States of America* 50, no. 6 (December 1963): 1194–1200, here 1195.

87. Ibid.

88. Ibid., 1196.

89. Ibid.

90. Ibid., 1198 ("external relations"); 1199.

91. Ibid., 1197.

92. Ibid., 1196

93. Ibid., 1200.

94. Ibid.

CHAPTER 3

1. Charles Olson, "The Chiasma; or, Lectures in the New Sciences of Man," *Olson: The Journal of the Charles Olson Archives* 10 (Fall 1978): 20–21, 65.

2. Olson echoes Nietzsche's title *Die fröhliche Wissenschaft*, meaning the joyful science or, as it has become known, *The Gay Science*, which might suggest a certain caution behind Olson's joy. Nietzsche argues that even the supposedly rigorous, antimetaphysical science of his day rests on faith and the will to truth. Olson doesn't pursue such radical doubt, however, leaving just a faint hint that his joy also derives from his belief in the possibility of a productive disenchantment with the certainties of modern knowledge. Emerson may also be on Olson's mind. As Robert Pippin observes, Ralph Waldo Emerson called himself

a "Professor of Joyous Science" in his journals. Robert B. Pippin, *Nietzsche, Psychology, and First Philosophy* (Chicago: University of Chicago Press, 2010), 34.

3. Catherine Seelye, ed., *Charles Olson and Ezra Pound: An Encounter at St. Elizabeths* (New York: Paragon House, 1975), 29.

4. Ibid. Olson's comment is very much aimed at the Pound of the 1940s. The younger Pound wrote about the importance of acknowledging scientific developments in a manner similar to Olson's statements in the 1950s: "we continue with thought forms and with language structures used by monolinear medieval logic, when the aptitudes of human mind developed in course of bio-chemical studies have long since outrun such simple devices." "Simplicities," *The Exile* I, 4 (1928), 3–4. Cited in Ian F. A. Bell, *Critic as Scientist: The Modernist Poetics of Ezra Pound* (London: Methuen, 1981), 210.

5. Hermann Weyl, *Space-Time-Matter*, trans. Henry L. Brose (London: Methuen, 1922), 187.

6. Olson, "The Heart is a clock," in *The Collected Poems of Charles Olson: Excluding the "Maximus" Poems*, ed. George F. Butterick (Berkeley: University of California Press, 1987), 641.

7. Donald Allen and Robert Creeley, *The New Writing in the USA* (Harmondsworth, Middx.: Penguin, 1967), 20.

8. Robert Duncan, "As an Introduction: Charles Olson's *Additional Prose*," in *A Selected Prose*, ed. Robert J. Bertholf (New York: New Directions, 1995), 151.

9. Robin Blaser, *The Fire: Collected Essays*, ed. Miriam Nichols (Berkeley: University of California Press, 2006), 204.

10. Wordsworth wrote: "The remotest discoveries of the Chemist, the Botanist, or Mineralogist, will be as proper objects of the Poet's art as any upon which it can be employed, if the time should ever come when these things shall be familiar to us, and the relations under which they are contemplated by the followers of these respective Sciences shall be manifestly and palpably material to us as enjoying and suffering beings. If the time should ever come when what is now called Science, thus familiarized to men, shall be ready to put on, as it were, a form of flesh and blood, the Poet will lend his divine spirit to aid the transfiguration, and will welcome the Being thus produced, as a dear and genuine inmate of the household of man." William Wordsworth and Samuel Taylor Coleridge, *Lyrical Ballads: 1798 and 1802*, ed. Fiona Stafford (Oxford: Oxford University Press, 2013), 107.

11. Blaser, *Fire*, 198.

12. George F. Butterick, *A Guide to the "Maximus" Poems of Charles Olson* (Berkeley: University of California Press, 1978), xl.

13. Ralph Maud, *Charles Olson's Reading: A Biography* (Carbondale: Southern Illinois University Press, 1996), 3.

14. Paul Christensen, *Charles Olson: Call Him Ishmael* (Austin: University of Texas Press, 1979), 7.

15. Sherman Paul, *Olson's Push: Origin, Black Mountain and Recent American Poetry* (Baton Rouge: Louisana State University Press, 1978), 114.

16. Robert von Hallberg, *Charles Olson: The Scholar's Art* (Cambridge, MA: Harvard University Press, 1978). Von Hallberg's book is chronologically located with Paul and Christensen and before Don Byrd's *Charles Olson's Maximus* (Urbana: University of Illinois Press,

1980) but is infused with a new and growing spirit of resistance to uncritical acceptance of the entirety of Olson's work.

17. Von Hallberg, *Charles Olson*, 211.

18. Andrew Ross, *The Failure of Modernism: Symptoms of American Poetry* (New York: Columbia University Press, 1986).

19. Tom Clark, *Charles Olson: The Allegory of a Poet's Life* (Berkeley: North Atlantic Books, 2000).

20. Susan Howe, "Since a Dialogue We Are," *Acts* 10 (1989): 166–73.

21. Charles Bernstein, "Undone Business," in *Content's Dream: Essays, 1975–1984* (Los Angeles: Sun and Moon Press, 1986), 328.

22. Howe, "Since a Dialogue We Are," 169.

23. Bernstein, "Undone Business," 328.

24. Libbie Rifkin, *Career Moves: Olson, Creeley, Zukofsky, Berrigan, and the American Avant-Garde* (Madison: University of Wisconsin Press, 2000), 65.

25. Heriberto Yepez, *The Empire of Neomemory*, trans. Jen Hofer, Christian Nagler, and Brian Whitener (Oakland, CA: Chainlinks, 2013), 233. There has been fierce, productive debate about this critique of Olson. See, e.g., "Il Gruppo responds to Rothenberg and Yepez," *Jacket2* (2013), accessed February 18, 2015, http://jacket2.org/commentary/il-gruppo -responds-jerome-rothenberg-and-heriberto-y%C3%A9pez.

26. See, e.g., David Herd, ed., *Contemporary Olson* (Manchester: Manchester University Press, 2015).

27. "For my part I do, qua lay physicist, believe in physical objects and not in Homer's gods; and I consider it a scientific error to believe otherwise." Willard Van Orman Quine, "Two Dogmas of Empiricism," in *From a Logical Point of View* (Cambridge, MA: Harvard University Press, 1961), 44. The essay began as a lecture in 1950 and was published the following year in *Philosophical Review* 60, no. 1 (January 1951): 20–43.

28. Thomas F. Merrill, *The Poetry of Charles Olson: A Primer* (East Brunswick, NJ: Associated University Presses, 1982), 46.

29. Michael André Bernstein, *The Tale of the Tribe: Ezra Pound and the Modern Verse Epic* (Princeton, NJ: Princeton University Press, 1980), 242.

30. Von Hallberg, *Charles Olson*, 86.

31. Alfred North Whitehead, *Process and Reality: An Essay in Cosmology* (Cambridge: Cambridge University Press, 1929), 314.

32. Charles Olson, *Collected Prose*, ed. Donald Allen and Benjamin Friedlander (Berkeley: University of California Press, 1997), 206.

33. Olson, "The Chiasma," 18.

34. Rolf Fjelde, "Mid-century American Poets," review of *Mid-century American Poets*, ed. John Ciardi, *Poetry New York* 3 (1950): 38–40, here 38. The poem advertising familiarity with test tubes was Richard Eberhart's "The Cancer Cells," in *Mid-century American Poets*, ed. John Ciardi (New York: Twayne, 1950), 238–39.

35. Wallace Stevens, *Collected Poetry and Prose* (New York: Library of America, 1997), 728.

36. Richard Eberhart, "Notes on Poetry," in Ciardi, *Mid-century American Poets*, 227.

37. Karl Shapiro, "Elegy for a Dead Soldier," in Ciardi, *Mid-century American Poets*, 96. Shapiro may be remembering the opening of Ezra Pound's *ABC of Reading*, where Pound announces, "We live in an age of science and of abundance." Ezra Pound, *ABC of Reading* (London: George Routledge and Sons, 1934), 1.

38. The other editors were Keith Botsford, Roger Shattuck (whose own interest in science and poetry led him to edit the 'Pataphysics issue of *Evergreen Review*), and Harvey Shapiro. All of them were recent Yale graduates.

39. Charles Olson, "Projective Verse," in *Collected Prose*, 240.

40. In one of the relatively rare instances of a poet revising a poem to take account of the accuracy with which it reports on a scientific issue, Williams altered this passage to read: "Release the Gamma rays that cure the cancer / . the cancer, usury." William Carlos Williams, *Paterson* (New York: New Directions, 1963), 182.

41. Daniel J. Kevles, *The Physicists: The History of a Scientific Community in Modern America* (Cambridge, MA: Harvard University Press, 1995), 31; Muriel Rukeyser, *Willard Gibbs* (Garden City, NY: Doubleday, Doran, 1942), 11; Muriel Rukeyser, *The Life of Poetry* (1949; Ashfield, MA: Paris Press, 1996), 163.

42. Rukeyser, *Life of Poetry*, 163. Rukeyser's argument is extremely compressed at this point: "When Baudelaire said the imagination is 'the most scientific of the faculties, because it alone understands the universal analogy,' he set the trap and sprung it in one phrase. The trap is the use of the discoveries of science instead of the methods of science. Woodrow Wilson was caught there, when he called for a Darwinian, rather than Newtonian, system of government." Her example only serves to obscure her important point that merely advertising new scientific discoveries does not make poetry, or any form of intellectual inquiry, scientific. Baudelaire, she means to say, is wrong to suggest that what makes the imagination scientific is the grasp of the "universal analogy." It is the imaginative reuse of scientific methods in other contexts that makes the imagination scientific. The jump from Baudelaire's defense of the imagination to the president's use of analogy does not bring clarity. One perception follows another too quickly.

43. William Carlos Williams, "The Poem as a Field of Action," in *Selected Essays* (New York: New Directions, 1954), 282.

44. Karl Jay Shapiro, *Essay on Rime* (New York: Reynal and Hitchcock, 1945), 24.

45. Fred A. Dudley, "The Impact of Science on Literature," *Science* 115, no. 2990 (1952): 412–15, here 414.

46. Richard Wilbur, "The Genie in the Bottle," in Ciardi, *Mid-century American Poets*, 6.

47. Randall Jarrell, "Answers to Questions," in Ciardi, *Mid-century American Poets*, 182.

48. Randall Jarrell, "Robert Lowell's Poetry," in Ciardi, *Mid-century American Poets*, 160. Jarrell's essay is a reprint of a review of *Lord Weary's Castle* that first appeared in the *Nation*. In an editor's note introducing the essay, Ciardi reports that Lowell wrote declining to offer a statement of poetics, saying, "I'm just getting back to writing and feel pretty numb about theorizing on old poems" (158).

49. Cited in Ian Hamilton, *Robert Lowell: A Biography* (London: Faber and Faber, 1983), 85.

50. R. S. Crane, "Cleanth Brooks; or, The Bankruptcy of Critical Monism," *Modern Philology*

45, no. 4 (1948): 226–45, here 243. Crane mentions I. A. Richards, Cleanth Brooks, and John Crowe Ransom as key participants in the defense of poetry.

51. These were not the only ones. Other studies of science and poetry of that period include Frederick William Conner, *Cosmic Evolution: A Study of the Interpretation of Evolution by American Poets from Emerson to Robinson* (Gainesville: University of Florida Press, 1949); Norman Holmes Pearson, "The American Poet in Relation to Science," *American Quarterly* 1, no. 2 (1949): 116–26; and Oscar Cargill, "Science and the Literary Imagination in the United States," *College English* 13, no. 2 (1951): 90–94.

52. Hyatt Howe Waggoner, *The Heel of Elohim: Science and Values in Modern American Poetry* (Norman: University of Oklahoma Press, 1950), 208, 16.

53. Douglas Bush, *Science and English Poetry: A Historical Sketch, 1590-1950* (Oxford: Oxford University Press, 1950), 151. Cleanth Brooks, *Modern Poetry and the Tradition* (Chapel Hill: University of North Carolina Press, 1939), 174.

54. Douglas Bush, *Science and English Poetry: A Historical Sketch, 1590-1950* (repr., Oxford: Oxford University Press, 1967), v.

55. Donald Allen, ed., *The New American Poetry, 1945-1960* (Berkeley: University of California Press, 1999), 417.

56. "U.S. Science Holds Its Biggest Powwow," *Life* 28, no. 2 (January 9, 1950): 17.

57. Lincoln Barnett, "The Meaning of Einstein's New Theory," *Life* 28, no. 2 (January 9, 1950): 22.

58. Bush hits this note hard: "Science offers a largely unexplored hinterland for the pioneer who has the tools for his task." For those wanting to find new frontiers, the sciences were the place to be. Bush argues that the rewards of such exploration for both the nation and the individual are great and that new knowledge that "can be obtained only through basic scientific research" is essential to the health, wealth, and security of the nation. Vannevar Bush, *Science, the Endless Frontier: A Report to the President by Vannevar Bush, Director of the Office of Scientific Research and Development, July 1945*, accessed July 1, 2011, http://www.nsf.gov/od/lpa/nsf50/vbush1945.htm.

59. Barnett, "Meaning of Einstein's New Theory," 22.

60. Charles Olson to Frances Boldereff, January 10, 1950, in Ralph Maud and Sharon Thesen, eds., *Charles Olson and Frances Boldereff: A Modern Correspondence* (Hanover, NH: Wesleyan University Press, 1999), 112.

61. Butterick, ed., *Collected Poems of Charles Olson*, 102.

62. Barnett, "Meaning of Einstein's New Theory," 22.

63. Charles Olson, "A Letter to the Faculty of Black Mountain College," *Olson: The Journal of the Charles Olson Archives* 8 (Fall 1977). See chap. 1, n. 10.

64. Charles Olson to Frances Boldereff, February 6, 1950, in Maud and Thesen, *Charles Olson and Frances Boldereff*, 149.

65. Olson, *Collected Prose*, 240.

66. Ibid.

67. Walter Isaacson, *Einstein: His Life and Universe* (New York: Simon and Schuster, 2008), 514.

68. Albert Einstein, "A Generalized Theory of Gravitation," *Reviews of Modern Physics* 20, no. 1 (January 1948): 35-39.

69. William L. Laurence, "New Einstein Theory Gives a Master Key to Universe," *New York Times*, December 27, 1949, 1.

70. James Gleick, *Genius: Richard Feynman and Modern Physics* (London: Little, Brown, 1992), 240.

71. William L. Laurence, "Schwinger States His Cosmic Theory," *New York Times*, June 25, 1948.

72. William L. Laurence, "New Guide Offered on Atom Research," *New York Times*, February 1, 1948.

73. Richard P. Feynman, *QED: The Strange Theory of Light and Matter* (Princeton, NJ: Princeton University Press, 1985), 9.

74. Charles Olson to Cid Corman, October 18, 1950, in Charles Olson, *Letters for Origin*, ed. Albert Glover (London: Cape Goliard, 1969), 5.

75. Ibid., 6.

76. Alan Golding, "Little Magazines and Alternative Canons: The Example of Origin," *American Literary History* 2, no. 4 (1990): 691-725, here 701.

77. Olson, *Collected Prose*, 155.

78. Henry Bugbee, *The Inward Morning: A Philosophical Exploration in Journal Form* (Athens: University of Georgia Press, 1999), 33. This philosophical journal was first published in 1958. Bugbee was one of Stanley Cavell's teachers.

79. F. J. Dyson, "The S Matrix in Quantum Electrodynamics," *Physical Review*, ser. 2, 75, no. 11 (June 1, 1949): 1736-55, here 1754.

80. Richard Feynman, "Space-Time Approach to Quantum Electrodynamics," *Physical Review*, ser. 2, 74, no. 6 (September 15, 1949): 769-89, here 772, 769.

81. J. R. Dunning, "Atomic Structure and Energy," *American Scientist* 37, no. 4 (October 1949): 505-27, here 505.

82. Selig Hecht, *Explaining the Atom* (New York: Viking Press, 1947), 110.

83. Charles Olson to Robert Creeley, April 7, 1951, in George F. Butterick ed., *Charles Olson and Robert Creeley: The Complete Correspondence*, ed. George F. Butterick, vol. 5 (Santa Barbara, CA: Black Sparrow Press, 1983), 129. Olson knew of Hecht through a Worcester friend, Clarence Graham, who was a colleague of Hecht's at Columbia. It seems likely that Olson is referring to Hecht's articles on animal vision rather than his book on the atom. An obituary describes Hecht as "a musician and painter, a devotee of literature and the arts," confirming Olson's appreciation of what the obituary calls Hecht's "elegant and lively style." Brian O'Brien, Harry Grundfest, and Emil Smith, "Selig Hecht: 1892-1947," *Science*, n.s., 107, no. 2770 (1948): 105-6.

84. Kai Bird and Martin J. Sherwin, *American Prometheus: The Triumph and Tragedy of J. Robert Oppenheimer* (London: Atlantic Books, 2008), 430.

85. E. Rabinowich, editorial, "It's Not What's Said It's Who Says It," *Bulletin of the Atomic Scientists* 6, no. 5 (May 1950): 130.

86. Herman Feshbach, Julian Schwinger, and John A. Harr, "Effect of Tensor Range in Nuclear

Two Body Problems" (November 1949), HUX-5 ACC0109, Atomic Energy Commission, 1–57, accessed December 29, 2010, http://www.osti.gov/cgi-bin/rd_accomplishments/display_biblio.cgi?id=ACC0109&numPages=57&fp=N.

87. Peter Galison, "Removing Knowledge," *Critical Inquiry* 31 (2004): 229–43, here 232.

88. Alvarez talks of how "unsuitable for the job" the cloud chamber was. Luis W. Alvarez, "Recent Developments in Particle Physics," *Science*, n.s., 165, no. 3898 (1969): 1071–91, here 1074. In his autobiography he explains how experimental physicists were frustrated by the cloud chamber's limitations for studying the new particles: "We could produce them with the Bevatron and wanted to study them, but none of the existing techniques used to track particles was well suited to studying the basic reactions of strange-particle physics." Alvarez went on to design the bubble chamber and record the discovery of many new particles. Luis W. Alvarez, *Adventures of a Physicist* (New York: Basic Books, 1987), 185.

89. Hans Albrecht Bethe, "Within the Atom," in *The Scientists Speak*, ed. Warren Weaver (New York: Boni and Gaer, 1947), 98.

90. Olson, "The Babe," in Butterick, *Collected Poems of Charles Olson*, 102.

91. Dunning, "Atomic Structure and Energy," 507.

92. Kragh, *Quantum Generations*, 312–313. Quotation from *Time* magazine taken from J. L. Heilbron, Robert W. Seidel, and Bruce R. Wheaton, *Lawrence and His Laboratory: A Historian's View of the Lawrence Years* (Berkeley, CA: Berkeley Lab, 1996), chap. 3, "Machine Made Mesons," accessed December 29, 2010, http://www.lbl.gov/Science-Articles/Research-Review/Magazine/1981/81fepi3.html.

93. Watson Davis, "From Now On: The Atom," *Science News-Letter* 58, no. 7 (August 12, 1950): 98.

94. Julian Schwinger, "Quantum Electrodynamics. I. A Covariant Formulation," *Physical Review*, ser. 2, 74, no. 10 (November 15, 1948): 1439–61, here 1439.

95. Feynman, *QED*, 9.

96. Dyson, "*S* Matrix in Quantum Electrodynamics," 1754.

97. Davis, "From Now On," 98.

98. George Johnson, *Strange Beauty: Murray Gell-Mann and the Revolution in Twentieth-Century Physics* (London: Vintage, 2001), 73; Eugene Feenberg, "Nuclear Shell Models," *Physical Review*, ser. 2, 77, no. 6 (March 15, 1950): 771–76, here 771. Feenberg develops the metaphorical image of the well through metaphors of depth, elevation, shallowness, and depression.

99. Evans Hayward, "The Use of Cloud Chambers with Pulsed Accelerators," *Science* 111 (April 7, 1950): 349–55, here 349.

100. Andrew Pickering, *The Mangle of Practice: Time, Agency, and Science* (Chicago: University of Chicago Press, 1995), 52.

101. Isidor I. Rabi, "The Atomic Nucleus," in Weaver, *Scientists Speak*, 102.

102. Richard C. Tolman, "A Survey of the Sciences," *Science*, n.s., 106, no. 2746 (August 15, 1947): 135–40, here 137.

103. Mary Poovey, *Genres of the Credit Economy: Mediating Value in Eighteenth- and Nineteenth-Century Britain* (Chicago: University of Chicago Press, 2008), 345.

104. The "fertility" of scientific metaphors is discussed by Ernan McMullin, "A Case for Scientific Realism," in *Scientific Realism*, ed. J. Leplin (Berkeley: University of California Press, 1984), 8–40. Reprinted in Yuri Balashov and Alex Rosenberg, eds., *Philosophy of Science: Contemporary Readings* (London: Routledge, 2002), 248–80. Thomas Kuhn discusses the "open-endedness" of metaphors for scientific research in "Metaphor in Science," in *Metaphor and Thought*, ed. Andrew Ortony (Cambridge: Cambridge University Press, 1993), 533–42.

105. Mary Hesse, *Forces and Fields: The Concept of Action at a Distance in the History of Physics* (London: Thomas Nelson and Sons, 1961), 27.

106. Martin H. Krieger, *Doing Physics: How Physicists Take Hold of the World* (Bloomington: University of Indiana Press, 1992), 6, 16.

107. Ibid., 24, 27.

108. Sarah Winter, *Freud and the Institution of Psychoanalytic Knowledge* (Stanford, CA: Stanford University Press, 1999), 223. She writes that Freud's "cultural theory gives evidence of his strategic supposition that the institutional resources attached to an academic discipline will become available to his 'new science' if, in part through the agency and example of his writings, it can win epistemological competitions with other fields." Her argument appears to assume that this is a zero-sum struggle for resources; my assumption is that these resources are expanding and at any one time elastic, since they include cultural, as well as financial, capital.

109. Norbert Wiener, *Cybernetics; or, Control and Communication in the Animal and the Machine* (New York: J. Wiley, 1948), 76.

110. Robert von Hallberg, *American Poetry and Culture, 1945–1980* (Cambridge, MA: Harvard University Press, 1985), 38. He suggests that avant-garde poets "subjected to scrutiny in verse" the cybernetic ideas of feedback and system.

111. Wiener, *Cybernetics*, 16.

112. See n. 26.

113. Wilfrid Sellars, *Empiricism and the Philosophy of Mind* (Cambridge, MA: Harvard University Press, 1997). The essay originated as a series of lectures presented in 1956 as "The Myth of the Given: Three Lectures on Empiricism and the Philosophy of Mind."

114. Quine, "Two Dogmas of Empiricism," 23, 42.

115. Ibid., 42.

116. Ibid., 45.

117. See, e.g., Duncan's essays "Towards an Open Universe" and "Rites of Participation." "Perhaps we recognize as never before in man's history that not only our own personal consciousness but also the inner structure of the universe itself has only this immediate event in which to be realized. Atomic physics has brought us to the threshold of such a—I know not whether to call it certainty or doubt." Robert Duncan, *A Selected Prose*, ed. Robert J. Bertholf (New York: New Directions, 1995), 12.

118. René Wellek and Austin Warren, *Theory of Literature* (New York: Harcourt, Brace, 1949), 298.

CHAPTER 4

1. George A. Lundberg, *Can Science Save Us?* (New York: Longmans, Green, 1947), 104.

2. Ibid., 104, 26.

3. Ibid., 87–88, 83, 89, 82, 80.

4. Charles Olson, *Collected Prose*, ed. Donald Allen and Benjamin Friedlander (Berkeley: University of California Press, 1997), 155.

5. Edward C. Tolman, "Kurt Lewin, 1890–1947," *Psychological Review* 55, no. 1 (1948): 1–4, here 4.

6. Everett M. Rogers, *A History of Communication Study* (New York: Free Press, 1997).

7. Rosmarie Waldrop, "Charles Olson: Process and Relationship," *Twentieth-Century Literature* 23, no. 4 (December 1977): 467–86.

8. Kurt J. Lewin, "Field Theory and Experiment in Social Psychology: Concepts and Methods," *American Journal of Sociology* 44, no. 6 (May 1939): 868–96, here 871.

9. Kurt J. Lewin, "Formalization and Progress in Psychology" (1940), in *Resolving Social Conflicts and Field Theory in Social Science* (Washington, DC: American Psychological Association, 1997), 169.

10. Wilbur Schramm, *The Beginnings of Communication Study in America: A Personal Memoir*, ed. Stephen H. Chaffee and Everett Mitchell Rogers (Thousand Oaks, CA: Sage, 1997), 67.

11. In addition to Murray, Selig Hecht is another possible link between Olson and Henderson, though a more tenuous one. Hecht studied with Lawrence Henderson in the early 1920s and led a reading group that closely analyzed Henderson's biological writing. Further information on Hecht can be found in George Wald, "Selig Hecht (1892–1947)" (1991), National Academy of Sciences, accessed January 13, 2014, http://www.nasonline.org/publications/biographical-memoirs/memoir-pdfs/hecht-selig.pdf.

12. Tom Clark, *Charles Olson: The Allegory of a Poet's Life* (Berkeley, CA: North Atlantic Books, 2000), 44.

13. Clyde Kluckhohn and Henry A. Murray, *Personality in Nature, Society, and Culture* (New York: Alfred A. Knopf, 1948), 23, 65.

14. Lewin, "Field Theory and Experiment in Social Psychology," 884.

15. For more on this history, see the excellent new study by Joel Isaac, *Working Knowledge: Making the Human Sciences from Parsons to Kuhn* (Cambridge, MA: Harvard University Press, 2012).

16. Lewin, "Field Theory and Experiment in Social Psychology," 890.

17. Ibid.

18. Charles Olson, *The Special View of History*, ed. Ann Charters (Berkeley, CA: Oyez, 1970), 27, 47.

19. Olson, *Collected Prose*, 155; Lewin, "Field Theory and Experiment in Social Psychology," 888.

20. For a discussion of why it can be valid to talk of a plurality of knowledges, see Peter Burke, *A Social History of Knowledge: From Gutenberg to Diderot* (Cambridge: Polity Press, 2000), chap. 1.

21. Muriel Rukeyser, *The Life of Poetry* (1949; Ashfield, MA: Paris Press, 1996), 132.

22. James Brock, "The Perils of a 'Poster Girl': Muriel Rukeyser, *Partisan Review*, and *Wake Island*," in *"How Shall We Tell Each Other of the Poet?": The Life and Writing of Muriel Rukeyser*, ed. Anne E. Herzog and Janet E. Kaufman (New York: St. Martin's Press, 1999), 256. See also unsigned editorial, "Grandeur and Misery of a Poster Girl," *Partisan Review* 10 (September–October 1943): 471–73; cited in Louise Kertesz, *The Poetic Vision of Muriel Rukeyser* (Baton Rouge: Louisiana State University Press, 1980), 179–80. According to Brock, although it was signed with the initials R.S.P. to indicate the editors Philip Rahv, William Phillips, and Delmore Schwartz, the editorial was written by Schwartz, who had a homophobic dislike for Rukeyser's poetry and ideas. He and others objected to what they saw as Rukeyser's opportunistic abandonment of her earlier left-wing sympathies to take advantage of the new wartime patriotic spirit, as well as the links she made between the noble resistance of the republicans against fascism in the civil war in Spain with the cynical war in the Pacific.

23. "Poetess in OWI Here Probed by U.S. as Red," *New York Times*, May 7, 1943. Cited in Jeanne Perreault, "Poetess Probed as Red: Muriel Rukeyser and the FBI," in *Modernism on File: Modern Writers, Artists, and the FBI, 1920–1950*, ed. Claire Culleton and Karen Kleick (New York: Palgrave Macmillan, 2008), 154.

24. FBI file, ref. 100-102441.

25. Rukeyser, *Life of Poetry*, 163. From her enthusiastic endorsement of the historian of science George Sarton it is evident that Rukeyser accepts his vision of the "unity of science." She wants to believe that the diverse sciences will converge on one unified science because she thinks that such a unified science would lead to a unification of humanity in a shared knowledge of the cosmos, assuming that scientific knowledge of a mind-independent universe has to be the same in every language and culture.

26. Theodor Adorno, *Aesthetic Theory*, trans. Robert Hullot-Kentor (Minneapolis: University of Minnesota Press, 1997). Simon Jarvis gives this a more intelligibly idiomatic translation: "the possibility that the non-existent could exist." Simon Jarvis, *Adorno: A Critical Introduction* (Cambridge: Polity Press, 1998), 115. Jarvis comments: "What makes a work of art more than the empirical world from which it distinguishes itself is something non-existent: 'that works of art are there, however, points to the possibility that the non-existent could exist.'" This is often taken to imply that art keeps alive hopes of change by its very contradictory nature, but Adorno's typically paradoxical formulation also relies on a key feature of the sorts of inquiry that nuclear physics represented for many intellectuals: its dramatic engagement in prospective speculation that could lead to the emergence of actualities from theories.

27. Rukeyser, *Life of Poetry*, 132. Louise Kertesz describes Rukeyser's attitude as "American 'meliorism.'" Kertesz, *Poetic Vision of Muriel Rukeyser*, 366.

28. Rukeyser, *Life of Poetry*, 17.

29. Ibid., 162, 163, 176. For interesting examples of poetry that does work with mostly very well understood discoveries of science, see the poetry in Kurt Brown's excellent anthology. As the editor explains in his introduction, "different fields of knowledge are explored for themes, images, metaphors, and language that might be used in the making of new and

unique poems" and "reveal for the rest of us unexpected meanings in the work of scientists and mathematicians." On the whole, epistemological questions about the role of the self, language, and methodology are set aside. Kurt Brown, *Verse and Universe: Poems about Science and Mathematics* (Minneapolis: Milkweed Editions, 1998), xiii.

30. Rukeyser, *Life of Poetry*, 13, 15, 163.

31. Ibid., 172.

32. Ibid., 173.

33. Olson continues: "Okay. Then the poem itself must, at all points, be a high energy-construct and, at all points, an energy-discharge. So: how is the poet to accomplish same energy, how is he, what is the process by which a poet gets in, at all points energy at least the equivalent of the energy which propelled him in the first place, yet an energy which is peculiar to verse alone and which will be, obviously, also different from the energy which the reader, because he is a third term, will take away?" (*Collected Prose*, 240).

34. Ibid., 240. Olson may be remembering Rukeyser's description of the reader relation as a "triadic relation," a formulation that draws on a discussion of C. S. Peirce's theory of the triadic sign; see Rukeyser, *Life of Poetry*, 187.

35. Olson, *Collected Prose*, 240.

36. Pippin continues: "Whatever sense I can make of anything, even in the most immediate and direct way, my sense of myself, if it is to be a making sense, must involve this aspect of normativity and thus the space of reasons, the public 'We' and the deep connection between my sense of others and any possible sense of myself (myself as always and necessarily, first of all, one among others)." Robert Pippin, *The Persistence of Subjectivity: On the Kantian Aftermath* (Cambridge: Cambridge University Press, 2005), 185.

37. Rukeyser, *Life of Poetry*, 19; Olson, *Collected Prose*, 240.

38. Ibid., 245.

39. Olson, *Collected Prose*, 240; Rukeyser, *Life of Poetry*, 40, 55.

40. Olson, *Collected Prose.*, 240.

41. Rukeyser, *Life of Poetry*, 77, 117.

42. I have discussed this issue of Olson's unacknowledged intellectual debts in my essay "Charles Olson: A Short History," *Parataxis: Modernism and Modern Writing* 10 (2001): 54–66.

43. Mary Beth Hinton, ed., *The Diaries of Marya Zaturenska, 1938–1944* (Syracuse: Syracuse University Press, 2001), 123.

44. She specifically thanks "Dr. Theodore Shedlovksy of the Rockefeller Institute, from whom I first learned about Willard Gibbs, and who offers the imaginative approach of a scientist who is alive to many purposes." In addition she thanks Norman Holmes Pearson and Donald Elder. Muriel Rukeyser, *Willard Gibbs* (Garden City, NY: Doubleday, Doran, 1942), 443.

45. Raymond M. Fuoss, "Theodore Shedlovsky, 1898–1976," in National Academy of Sciences, *Biographical Memoirs*, vol. 52 (Washington, DC: National Academy of Sciences, 1980), 381.

46. Rukeyser, *Willard Gibbs*, 74.

47. Catherine Gander, *Muriel Rukeyser and Documentary: The Poetics of Connection* (Edinburgh: Edinburgh University Press, 2013), 147. See also the discussions of Willard Gibbs in Kertesz, *Poetic Vision of Muriel Rukeyser*; and in Reginald Gibbons, "Fullness, Not War: On Muriel Rukeyser," in Herzog and Kaufman, *"How Shall We Tell Each Other of the Poet?,"* 101–9.

48. Kertesz, *Poetic Vision of Muriel Rukeyser*, 183.

49. F. O. Matthiessen was also an influence on Rukeyser. See Gander, *Muriel Rukeyser and Documentary*, 147.

50. Rukeyser, *Willard Gibbs*, 50.

51. Ibid., 88.

52. Ibid., 65.

53. Ibid., 62.

54. Ibid., 185.

55. Ibid., 176.

56. Ibid., 117.

57. Hinton, *Diaries of Marya Zaturenska*, 187.

58. Rukeyser, *Williard Gibbs*, 443.

59. Muriel Rukeyser, *Collected Poems* (New York: McGraw-Hill, 1978), 184.

60. Rukeyser, *Willard Gibbs*, 270; Milton Friedman, *Money Mischief: Episodes in Monetary History* (New York: Houghton Mifflin Harcourt, 1994), 37.

61. Philip Mirowski, *More Heat than Light: Economics as Social Physics, Physics as Nature's Economics* (Cambridge: Cambridge University Press, 1989), 251.

62. Rukeyser, *Willard Gibbs*, 418.

63. Joel Isaac, *Working Knowledge: Making the Human Sciences from Parsons to Kuhn* (Cambridge, MA: Harvard University Press, 2012), 68.

64. Bernard Barber, ed., *L. J. Henderson on the Social System* (Chicago: University of Chicago Press, 1970), 88. Cited in Isaac, *Working Knowledge*, 68.

65. Rukeyser, *Willard Gibbs*, 418.

66. Lawrence J. Henderson, *Pareto's General Sociology: A Physiologist's Interpretation* (Cambridge, MA: Harvard University Press, 1935), 15.

67. Rukeyser, *Willard Gibbs*, 417–18.

68. Isaac, *Working Knowledge*, 61.

69. Henry Murray is also known for his alleged responsibility in permanently warping the mind of Ted Kaczynski, the so-called Unabomber, when Kaczynski was a volunteer for psychological experimentation in stress interviews, research undertaken for the CIA.

70. Ibid., 164.

71. Ibid., 180.

72. Ibid., 75.

73. Clyde Kluckhohn, "A Navaho Personal Document with a Brief Paretian Analysis," *Southwest Journal of Anthropology* 1 (1945): 260–83, here 260. Kluckhohn gives as his institutional location the Pentagon.

74. Kluckhohn to Parsons, October 25, 1953, Talcott Parsons Papers, Harvard University,

HUG/FP 42.8.4, box 13. Cited in Willow Roberts Powers, "The Harvard Study of Values: Mirror for Postwar Anthropology," *Journal of the History of the Behavioral Sciences* 36, no. 1 (Winter 2000): 15–29, here 21.

75. Clark, *Charles Olson*, 47.

76. John Morion Blum, "A Celebration of Frederick Merk (1887–1977), *Virginia Quarterly Review*, Summer 1978, accessed January 15, 2014, http://www.vqronline.org/articles/1978/summer/blum-celebration/.

77. Frederick Merk, review of *Frontier Folkways*, by James G. Leyburn, *New England Quarterly* 9, no. 3 (September 1936): 542–44, here 542.

78. Henderson, *Pareto's General Sociology*, 15.

79. Ibid., 8.

80. Donald Davidson, "On the Very Idea of a Conceptual Scheme," in *Inquiries into Truth and Interpretation* (New York: Oxford University Press, 1984), 191. Throughout *The Structure of Scientific Revolutions* Kuhn uses the phrase "conceptual categories" rather than "conceptual scheme." The quotation from Quine comes from Willard Van Orman Quine, "Two Dogmas of Empiricism," in *From a Logical Point of View* (Cambridge, MA: Harvard University Press, 1961), 44.

81. Talcott Parsons, *Essays in Sociological Theory Pure and Applied* (Glencoe, IL: Free Press, 1949), 44.

82. Rukeyser, *Willard Gibbs*, 81.

83. Ibid.

84. Rukeyser, *Life of Poetry*, 163.

85. Ibid., 19.

86. Ibid.

87. David D. Nolte, "The Tangled Tale of Phase Space," *Physics Today* 63, no. 4 (April 2010): 33–38, here 33. Nolte's quotation of Gleick comes from James Gleick, *Chaos: Making a New Science* (New York: Viking, 1987). Nolte argues that although Joseph Liouville is generally credited with having first conceived the mathematics of a phase space, although not the terminology, Ludwig Boltzmann probably deserves credit for originating the mathematical idea. Nolte points out that Gibbs himself is cautious about the idea of a phase space, and the closest he comes to naming it is to say, "If we regard a phase as represented by a point in space of 2n dimensions, the changes which take place in the course of time in our ensemble of systems will be represented by a current in such space"; quoting from Josiah Willard Gibbs, *Elementary Principles in Statistical Mechanics: The Rational Foundation of Thermodynamics* (New York: Charles Scribner's Sons, 1902). Nevertheless, Gibbs is generally seen as one of the architects of the idea and was certainly thought of as such in the 1930s.

88. Rukeyser, *Life of Poetry*, 13.

89. Ibid., 178.

90. Rukeyser, *Willard Gibbs*, 339.

91. Gibbs, *Elementary Principles*, x.

92. Catherine Gander suggests that there is a similarity between Rukeyser's thinking about

structure and the concept of the rhizome in the work of Gilles Deleuze and Félix Guattari. Gander bases this partly on the similar ideas of an uncentered network in constant transformation. Gander, *Muriel Rukeyser and Documentary*, 183.

93. Norbert Weiner, *Cybernetics; or, Control and Communication in the Animal and the Machine* (New York: John Wiley and Sons, 1948), 110.

94. Pierre-Simon Laplace is credited with this thought experiment: if we knew the exact state of every particle of matter in the universe, could we predict the whole future of the universe from that knowledge?

95. William Packard, ed., *The Craft of Poetry Interviews from the "New York Quarterly"* (New York: Doubleday, 1974), 130–31. Cited in Lorrie Goldensohn, "Our Mother Muriel," in Herzog and Kaufman, *"How Shall We Tell Each Other of the Poet?,"* 131.

96. Jay Brassier, *Nihil Unbound: Enlightenment and Extinction* (New York: Palgrave Macmillan, 2007), 50. He is discussing modern philosophers rather than writers, but the judgment could apply to intellectuals in many humanities disciplines.

97. Leo Marx, "On Recovering the 'Ur' Theory of American Studies," *American Literary History* 17, no. 1 (Spring 2005): 118–34, here 119.

98. Charles Olson, *Call Me Ishmael: A Study of Melville* (London: Jonathan Cape, 1967), 15, 67, 68, 66.

99. Ibid., 19, 20, 68.

100. There is now a long-running argument about Olson's originality, or lack of it, in "Projective Verse." Marjorie Perloff showed in a key article in 1973 that many ideas attributed to him on the basis of "Projective Verse" were actually derived from unacknowledged modernist sources, though she does not mention Rukeyser. Marjorie Perloff, "Charles Olson and the 'Inferior Predecessors': 'Projective Verse' Revisited," *ELH* 40 (1973): 285–306. In a recent rejoinder, Ralph Maud takes issue with her, largely on the grounds that either Olson could not have read or there is no direct evidence that he did read the texts that she cites as sources and that Olson's language is different from the sources cited, rather than on the more central issue of the genealogy of the poetic concepts. Ralph Maud, *Charles Olson at the Harbour* (Vancouver: Talonbooks, 2008), 168–72.

101. Martin Puchner, *Poetry of the Revolution: Marx, Manifestos, and the Avant-Gardes* (Princeton, NJ: Princeton University Press, 2006), 151.

102. H. S. M. Coxeter, *Non-Euclidean Geometry* (Toronto: University of Toronto Press, 1942).

103. Alfred North Whitehead, *Adventures of Ideas* (Cambridge: Cambridge University Press, 1933), 242–43, 176.

104. Werner Brock, introduction to Martin Heidegger, *Existence and Being* (London: Vision Press, 2004), 100, 51.

105. Martin Heidegger, *Being and Time*, trans. John Macquarrie and Edward Robinson (Oxford: Basil Blackwell, 1962), 185.

106. Miles Groth, *Translating Heidegger* (Amherst, NY: Humanity Books, 2004), 100.

107. The leading historian of the modern social sciences Dorothy Ross argues that the term that best encapsulates how historians "understand the process of disciplinary formation is *project*." She continues: "A project is, on the one hand, a shared idea, aspiration, plan

or blueprint. . . . But, as [Bernard Yack] continues, in a usage taken from the philosopher Martin Heidegger, understanding 'always involves the projection of a world of possibilities within which things gain their meaning.'" Dorothy Ross, "Changing Contours of the Social Science Disciplines," in *The Modern Social Sciences*, ed. Theodore M. Porter and Dorothy Ross (Cambridge: Cambridge University Press, 2003), 206–7. She cites Bernard Yack, *The Fetishism of Modernities: Epochal Self-Consciousness in Contemporary Social and Political Thought* (South Bend, IN: University of Notre Dame Press, 1997), 116–17.

108. I. A. Richards, *Principles of Literary Criticism*, 2nd ed. (London: K. Paul, Trench, Trubner, 1926), 143, 185, 241, 29, 70.

109. William Carlos Williams, "The Poem as a Field of Action," in *Selected Essays* (New York: New Directions, 1954), 281. On the field as analogy, see Michael Golston, *Rhythm and Race in Modernist Poetry and Science* (New York: Columbia University Press, 2008), 215–223.

110. Ralph Maud points out that the lecture was not printed until 1954 in the *Selected Essays*. It does seem plausible that Olson might have heard of the lecture either from Williams or friends. What the coincidence of the two essays demonstrates is the attractiveness of developing a literary concept of the poem as a field given the increasing prominence of physics.

111. Stephen Fredman points out that in one sense "the 'field' spoken of by poets like Pound, Williams, Olson, Duncan, and Creeley is an alternate ground to the stanzaic measures provided by traditional poetics." Stephen Fredman, *The Grounding of American Poetry: Charles Olson and the Emersonian Tradition* (Cambridge: Cambridge University Press, 1993), 147.

112. The concept of the field in Olson's poetics has been given widely varying interpretations by literary critics. Charles Altieri makes extensive use of the concept of a poetic field to describe an otherwise-fluid semantic space of thinking and feeling generated by a poem, notably in *Self and Sensibility in Contemporary American Poetry* (Cambridge: Cambridge University Press, 1984). Daniel Belgrad has shown that some influential 1940s theories of society drew on analogies with energy fields. He argues that "as opposed to the atomistic individualism of classical liberalism, the energy field model of human experience defined it as emerging from a 'field of force' that was prior to any human identity." Daniel Belgrad, *The Culture of Spontaneity: Improvisation and the Arts in Postwar America* (Chicago: University of Chicago Press, 1998), 120. Anne Dewey takes such arguments further in her discussion of Olson and his contemporaries. She argues that "since the nineteenth century, writers have used the scientific model of the force field to describe collective economic, political, and ethnic forces as the dominant agents of history" and that Olson's field poetics involves "the poem tracking multiple forces in the force field" of history. Anne Day Dewey, *Beyond Maximus: The Construction of Public Voice in Black Mountain Poetry* (Stanford: Stanford University Press, 2007), 207. In these and most other cases, the emphasis falls on the metaphoric elaboration of the idea rather than its more direct historical interrelations, epistemic and rhetorical, with the discourses of contemporary physics and the social sciences.

CHAPTER 5

1. Catherine Gander, *Muriel Rukeyser and Documentary: The Poetics of Connection* (Edinburgh: Edinburgh University Press, 2013), 156. A third coeditor, Lloyd Mallan, was based at the proposed publisher Fawcett Books.

2. Ibid., 155-56. See also Gander's account of Rukeyser's debts to Waldo Frank and the list of potential contributors.

3. Ibid., 157.

4. Cited in Adam Bernstein, "Physicist, Educator Philip Morrison Dies," *Washington Post*, April 26, 2005, accessed January 13, 2014, http://www.washingtonpost.com/wp-dyn/content/article/2005/04/25/AR2005042501456.html.

5. Philip Morrison, "Blackett's Analysis of the Issues," *Bulletin of the Atomic Scientists* 5, no. 2 (February 1949): 37-40, here 37.

6. Muriel Rukeyser, *The Collected Poems of Muriel Rukeyser*, ed. Janet E. Kaufman and Anne Herzog (Pittsburgh, PA: University of Pittsburgh Press, 2005), 467; Muriel Rukeyser, *The Speed of Darkness* (New York: Random House, 1968).

7. Muriel Rukeyser, *Collected Poems* (New York: McGraw-Hill, 1978), v.

8. Rukeyser, "Orpheus," in *Collected Poems*, 293. First published in Muriel Rukeyser, *Selected Poems* (New York: New Directions, 1951).

9. Muriel Rukeyser, "The Genesis of Orpheus," in *Mid-century American Poets*, ed. John Ciardi (New York: Twayne, 1950), 50-54.

10. Lorrie Goldensohn, "Our Mother Muriel," in *"How Shall We Tell Each Other of the Poet?"*: *The Life and Writing of Muriel Rukeyser*, ed. Anne E. Herzog and Janet E. Kaufman (New York: St. Martin's Press, 1999), 122.

11. Rukeyser, "Orpheus," in *Collected Poems*, 292.

12. Ibid., 295.

13. I. L. Salomon, "From Union Square to Parnassus," *Poetry*, April 1952, 53, 56, 57.

14. Michael Davidson, *Ghostlier Demarcations: Modern Poetry and the Material World* (Berkeley: University of California Press, 1997), 168-69.

15. Louise Kertesz, *The Poetic Vision of Muriel Rukeyser* (Baton Rouge: Louisiana University Press, 1980), 256.

16. Muriel Rukeyser Papers, Library of Congress, box 1.23, dated New York, June 18, 1967. Cited in Gander, *Muriel Rukeyser and Documentary*, 107.

17. Muriel Rukeyser, *U.S. 1* (New York: Covici Friede, 1938).

18. Muriel Rukeyser, *One Life* (New York: Simon and Schuster, 1957), 315.

19. Ibid., 330.

20. Ibid., 321.

21. In fact, as Catherine Gander documents, initial reviews were varied; some reviewers were fairly positive, notably Richard Eberhart. Gander, *Muriel Rukeyser and Documentary*, 99-100.

22. Rukeyser, *Collected Poems*, 413; Florence Howe and Ellen Bass, eds., *No More Masks: An Anthology of Poems by Women* (Garden City, NY: Anchor Books, 1973).

23. Louise Bernikow, *The World Split Open: Four Centuries of Women Poets in England and America, 1552-1950* (New York: Random House, 1974).

24. It is also significant that one of our most astute anthologists, Eliot Weinberger, chooses "The Speed of Darkness" as one of three Rukeyser poems for his anthology of American poet "innovators and outsiders" from midcentury onward. Eliot Weinberger, *American Poetry Since 1950: Innovators and Outsiders* (New York: Marsilio, 1993).

25. Charles Olson, *The Collected Poems of Charles Olson: Excluding the "Maximus" Poems*, ed. George F. Butterick (Berkeley: University of California Press, 1987), 90. My discussion of "The Kingfishers" is informed by Ralph Maud's thorough gloss on the poem. See Ralph Maud, *What Does Not Change: The Significance of Charles Olson's "The Kingfishers"* (Madison, WI: Fairleigh Dickinson University Press, 1998).

26. Norbert Wiener, *Cybernetics; or, Control and Communication in the Animal and the Machine* (New York: J. Wiley, 1948), 16.

27. "Very early in the year it prepares its nest, which is at the end of a tunnel bored by itself in a bank, and therein the six or eight white, glossy, translucent eggs are laid, sometimes on the bare soil, but often on the fishbones which, being indigestible, are thrown up in pellets by the birds; and, in any case, before incubation is completed these *rejectamenta* accumulate so as to form a pretty cup-shaped structure that increases in bulk after the young are hatched, but, mixed with their fluid excretions and with decaying fishes brought for their support, soon becomes a dripping fetid mass." *Encyclopaedia Britannica*, 11th ed., s.v. "kingfisher."

28. I. A. Richards, *Principles of Literary Criticism*, 2nd ed. (London: Routledge and Kegan Paul, 1926), 254.

29. George Steinmetz, "American Sociology before and after World War II: The (Temporary) Settling of a Disciplinary Field," in *Sociology in America: A History*, ed. Craig Calhoun (Chicago: University of Chicago Press, 2007), 358. Citing Paul Lazarsfeld and Allen H. Barton, "Qualitative Measurement in the Social Sciences: Classification, Typologies, and Indices," in *The Policy Sciences: Recent Developments in Scope and Method*, ed. Daniel Lerner (Stanford, CA: Stanford University Press, 1951), 156.

30. Lazarsfeld and Barton, "Qualitative Measurement in the Social Sciences," 159.

31. Charles Olson, *The Maximus Poems*, ed. George F. Butterick (Berkeley: University of California Press, 1983), 163, 149, 132, 121, 65, 59.

32. In his major study of the philosophy of history, Aviezer Tucker includes a chapter on scientific historiography, which he defines as "historiography that generates probable knowledge of the past." Because history belongs to a group of human sciences that are "systematically interested in common cause tokens, in historical events and processes," it differs from the natural sciences. Aviezer Tucker, *Our Knowledge of the Past: A Philosophy of Historiography* (Cambridge: Cambridge University Press, 2004), 1, 100.

33. Ernest Nagel, *The Structure of Science: Problems in the Logic of Scientific Explanation* (London: Kegan Paul, 1961), 550, 570, 606. The American edition was published by Harcourt, Brace and World also in 1961.

34. Charles Olson, "The Chiasma; or, Lectures in the New Sciences of Man," *Olson: The Journal of the Charles Olson Archives* 10 (Fall 1978): 70–71.

35. Ibid.

36. Charles Olson, "Maximus Letter #28," in "Early Unpublished Maximus Poems, 1953–1957," *Olson: The Journal of the Charles Olson Archives* 6 (Fall 1976): 15. Several different

poems in this presentation of archive material are all given the number 28, indicating how uncertain Olson was about how to continue the sequence.

37. George Butterick, notes, *Olson: The Journal of the Charles Olson Archives* 6 (Fall 1976): 1.

38. Charles Olson, *Collected Prose*, ed. Donald Allen and Benjamin Friedlander (Berkeley: University of California Press, 1997), 155.

39. Henry Adams, *The Education of Henry Adams* (Oxford: Oxford University Press, 2008), 318, 319.

40. Charles Olson, "Maximus Letter #28," *Olson: The Journal of the Charles Olson Archives* 6 (Fall 1976): 17.

41. Adams, *Education of Henry Adams*, 319.

42. Olson, "Maximus Letter #28," 18.

43. In saying this I am generalizing about just one feature of a number of insightful readings of Olson from which I have learned a lot. These include valuable discussions of this poem in the following: Don Byrd, *Charles Olson's Maximus* (Urbana: University of Illinois Press, 1980); Anne Day Dewey, *Beyond Maximus: The Construction of Public Voice in Black Mountain Poetry* (Stanford, CA: Stanford University Press, 2007); Miriam Nichols, *Radical Affections: Essays on the Poetics of Outside* (Tuscaloosa: University of Alabama Press, 2010).

44. Shahar Bram, *Charles Olson and Alfred North Whitehead: An Essay on Poetry*, trans. Batya Stein (Lewisburg: Bucknell University Press, 2004), 46.

45. Joshua Hoeynck, "Deep Time and Process Philosophy in the Charles Olson and Robert Duncan Correspondence," *Contemporary Literature* 55, no. 2 (Summer 2014): 336–68, here 350.

46. Ralph Maud, ed., *A Charles Olson Reader* (Manchester: Carcanet, 2005), 67.

47. Whitehead recognizes that "the word '*perceive*' is, in our common usage, shot through and through with the notion of cognitive apprehension. . . . I will use the word 'prehension' for *uncognitive apprehension*: by this I mean *apprehension* which may or may not be cognitive." Although the final definition of "apprehension" appears to be ambiguous, I take Whitehead to mean that some apprehensions are prehensions—and therefore noncognitive or nonconceptual—and some are not prehensions because they are cognitive or conceptual. These issues have returned to the fore in John McDowell's work, which is sometimes labeled "new pragmatism." Alfred North Whitehead, *Science and the Modern World* (1926; Cambridge: Cambridge University Press, 1932), 86.

48. Alfred North Whitehead, *Science and the Modern World* (1926; Cambridge: Cambridge University Press, 1932), 194.

49. Isabelle Stengers, *Thinking with Whitehead: A Free and Wild Creation of Concepts*, trans. Michael Chase (Cambridge, MA: Harvard University Press, 2011), 7, 124.

50. Steven Shaviro, *Without Criteria: Kant, Whitehead, Deleuze, and Aesthetics* (Cambridge, MA: MIT Press, 2009), 45.

51. Miriam Nichols, *Radical Affections: Essays on the Poetics of Outside* (Tuscaloosa: University of Alabama Press, 2010).

52. Hoeynck, "Deep Time and Process Philosophy," 342.

53. Whitehead, *Science and the Modern World*, 194.

54. Alfred North Whitehead, *Adventures of Ideas* (Cambridge: Cambridge University Press, 1933), 242.

55. Michael North identifies Whitehead, along with William James and Henri Bergson, as philosophers who "identify novelty as one of the great unsolved problems in modern thought." North's discussions of Wiener, Arendt, Pound, and other modern thinkers provide a valuable context for the study of the recent history of poetry and science. Michael North, *Novelty: A History of the New* (Chicago: University of Chicago Press, 2013), 6.

56. Whitehead, *Adventures of Ideas*, 242.

57. Ibid., 243.

58. Ibid.

59. The battle is also mentioned in "Letter 22." Butterick provides detailed information on the allusions to Graves and Whitehead. George Butterick, *A Guide to the Maximus Poems of Charles Olson* (Berkeley: University of California Press, 1978), 140, 262–64. I think it is worth noting that according to Graves, the poets stop the battle because they admire the "valor" of the combatants and want to commemorate the battle. Olson is not dismissing either those who have argued for the soul or those who have insisted on a severe, reductive materialism. Robert Graves, *The White Goddess: A Historic Grammar of Poetic Myth*, enl. ed. (New York: Farrar, Straus and Giroux, 1966), 22.

60. The spacing of these lines is only a rough indicative approximation of Olson's highly visual page presentation.

61. T. S. Eliot, "East Coker" II, *Collected Poems 1909–1962* (London: Faber, 2002), 184.

62. Hoeynck, "Deep Time and Process Philosophy," 349.

63. Ibid., 346.

64. Whitehead, *Adventures of Ideas*, 242.

65. Shaviro, *Without Criteria*, 31.

66. Joshua Hoeynck, *Poetic Cosmologies: Black Mountain Poetry and Process Philosophy* (Ann Arbor, MI: Proquest, 2008), 108–9. He is citing Alfred North Whitehead, *Process and Reality: An Essay in Cosmology* (Cambridge: Cambridge University Press, 1929), 112.

67. Nichols, *Radical Affections*, 57.

68. Olson, *Collected Prose*, 155.

69. A collection of new readings of Olson by twenty-four contributors (including myself), *Contemporary Olson*, indicates that Olson's poetry is generating new interest still. It also shows that Whitehead continues to have supporters. In an excellent essay on Olson's attempt to challenge subject-object dualism, Miriam Nichols offers a positive endorsement of Whitehead's influence on Olson, in which she argues that Olson's concentration on method displaces explicitly ethical preoccupations, and that Whitehead encourages Olson to turn away from epistemology. Her defense of Whitehead leads her to argue that Whitehead helped Olson discard dualist legacies of thought: "Olson displaces objectivity with shareable cultural paradigms and geohistories." In my account of Olson, conceptual schemes were already a self-conscious manipulation of such paradigms and histories. See Miriam Nichols, "Myth and Document in Charles Olson's Maximus Poems," in *Contemporary Olson*, ed. David Herd (Manchester: Manchester University Press, 2015), 26.

70. Charles Olson, *Maximus Poems*, 175. "Letter, May 2, 1959" is on p. 150.

71. Thomas Kuhn, *Structure of Scientific Revolutions*, 2nd ed. (Chicago: University of Chicago Press, 1970), 50. In a sometimes-impatient response to critics and commentators, Kuhn insists that paradigms are "concrete problem solutions, the exemplary objects of an ostension" for specific communities of researchers (perhaps identifying "an individual specialists' group"). Thomas Kuhn, "Reflections on My Critics," in Thomas Kuhn, *The Road since Structure*, ed. James Conant and John Haugeland (Chicago: University of Chicago Press, 2000), 167-68.

72. Wilfrid Sellars, *In the Space of Reasons: Selected Essays of Wilfrid Sellars*, ed. Kevin Scharp and Robert B. Brandom (Cambridge, MA: Harvard University Press, 2007), 373.

73. Ibid., 375.

74. Bas C. Van Fraassen, *The Scientific Image* (Oxford: Clarendon Press, 1980), 4.

75. Robert Duncan, "As an Introduction: Charles Olson's *Additional Prose*," in *A Selected Prose*, ed. Robert J. Bertholf (New York: New Directions, 1995), 150.

76. Ibid., 148.

77. Ibid., 151.

78. Robert Duncan, "Eleven Letters," ed. Robert J. Bertholf, *Sulfur* 35 (Fall 1994): 87-118, here 104-6. Cited in Peter O'Leary, *Gnostic Contagion: Robert Duncan and the Poetry of Illness* (Middletown, CT: Wesleyan University Press, 2002), 106-7.

79. Robert Duncan, "Apprehensions," *Roots and Branches* (New York: New Directions, 1964), 31-43, here 40.

80. Duncan, "Eleven Letters," 105.

81. Robert Duncan, "The Fire: Passages 13," *Bending the Bow* (New York: New Directions, 1968), 40-45, here 43.

82. Lisa Jarnot, *Robert Duncan, the Ambassador from Venus: A Biography* (Berkeley: University of California Press, 2012), xxi.

83. George Bowering and Robert Hogg, *Robert Duncan: An Interview, April 19, 1969* (Toronto: Coach House Press, Beaver Kosmos Folio, 1971), no page numbers. For the convenience of readers with access to this text, I shall give my page estimate: this quotation comes from p. 5. An extract from this interview, which does not contain the discussion of science, appears in Christopher Wagstaff, ed., *A Poet's Mind: Collected Interviews with Robert Duncan, 1960-1985* (Berkeley: North Atlantic Books, 2012).

84. Ibid.

85. "Science and the Citizen," *Scientific American* 219 (December 1968): 48-56, here 48.

86. Luis W. Alvarez won the Nobel Prize for Physics in 1968 for, in the words of the Nobel committee, "his decisive contributions to elementary particle physics, in particular the discovery of a large number of resonance states, made possible through his development of the technique of using hydrogen bubble chamber and data analysis." Mark Johnson is, I think, mistaken to identify the physicist that Duncan refers to as Murray Gell-Mann. Gell-Mann was not awarded the Nobel Prize until October 1969, sometime after this interview, and was notorious for having writer's block during this period of his life, which meant that he never wrote a Nobel lecture and rarely wrote essays. Mark Andrew Johnson, *Robert Duncan* (Farmington Hills, MI: Twayne, 1988), 110.

87. Luis W. Alvarez, "Recent Developments in Particle Physics," *Science*, n.s.,165, no. 3898 (September 12, 1969): 1071–91, here 1071.

88. Walter Sullivan, "Cosmic Rays Fail to Discover Pharaoh's Chamber in Pyramid," *New York Times*, May 1, 1969.

89. Editorial, "Luis Walter Alvarez," *New York Times*, October 31, 1968.

90. Bowering and Hogg, *Robert Duncan: An Interview*, [6].

91. Robert Duncan, *The H.D. Book*, ed. Michael Boughn and Victor Coleman (Berkeley: University of California Press, 2011), 429.

92. Wagstaff, *A Poet's Mind*, 41.

93. Graça Capinha, "Robert Duncan and the Question of Law: Ernst Kantorowicz and the Poet's Two Bodies," in *Robert Duncan and Denise Levertov: The Poetry of Politics, the Politics of Poetry*, ed. Albert Gelpi and Robert Bertholf (Stanford, CA: Stanford University Press, 2006), 19.

94. Robert Duncan, "Towards an Open Universe," in *A Selected Prose*, 3.

95. Duncan, *H.D. Book*, 142.

96. Richard Boyd, Philip Gasper, and J. D. Trout, eds., *The Philosophy of Science* (Cambridge, MA: MIT Press, 1991), 778. Cited in Hilary Putnam, *Philosophy in an Age of Science: Physics, Mathematics, and Skepticism*, ed. Mario De Caro and David Macarthur (Cambridge, MA: Harvard University Press, 2012), 110. Putnam is scornful of this definition, saying in effect that it is incoherent. He points out that any activity, including writing a text, does not violate natural laws and then that the second part of the definition ignores too many questions, such as whether the interpretation of texts is a form of inquiry, and whether there might be one science, say physics, that would claim to be able to explain even the act of textual interpretation. Nevertheless, I think the definition does capture, however crudely, a prevailing twentieth-century overconfidence in the potential of natural science to eventually provide explanations for virtually everything.

97. Kevin Johnston, *Precipitations: Contemporary American Poetry as Occult Practice* (Middletown, CT: Wesleyan University Press, 2002), 50.

98. Putnam, *Philosophy in an Age of Science*, 112.

99. Peter O'Leary, *Gnostic Contagion: Robert Duncan and the Poetry of Illness* (Middletown, CT: Wesleyan University Press, 2002), 73.

100. Ibid., 110.

101. Box 26, Notebook 28 (August 8, 1959), Poetry / Rare Book Collection, University Libraries, State University of New York, Buffalo, 52. Cited in Johnston, *Precipitations*, 83.

102. Robert Duncan, *Roots and Branches* (New York: New Directions, 1964), 30.

103. The final section is on pp. 42–43 of *Roots and Branches*.

104. Others have noticed this similarity. Mark Johnson says confidently that "this matrix sends the reader to nuclear physics, for it evokes the matrices with which field physicists represent three-dimensional electron patterns in a two-dimensional mathematical grid." For the reasons given in the text, and because matrices had wide use in many areas of science at the time he was writing, I think we should be cautious about making this assumption. Johnson, *Robert Duncan*, 109.

105. Rukeyser, *One Life*, 314.

CHAPTER 6

1. Charles Olson, "The Chiasma; or, Lectures in the New Sciences of Man," *Olson: The Journal of the Charles Olson Archives* 10 (Fall 1978): 70.

2. Fred W. Decker, "Scientific Communications Should Be Improved," *Science*, n.s., 125, no. 3238 (January 18, 1957): 101–5, here 101, 103, 101, 104.

3. Alan Waterman, "National Science Foundation: A Ten Year Résumé," *Science*, n.s., 131, no. 3410 (May 6, 1960): 1341–54, here 1349.

4. Cited in John Troan, "Science Reporting—Today and Tomorrow," *Science*, n.s., 131, no. 3408 (April 22, 1960): 1193–96, here 1193.

5. Gerard Piel, "Need for Public Understanding of Science," *Science* 121, no. 3140 (March 4, 1955): 317–22.

6. Glenn Seaborg, "Goals in Understanding Science," *Science News* 90, no. 18 (October 29, 1966): 354–56, here 356.

7. Gerard Piel, *The Age of Science: What Scientists Learned in the 20th Century* (Oxford: Perseus Press, 2001), xii.

8. Bruce V. Lewenstein, "The Meaning of 'Public Understanding of Science' in the United States after World War II," *Public Understanding of Science* 1, no. 1 (January 1992): 45–68, here 50.

9. Dennis Flanagan and Gerard Piel, "The Sciences: A Prospectus in the Form of a Dialogue" (1947), *Scientific American* archives. Cited in Lewenstein, "Meaning of 'Public Understanding of Science,'" 51.

10. Lewenstein, "Meaning of 'Public Understanding of Science,'" 51.

11. Piel, *Age of Science*, xix.

12. Rae Armantrout, "The Lyric," in *Grand Piano*, pt. 8 (Detroit, MI: Mode A, 2009), 36. *Grand Piano* is "an experiment in collective autobiography" by poets active in the San Francisco area in the late 1970s and is published in ten parts.

13. Rae Armantrout, "Metaphor or More," in "Like a Metaphor," ed. Gilbert Adair, *Jacket* 2 (2012), accessed January 23, 2014, http://jacket2.org/interviews/metaphor-or-more. This series of informal exchanges, in which I took part, provides a useful insight into the assumptions made by poets about their engagements with science. One striking feature is how clear the majority of poets are about their own responsibility to respect the epistemic claims of scientific method.

14. Lydia Davis, "'Why Stop with a Barnacle?,'" in *A Wild Salience: The Writing of Rae Armantrout*, ed. Tom Beckett (Cleveland, OH: Burning Press, 1999), 82.

15. Rae Armantrout, "Natural History," in *The Invention of Hunger* (San Francisco: Tuumba Press, 1979), no page numbers (her ellipses). Reprinted in *Veil: New and Selected Poems* (Middletown, CT: Wesleyan University Press, 2001). Tuumba Press was run by Lyn Hejinian, who printed the magazine in a printshop near her home in Northern California. For an excellent, detailed account of the press and its significance, see Ann Vickery, *Leaving Lines of Gender: A Feminist Genealogy of Language Writing* (Hanover, NH: Wesleyan University Press, 2000), chap. 4.

16. Steven Vogel, "Organisms That Capture Currents," *Scientific American* 239 (August 1978): 128–39.

17. The analogy with air-conditioning is taken from an earlier *Scientific American* article whose title Vogel cites: Martin Lüscher, "Air-Conditioned Termite Nests," *Scientific American* 205 (July 1961): 138–45.

18. Vogel, "Organisms That Capture Currents," 139.

19. Berthold K. Hölldobler and Edward O. Wilson, "Weaver Ants," *Scientific American* 237, no. 6 (December 1977): 146–54; Edward O. Wilson, "Slavery in Ants," *Scientific American* 232, no. 6 (June 1975): 32–36.

20. Wilson, "Slavery in Ants," 35.

21. Ibid., 36.

22. Richard E. Dickerson, "Chemical Evolution and the Origin of Life," *Scientific American* 239 (September 1978):70–86, here 85.

23. Kinsey A. Anderson, "Solar Particles and Cosmic Rays," *Scientific American* 202, no. 6 (June 1960): 64–71, here 64.

24. Jackson Mac Low, "Solar Particles and Cosmic Rays," *Stanzas for Iris Lezak* (Millerton, NY: Something Else Press, 1971), 122–26. Mac Low's work is usefully discussed in *Paper Air* 2, no. 3 (1980), which is also available at http://english.utah.edu/eclipse/projects /PAPER/PA2.3/.

25. Mac Low, *Stanzas*, 405. Mac Low drew extensively on articles in the June, July, and August issues and then revisited the October issue to write the poem "Asymmetry from *Scientific American* October 1960 *The New York Times Book Review* Section 25 September 1960 — (5 Oct. 1960)." It is hard to be exact about how many articles he used overall because this complex assemblage poem and some of the Asymmetries write across several issues of the magazine.

26. Mac Low, *Stanzas*, 399.

27. I am thinking here of the sorts of arguments made by Andrew Epstein and Lytle Shaw about Frank O'Hara, John Ashbery, and others and also about the arguments developed by Libbie Rifkin, Anne Dewey, and their collaborators about other poets and poetry communities.

28. Lytle Shaw, *Frank O'Hara: The Poetics of Coterie* (Iowa City: University of Iowa Press, 2006), 37.

29. Andrew Epstein, *Beautiful Enemies: Friendship and Postwar American Poetry* (New York: Oxford University Press, 2006), 57.

30. Brian McHale, "Poetry as Prosthesis," *Poetics Today* 21, no. 1 (Spring 2000): 1–32, here 26.

31. Ellen Zweig, "Jackson Mac Low: The Limits of Formalism," *Poetics Today* 3, no. 3 (Summer 1982): 79–86, here 82.

32. Craig Dworkin, *Reading the Illegible* (Urbana, IL: Northwestern University Press, 2003), passim.

33. A riometer is a relative ionospheric opacity meter, used to record the absorption of electromagnetic radiation by the ionosphere.

34. Marcel Duchamp, *The Essential Writings of Marcel Duchamp: Marchand du sel — Salt Seller*, ed. Michel Sanouillet and Elmer M. Peterson (London: Thames and Hudson, 1975), 71. Cited in Marjorie Perloff, *21st Century Modernism: The "New" Poetics* (Oxford: Blackwell, 2002), 87.

35. Mac Low, *Stanzas for Iris Lezak*, 122. Further refs in text as *SFIL*.

36. Anderson, "Solar Particles and Cosmic Rays," 64.

37. Joseph E. Harmon and Alan G. Gross, *The Scientific Literature: A Guided Tour* (Chicago: University of Chicago Press, 2007), 221–22.

38. Robert Duncan, "Osiris and Set," in *Roots and Branches* (New York: New Directions, 1964), 68.

39. Robert Duncan, "Structure of Rime XVIII," in *Roots and Branches*, 67.

40. Duncan, "Osiris and Set," 68.

41. Sherwood L. Washburn, "Tools and Human Evolution," *Scientific American* 203, no. 3 (September 1960): 63–75, here 69.

42. Herbert Butterfield, "The Scientific Revolution," *Scientific American* 203, no. 3 (September 1960): 173–92, here 192. Butterfield was a historian now better known for his essay "The Whig Interpretation of History" and his work on historiography and international relations than for his work on the history of science, but his research on the early development of science was influential in its day.

43. "Science and the Citizen," *Scientific American* 203, no. 3 (September 1960): 98–108, here 98.

44. Marshall Sahlins, "The Origin of Society," *Scientific American* 203, no. 3 (September 1960): 76–87, here 78, 86.

45. Ibid., 82.

46. Theodosius Dobzhansky, "The Present Evolution of Man," *Scientific American* 203, no. 3 (September 1960): 206–17, here 212.

47. Ibid., 212.

48. In 1957 the American army had been divided into five divisions to respond to the possibility of having to fight on a nuclear battlefield.

49. *Scientific American* 203, no. 3 (September 1960): Hughes, 228–29; Space Technology Laboratories, 227; Kollmorgen, 215; EAGLE, 211; Avco, 175; Laboratory for Electronics, 57; Martin, 168; RCA, 170–71.

50. Ibid., 191.

51. Dwight D. Eisenhower, Farewell Address, reading copy of the speech, Dwight D. Eisenhower Presidential Library, accessed January 29, 2015, http://www.eisenhower.archives .gov/research/online_documents/farewell_address/Reading_Copy.pdf.

52. Robert Duncan, "Up Rising: Passages 25," in *Bending the Bow* (New York: New Directions, 1968), 82.

53. *Scientific American* 203, no. 3 (September 1960): 59, 193.

54. Ibid., 19.

55. Haeckel's ideas on eugenics are alleged to have provided elements of Nazi ideology.

56. The Rand Corporation of Santa Monica, California, presents itself in small print at the bottom of the page in terms that cleverly suggest its scientific research is similarly Humboldtian to that of a university, only slipping in the reference to what it carefully calls "national security" at the end: "A nonprofit organization engaged in a program of research in the physical sciences, economics, mathematics, and the social sciences. These diverse skills are joined in the analysis of complex problems related to national security and the public interest."

57. Paul Weindling defends Haeckel against historians who treat his views as a precursor of the Nazi euthanasia program, saying that Haeckel saw practices such as the killing of weak children by the Spartans as evidence of their lack of civilization. Paul Weindling, "Genetics, Eugenics and the Holocaust," in *Biology and Ideology from Descartes to Dawkins*, ed. Denis R. Alexander and Ronald L. Numbers (Chicago: University of Chicago Press, 2010), 198.

58. Peter O'Leary, *Gnostic Contagion: Robert Duncan and the Poetry of Illness* (Middletown, CT: Wesleyan University Press, 2002), 66.

CHAPTER 7

1. George Oppen, *Selected Prose, Daybooks, and Papers*, ed. Stephen Cope (Berkeley: University of California Press, 2007), 167.

2. George Oppen, letter to Paul Vangelisti, no date, in Paul Vangelisti, *Communion* (Fairfax, CA: Red Hill Press, 1970), no page numbers.

3. J. R. Oppenheimer, "The Age of Science: 1900–1950," *Scientific American* 183, no. 3 (September 1950): 20–23, here 20, 23.

4. Gary Snyder, *Turtle Island* (New York: New Directions, 1974), 101, 97. The second part of the book is composed of essays about the politics of the environment.

5. Burt Bolin, "The Carbon Cycle," *Scientific American* 223, no. 3 (September 1970): 125–32, here 132.

6. Snyder, *Turtle Island*, 97, 94, 101.

7. Chauncey Starr, "Energy and Power," *Scientific American* 225, no. 3 (September 1971): 37–49, here 37, 44.

8. Roy A. Rappaport, "The Flow of Energy in an Agricultural Society," *Scientific American* 225, no. 3 (September 1971): 116–32, here 132. This much-cited article was widely circulated as a special reprint, and it continues to be recommended to students today. It was based on the research that underlay his brilliant study of New Guinea highlanders, *Pigs for the Ancestors: Ritual in the Economy of a New Guinea People* (New Haven, CT: Yale University Press, 1968).

9. The quotation in the text is the subtitle to the introductory article by Kingsley Davis, "The Urbanization of the Human Population," *Scientific American* 213, no. 3 (September, 1965): 41–53, here 41.

10. L. S. Dembo and George Oppen, "George Oppen," *Contemporary Literature* 10, no. 2 (Spring 1969): 159–77, here 162.

11. Gillian Rose, *Hegel contra Sociology* (London: Athlone Press, 1981), 22.

12. Charles Altieri, *The Particulars of Rapture: An Aesthetics of the Affects* (Ithaca, NY: Cornell University Press, 2003), 198, 200.

13. George Oppen, "Of Being Numerous," in *New Collected Poems*, ed. Michael Davidson (New York: New Directions, 2008), 183–84. Further page references to "Of Being Numerous" will be inserted in the text after *NCP*.

14. Altieri, *The Particulars of Rapture*, 200.

15. Kingsley Davis and Wilbert E. Moore, "Some Principles of Stratification," *American Sociological Review* 10, no. 2 (1944): 242–49, here 242.

16. Hamilton Cravens, "Column Right, March! Nationalism, Scientific Positivism, and the Conservative Turn of the American Social Sciences in the Cold War Era," in *Cold War Social Science: Knowledge Production, Liberal Democracy, and Human Nature*, ed. Mark Solovey and Hamilton Cravens (New York: Palgrave Macmillan, 2012), 120.

17. Davis and Moore, "Some Principles of Stratification," 248, 243.

18. Ibid., 246.

19. Kingsley Davis, "The Myth of Functional Analysis as a Special Method in Sociology and Anthropology," *American Sociological Review* 24, no. 6 (December 1959): 757-72, here 758-59, 762n24, 763.

20. The phrase "suspension of the ethical" comes from the translation of Kierkegaard's *Fear and Trembling* that Oppen used as the source of section 16. Henry Weinfield argues that Oppen was very attentive to Kierkegaard's representation of the singularity of Abraham and the troubling implications of the suspension of the ethical involved in his willingness to sacrifice his son. Henry Weinfield, *The Music of Thought in the Poetry of George Oppen and William Bronk* (Iowa City: University of Iowa Press, 2009), 66-67.

21. Immanuel Wallenstein, "The Culture of Sociology in Disarray: The Impact of 1968 on U.S. Sociologists," in *Sociology in America: A History*, ed. Craig Calhoun (Chicago: University of Chicago Press, 2007), 430.

22. Craig Calhoun and Jonathan VanAntwerpen, "Orthodoxy, Heterodoxy, and Hierarchy: 'Mainstream' Sociology and Its Challengers," in Calhoun, *Sociology in America*, 388.

23. Wallenstein, "Culture of Sociology in Disarray," 428.

24. Peter Nicholls, *George Oppen and the Fate of Modernism* (Oxford: Oxford University Press, 2007), 105. Nicholls is citing from the Oppen archive: George Oppen Papers, Mandeville Special Collections, University of California at San Diego, MSS 16, 17, 2.

25. Abel Wolman, "The Metabolism of Cities," *Scientific American* 213, no. 3 (September 1965): 178-90, here 179.

26. Nathan Glazer, "The Renewal of Cities," *Scientific American* 213, no. 3 (September 1965): 194-204, here 197, 204.

27. Benjamin Chinitz, "New York: A Metropolitan Region," *Scientific American* 213, no. 3 (September 1965): 134-48, here 136.

28. Davis, "Urbanization of the Human Population," 42.

29. Chinitz, "New York," 136.

30. Robert A. Caro, *The Power Broker: Robert Moses and the Fall of New York* (New York: Random House, 2004), 4.

31. Chinitz, "New York," 148.

32. Hans Blumenfeld, "The Modern Metropolis," *Scientific American* 213, no. 3 (September 1965): 64-74, here 64, 68, 71, 72, 74.

33. George Oppen, *Selected Letters of George Oppen*, ed. Rachel Blau Du Plessis (Durham, NC: Duke University Press, 1990), 135.

34. Martin Heidegger, *Essays in Metaphysics: Identity and Difference*, trans. Kurt F. Liedecker (New York: Philosophical Library, 1960), 26; Peter Nicholls, *George Oppen and the Fate of Modernism* (Oxford: Oxford University Press, 2007), 77-81.

35. Oppen, *Selected Letters*, 135.

36. The 1969 translation by Joan Stambaugh makes the point even more sharply: "our whole human existence everywhere sees itself challenged . . . to devote itself to planning and calculating everything." Martin Heidegger, *Identity and Difference*, trans. Joan Stambaugh (New York: Harper and Row, 1969), 34–35.

37. An obvious case of this occurred when Robert Duncan played with lines from Denise Levertov and she thought he was mocking her.

38. I have discussed this practice in Creeley's work in my essay "Scenes of Instruction: Creeley's Reflexive Poetics," in *Form, Power, and Person in Robert Creeley's Life and Work*, ed. Stephen Fredman and Steve McCaffery (Iowa City: University of Iowa Press, 2010), 159–80.

39. Weinfield, *Music of Thought*, 38.

40. Ibid., 40.

41. Oren Izenberg, *Being Numerous: Poetry and the Ground of Social Life* (Princeton, NJ: Princeton University Press, 2011), 84.

42. Nicholls, *George Oppen*, 83, 84, 85, 94. The quotations from Oppen are taken from the Oppen archive, George Oppen Papers, Mandeville Special Collections, University of California at San Diego, MSS 16, 14, 5; and from L. S. Dembo, "The 'Objectivist' Poet: Four Interviews," *Contemporary Literature* 10, no. 2 (Spring 1969): 155–77, here 162. In the interview, Oppen is trying to make clear that he is not debating methodological concepts but whether humanity is real enough to be what he calls a "substantive" or "a thing."

43. Susan Thackrey, *George Oppen: A Radical Practice* (San Francisco: O Books, 2001), 20.

44. Ibid., 39; Oppen, *Selected Letters*, 133.

45. The original quotation is "In this explanation it is presumed that an experiencing subject is one occasion of a sensitive reaction to an actual world." Alfred North Whitehead, *Process and Reality: An Essay in Cosmology* (Cambridge: Cambridge University Press, 1929), 22. Whitehead is discussing the origin of both religion and science in the need for "an intellectual justification of brute experience" (21), or in other words, he is not talking directly about his own philosophy but about a whole history of attempts to explain the meaning of things.

46. Thomas Crow, *The Intelligence of Art* (Chapel Hill: University of North Carolina Press, 1999), 5.

47. The song was written by Alex Kramer and Joan Whitney and recorded by Louis Jordan and his Tympany Five. Jordan's delivery makes it a curiously subversive song, a lyric of double-consciousness, speaking to both the white boss and fellow African Americans.

48. J. A. Cuddon, *The Owl's Watchsong: A Study of Istanbul* (London: Century Hutchinson, 1986), 28. The book was first published in London in 1960 and then in New York in 1962.

49. Allen Ginsberg, "Howl," in *Howl and Other Poems* (San Francisco: City Lights, 1959), 9.

50. Robert Duncan, *The Opening of the Field* (New York: New Directions, 1960), 7.

51. William Carlos Williams, *Paterson* (New York: New Directions, 1992), 165.

52. Kevin Lynch, "The City as Environment," *Scientific American* 213, no. 3 (September 1965): 209–19, here 219.

53. Kevin Lynch, *The Image of the City* (Cambridge, MA: MIT Press, 1960), 1.

54. Ibid., 119.

55. Lynch, *Image of the City*, 119.

56. Ibid., 2.

57. Nicholls, *George Oppen*, 111–13.

58. Lynch, *Image of the City*, 4.

59. On images, see Michael Heller, *Speaking the Estranged* (Bristol: Shearsman, 2012), 54.

60. Ibid., 7, 81.

61. Ibid., 5.

62. Ibid., 10; Lynch, "The City as Environment," 210.

63. Lynch, "City as Environment," 219.

64. Ibid., 209.

65. Ibid., 213, 219.

66. Lynch, *Image of the City*, 127.

67. Lynch, "City as Environment," 209.

68. Ibid., 214.

69. Ibid., 219.

70. LeRoi Jones [Amiri Baraka], "The People Burning," in *Black Magic: Collected Poetry, 1961–1967* (Indianapolis: Bobbs-Merrill, 1969), 11.

71. Nathan S. Caplan and Jeffery M. Paige, "A Study of Ghetto Rioters," *Scientific American* 219, no. 2 (August 1968): 15–21.

72. Ibid., 21.

73. Amiri Baraka, "Newark Later," in *Black Magic*, 145.

74. Amiri Baraka, *The LeRoi Jones / Amiri Baraka Reader*, ed. William J. Harris (New York: Thunder's Mouth Press, 1991), 397.

75. LeRoi Jones, "An Explanation of the Work," in *Black Magic*, n.p. (front matter).

76. Ishmael Reed, "badman of the guest professor," in *The Black Poets*, ed. Dudley Randall (New York: Bantam Books, 1971), 287.

77. Amiri Baraka, "Ka 'Ba," in *Black Magic*, 146. Further page refs. inserted in text after *BM*.

78. Letters, *Scientific American* 224, no. 1 (January 1971): 6–8, here 6.

79. Andrew M. Greeley and Paul B. Sheatsley, "Attitudes toward Racial Integration," *Scientific American* 225, no. 6 (December 1971): 13–19, here 18.

80. Curt Stern, "The Biology of the Negro," *Scientific American* 191, no. 4 (October 1954): 80–85, here 81, 82, 85.

81. James P. Comer, "The Social Power of the Negro," *Scientific American* 216, no. 4 (April 1967): 21–27, here 26.

82. Richard F. America, "The Case of the Racist Researcher," in *The Death of White Sociology*, ed. Joyce A. Ladner (New York: Random House, 1973), 457.

83. As the evasive publication details attest—no publisher given and only a Post Office box number as an address for the distributor—George's book had a dubious, covert status. Wesley Critz George, *The Biology of the Race Problem*, Report Prepared by Commission of the Governor of Alabama (New York: distributed by the National Putnam Letters Committee, 1962), 33; Judson Herrick, *The Evolution of Human Nature* (Austin: University of Texas Press, 1956). The sheer number of postwar publications from scientifically respect-

able presses and in authoritative science journals from which George is able dig out support for his thesis is still disturbing.

84. George, *Biology of the Race Problem*, 75.

85. Ibid., 74. A footnote reads: "James G. Needham 1950 *About Ourselves*. George Allen and Unwin; London." A reader would have to do library research to establish that this was the work of an elderly writer out of the mainstream of his field and was first published in 1941. Needham's text reads: "The direct road to racial deterioration runs by way of continued breeding from inferior stock." Losing the adjective "direct" further enforces Wesley George's thesis. James G. Needham, *About Ourselves: Man's Development and Behavior from the Zoological Standpoint* (Lancaster, PA: Jacques Cattell Press, 1941), 147.

86. George, *Biology of the Race Problem*, 65.

87. Ibid.

88. Robert J. Braidwood, "The Agricultural Revolution," *Scientific American* 203, no. 3 (September 1960): 131–48, here 148.

89. Dwight J. Ingle, "Racial Differences and the Future," *Science*, n.s., 146, no. 3642 (October 16, 1964): 375–79, here 379.

90. Ibid., 375, 376.

91. Ibid., 376.

92. Ibid.

93. Ibid., 378.

94. Morton Fried et al., "Race, Science, and Social Policy," *Science*, n.s., 146, no. 3651 (December 18, 1964): 1526–30, here 1530.

95. James Richard Jacquith et al., "Race, Science, and Social Policy," *Science*, n.s. 146, no. 3650 (December 11, 1964): 1415–18, here 1417, 1418.

96. Ibid., 1418.

97. Ibid., 1415.

98. Fried et al., "Race, Science, and Social Policy," 1526.

99. Caplan and Paige, "A Study of Ghetto Rioters," 15.

100. Ibid., 21.

101. Ibid.

102. Ibid.

103. Le Roi Jones [Amiri Baraka], *Black Music* (New York: William Morrow, 1967), 247.

104. Letter from Ralph D. Lausa, MD, Department of Health, State of Ohio, in *Scientific American* 219 (December 1968): 6–7.

105. Caplan and Paige, "A Study of Ghetto Rioters," 19–20.

106. Nikki Giovanni, "Beautiful Black Men," in Randall, *Black Poets*, 320.

107. Lerone Bennett, *The Challenge of Blackness* (Chicago: Johnson Publishing, 1972), 36. These sentences are used as part of an epigraph at the start of Joyce A. Ladner, ed., *The Death of White Sociology* (New York: Random House, 1973), xiii.

108. Wallenstein, "Culture of Sociology in Disarray," 435.

109. Robert Staples, "What Is Black Sociology?," in Ladner, *Death of White Sociology*, 167.

110. Joseph Scott, "Black Science and Nation-Building," in Ladner, *Death of White Sociology*, 290-91. This milestone essay is reprinted in John P. Jackson and Nadine M. Weidman,

eds., *Race, Racism, and Science: Social Impact and Interaction* (New Brunswick, NJ: Rutgers University Press, 2006).

111. Kimberley W. Benston and Amiri Baraka, "Amiri Baraka: An Interview," *boundary 2* 6, no. 2 (Winter 1978): 303–18, here 314.

112. Amiri Baraka and Kalamu ya Salaam, "Amiri Baraka Analyzes How He Writes," *African American Review* 37, no. 2/3 (Summer–Autumn 2003): 211–36, here 234.

113. Baraka, *LeRoi Jones / Amiri Baraka Reader*, 475, 552.

114. Amiri Baraka, *Selected Poetry of Amiri Baraka / LeRoi Jones* (New York: William Morrow, 1979), 245.

CODA

1. "Doomsday: Tinkering with Life," *Time*, April 18, 1977.

2. Michael Rogers, "The Pandora's Box Congress: 140 Scientists Ask: Now That We Can Rewrite the Genetic Code What Are We Going to Say?," *Rolling Stone* 189 (June 19, 1975): 37–42 and 74–82, here 39.

3. Shane Crotty, *Ahead of the Curve: David Baltimore's Life in Science* (Berkeley: University of California Press, 2003), 112.

4. Rogers, "Pandora's Box Congress."

5. "In this book are instructions, in a curious and wonderful code, for making a human being." Sinsheimer quoted in Lily Kay, *Who Wrote the Book of Life? A History of the Genetic Code* (Stanford, CA: Stanford University Press, 2000), 34.

6. Rogers, "Pandora's Box Congress," 37.

7. John Maynard Smith, "The Concept of Information in Biology," *Philosophy of Science* 67, no. 2 (2000): 177–94, here 178.

8. Kay, *Who Wrote the Book of Life?*, 3.

9. "And thus we say 'writing' for all that gives rise to an inscription in general. . . . It is also in this sense that the contemporary biologist speaks of writing and pro-gram in relation to the most elementary processes of information within the living cell." Jacques Derrida, *Of Grammatology*, trans. Gayatri Chakravorty Spivak (Baltimore, MD: Johns Hopkins University Press, 1976), 9.

10. Ron Silliman, *The New Sentence* (New York: Roof Books, 1987), 63–93.

11. In several essays I have explored in greater detail the possible connections between Language Writing and the new biology: "'Particulars of Time and Space of Which One Is a Given Instance': Poetry, Science and Politics in the Sixties," *Open Letter: Poetry of the 1960s*, 11th ser., no. 1 (2001): 42–65; "Junk DNA in Recent American Poetry," in *Another Language: Poetic Experiments in Britain and North America*, ed. Kornelia Freitag and Katharina Vester (Berlin: Lit Verlag Dr. W. Hopf, 2008), 27–42; "Cutting and Pasting: Language Poetry and Molecular Biology," in *Science in Modern Poetry: New Directions*, ed. John Holmes (Liverpool: Liverpool University Press, 2012), 38–54.

12. Lyn Hejinian, "Note to 'The Person,'" *Mirage* 3 (1984): 24.

INDEX